MW00973792

Aerospace Science 300:

EXPLORING SPACE: THE HIGH FRONTIER

Second Edition

C^2 Technologies, Inc.
1921 Gallows Road, Suite 1000
Vienna, VA 22182-3900
703-448-7900
www.c2ti.com

Editorial Credits
Highera Consulting, LLC, Editorial oversight: Erin Kelmereit
Department of the Air Force Editors: Michael Wetzel

Production Credits
Chief Executive Officer: Dolly Oberoi
President: Curtis Cox
VP, Operations: James E. Threlfall
VP, Services: Jimmy Ruth
Program Manager: Gayla Thompson
Photo Researcher: Rodger Stuffel
Cover Design: Rodger Stuffel and Michael Wetzel

Cover Images: Rodger Stuffel

Printing and Binding: Chroma Graphics, Inc.

Some images in this book feature models. These models do not necessarily endorse, represent, or participate in the activities represented in the images.

Jeanne M. Holm Center for Officer Accessions and Citizen Development Credits
Director of Academic Affairs: James Wiggins
Chief, AFJROTC Curriculum: Vickie Helms, M.Ed.
Curriculum Instructional Systems Specialist: Michael Wetzel

Library of Congress Cataloging-in-Publication Data on file
ISBN 978-0-9984993-4-5

Printed by a United States of America–Based Company
15 14 13 12 11 10 9 8 7 6 5 4 3 2

Contents

CHAPTER 1 The History of Astronomy

CHAPTER 2 The Solar System 50

Preface

Exploring Space: The High Frontier 2nd Edition is typically the third/fourth-year science course in the high school sequence of Aerospace Science courses for the Air Force Junior ROTC. This course has been completely rewritten to include the latest information and teaching philosophies, incorporating 21st century learning strategies. This new course provides students with the latest information on exploring space and an introduction to cybersecurity and technology. The textbook will begin with early astronomy and the basic interest in the universe from the Greeks through the Renaissance and Enlightenment ages. Students will be provided an in-depth view of the solar system, including Earth, the Sun, the Moon, and planets. The text also discussed the history of space travel and more modern space probes and robotics. Students will examine the effects of space on the human body. The text also investigates the history of rockets, launch vehicles, and the coordinated systems required for a successful launch into space. Finally, the text will offer a cybersecurity chapter that outlines the importance of cybersecurity in space and in daily life.

The text is intended for high school students and complements the material taught in high school math, physics, and other STEM-related courses. This textbook focuses on the basics of spaceflight and includes full-color images and diagrams to enhance student learning. Other features of the text, include chapter quotes, "Quick Writes" that can be used for writing assignments, "Learn Abouts" that provide an overview of the lesson, a list of vocabulary words for each lesson, "Did You Know" call out boxes with interesting facts, and "Right Stuff" vignettes with additional details on important persons or events from spaceflight and cyber history. Each lesson closes with "Checkpoint Review Questions" that allow the student to review and reinforce what they have learned.

The text has eight chapter which are divided into lessons.

CHAPTER 1: THE HISTORY OF ASTRONOMY discusses the models of the universe throughout history. This section focuses on early astronomers and the theories they developed by observing the movements of the stars and planets. Lesson one describes the prehistoric and classical astronomy models presented by ancient Greek astronomers, Aristotle, and Ptolemy. Lesson two outlines the models of astronomy discovered during the renaissance period, including Copernicus's heliocentric system and Kepler's laws of planetary motion. Lesson three focuses on models of astronomy from the enlightenment era and modern astronomy models, such as Galileo's observations, Newton's laws of motion and gravity, and Einstein's theories.

CHAPTER 2: THE SOLAR SYSTEM explores the elements of the solar system. Lesson one focuses on the Earth and the Moon and explores the components of each and their relationship to each other. Lesson two analyzes the sun's energy and investigates how the solar system was formed. Lesson three examines the planets included in the solar system. In addition, in exploring each planet in the solar system, this lesson examines dwarf planets, comets, asteroids, meteors, and the Kuiper belt. Lesson four completes the discussion by exploring deep space.

CHAPTER 3: SPACE EXPLORATION examines components required to explore space and the current NASA strategic plan for space exploration. Lesson one analyzes the benefits of space exploration, NASA's plans for space exploration, and the role of private industry in space exploration. Lesson two discusses the components necessary for a successful space mission, including the qualifications and training of astronauts. Lesson three explores the hazards for spacecraft in space, such as radiation, impact damage, surface landing threats, and fire in space.

CHAPTER 4: SPACE PROGRAMS discusses past space programs and the effect of space on the human body. Lesson one examines space programs throughout the world, including the U.S. space program, Russian space program, and Chinese space program. The history and future goals of each program are examined. Lesson two provides a detailed explanation of the U.S. manned space program, including Project Mercury, Project Gemini, Project Apollo, and the Space Shuttle program. Lesson three focuses on methods that can be explored to make space people-friendly. This lesson examines the effect of space on the human body and the study of space biomedicine.

CHAPTER 5: SPACE STATIONS AND BEYOND examines the history of space stations and the future of space travel. Lesson one identifies key accomplishments of space stations throughout history, including Salyut, Skylab, Mir, and the International Space Station. Lesson two examines the future of U.S. space travel, including the potential of asteroid mining, lunar exploration, Mars missions, and space tourism. Lesson three explores the effects of space technology on daily lives. This lesson explores the technology used everyday on Earth that uses space technology, such as satellites and GPS.

CHAPTER 6: SPACE PROBES AND ROBOTICS discusses the role of space probes and robotics in space exploration. Lesson one explores key space probe missions to the Sun, Moon, Venus, and Mars. Lesson two analyzes the current and future use of robotics in space. And lesson three dives into the Mars Rover expeditions and looks ahead at Mars Rover 2020.

CHAPTER 7: HOW IT WORKS: ORBITING, SPACE TRAVEL, AND ROCKETS evaluates the science and technology required for space travel. Lesson one begins by examining how orbits work and identifying the different types of orbits. Lesson two explores trajectories in space travel, including how spacecraft navigates in space. Lesson three investigates rockets and launch vehicles. The evolution of rockets and the different types of rockets are analyzed. In addition, the text examines launch vehicles throughout history and the factors involved in a successful launch.

CHAPTER 8: CYBERSECURITY examines the concept of cybersecurity with regards to space assets as well as personal cybersecurity. Lesson one begins by defining cybersecurity and examining cyber threats to U.S. space assets. The text will explore strategies for protecting U.S. assets. Lesson two examines the principles of cybersecurity to individuals and explores basic methods for protecting yourself from cybersecurity attacks. Lesson three examines cybersecurity policies of the US military and government, including the Homeland Security and DoD Cyber Strategy.

At the end of the text, you will find a glossary of vocabulary terms, an index of key topics throughout the book, and a list of references.

This book has been prepared especially for cadets to introduce them to the science of space exploration and the importance of cybersecurity. It is our hopes that the text sparks an interest in space exploration and helps fuel the next space race amongst the next generation. Space exploration and cybersecurity requires scientists, mathematicians, engineers, technologists, and technicians. These careers can all lead you to a future in space or as cyber professionals. As you embark in your future endeavors, remember that when using technology, there may be someone ready and willing to compromise your identity. Being prepared is the best defense against a cyberattack. Finally, don't forget to look up at the night sky and enjoy the wonders of the universe that you have learned about in this text, you are the future of space exploration.

Acknowledgements

Aeronautical Science 300: Exploring Space: The High Frontier, Second Edition, was developed based in part on the recommendations from AFJROTC instructors, who are responsible for delivering this curriculum.

The Jeanne M. Holm Center for Officer Accessions and Citizen Development (Holm Center) Academic Affairs Directorate (DE) team involved in the production effort was under the direction of Mr. Jim Wiggins, Dean and Director of Academic Affairs for the Holm Center at Maxwell Air Force Base, Alabama, and Ms. Vickie Helms, M.Ed., Chief, AFJROTC Curriculum (DEJ). Special acknowledgment goes to Michael Wetzel, M.Ed., USAF, instructional systems designer for the Holm Center Academic Affairs, who was the Air Force researcher, editor, reviewer, and significant contributor for the AS 300 textbook. We commend Michael for his persistent efforts, commitment, and thorough review in producing the best academic materials possible for AFJROTC cadets worldwide.

We would also like to express our gratitude to the C2 Technologies team for all its hard work in publishing this book, especially project manager Gayla Thompson and graphic designer Rodger Stuffel. We would like to express our gratitude to the Highera Consulting, LLC writing team. That team consisted of contractors at Highera Consulting, LLC - Erin Kelmereit, Heidi Guthrie, Brandy Horvath, and Sandra Fugate.

Special thanks given to Astrobotic Technology, Inc. for granting permissions for the use of the CubeRover image.

All the people identified above came together on this project and combined their efforts to form one great team, providing 21st century learning materials for all our units. We believe this curriculum will continue the precedent of providing "world class" curriculum materials. Our goal is to create materials that provide a solid foundation for developing citizens of character dedicated to serving their nation and community.

CHAPTER 1

The Night Sky
Nuamfolio/Shutterstock

The History of Astronomy

Chapter Outline

LESSON 1

Prehistoric and Classical Astronomy

LESSON 2

Astronomy and the Renaissance

LESSON 3

The Enlightenment and Modern Astronomy

> "We revolve around the sun like any other planet."
>
> Nicolaus Copernicus

LESSON 1

Prehistoric and Classical Astronomy

Quick Write

After reading the vignette about the Mayan civilization use of astronomy, do you think their civilization placed too much importance on the movement of planets and stars?

Learn About

- the celestial sphere
- the Greek Earth-centered model
- Ptolemy's model

The Maya civilization lasted for over 2,000 years, however the period between 300 A.D. and 900 A.D. was its most successful. During that time, the Mayans developed a complex understanding of astronomy. The origin of astronomy can date back to the Maya people. The ancient Mayans were keen astronomers who methodically observed the sky. Mayans believed that the actions and wills of the gods could be read in the stars, strongly influencing their daily life. However, the Maya reading of planets and stars was more astrology then astronomy. **Astrology** is *the interpretation of the influence of the heavenly bodies on human affairs*.

The Mayans even constructed their temples and other structures with astronomy in mind. The temples were built so that the sun, Moon, stars, and planets were all visible from the top of the temple. Windows were even strategically placed in the temples so that celestial objects could be seen at various times throughout the year.

Mayans built observatories at many of their cities, aligning structures with the movement of stars and planets. Celestial events defined the rituals of Maya rulers. Transfers of royal power, for example, seem to have been timed by the summer solstice at certain cities. **Solstice** *occurs twice a year when the sun reaches the greatest distance from the equator*. At the Maya city of Palenque, located in southern Mexico, an inscription notes that King Chan-Bahlum dedicated Temples of the Cross Group on July 23, 690—timed to coincide with the conjunction of Jupiter, Saturn, Mars, and the Moon.

To the Mayans, this event may have represented life before the birth of the three ancestor gods of the Palenque dynasty with the First Mother (the Moon).

The planet Venus was particularly significant to the Maya; the important feathered serpent god Quetzalcoatl, (ket-sahl-koh-aht-l), for example, is identified with Venus. The Dresden Codex, one of four surviving Maya chronicles, *a factual written account of important or historical events in the order of their occurrence*, contains an extensive tabulation of the appearances of Venus, and was used to predict the future. The Maya also used the stars to help determine the best time to go to war, again triggered by the planet Venus. Images of Venus war clothing and ornaments is seen on upright stone slabs, columns, and other carvings, and raids and captures were timed by appearances of Venus, particularly as an evening "star."

Astronomy was essential to the Maya civilization, as it was used to develop two calendars. Using their knowledge of astronomy and mathematics, the ancient Mayans developed one of the most accurate calendar systems in human history. The ancient Mayans had a fascination with cycles of time. The first was a cyclical, *occurring in cycles or recurrent*, Haab (häb) calendar of 365 days. This calendar is composed of 18 months made of 20 days, and one month with five days. The Maya farmers conducted offerings and ceremonies in the same months every year, following a 365-day Haab cycle.

Vocabulary

- astrology
- solstice
- chronicles
- cyclical
- astronomy
- constellation
- concentric
- celestial sphere
- north celestial pole
- south celestial pole
- elliptic
- ecliptic
- eclipse
- solar eclipse
- lunar eclipse
- planet
- zodiac
- cosmological
- parallax
- epicycle

The Mayans also developed the Long Count calendar to chronologically date mythical and historical events. The 13 baktun, (bak-tun), cycle of the Maya Long Count calendar measures 1,872,000 days or 5,125.366 tropical years. This is one of the longest cycles found in the Maya calendar system. This cycle ended on the winter solstice, December 21, 2012.

This image shows Temples of the Cross Group at Mayan ruins of Palenque - Chiapas, Mexico.

Diego Grandi/Shutterstock

The Celestial Sphere

What do you see when you look at the stars? Do you notice how the stars rise? What about the movement of stars in the sky? Throughout history, civilizations have studied the stars and their movements in the sky. Astronomy is *the study of the universe beyond the earth's atmosphere*. Stars are included in the study of astronomy. If you observe the stars you will notice that some stars rise in the east and set in the west. You'll also notice that depending on the season the location of the stars in the sky may differ. For example, the Big Dipper is an easy constellation to pick out in the night sky. A constellation is *a group of stars forming a recognizable pattern named after its apparent form*. In the fall, the Big Dipper is visible on the horizon in the US and Canada. But in the summer, the Big Dipper will be higher in the sky and a bit more to the northwest.

Did You Know?

The Big Dipper is not actually classified as a constellation. The Big Dipper is, in fact, a part of the constellation Ursa Major, or the Greater Bear.

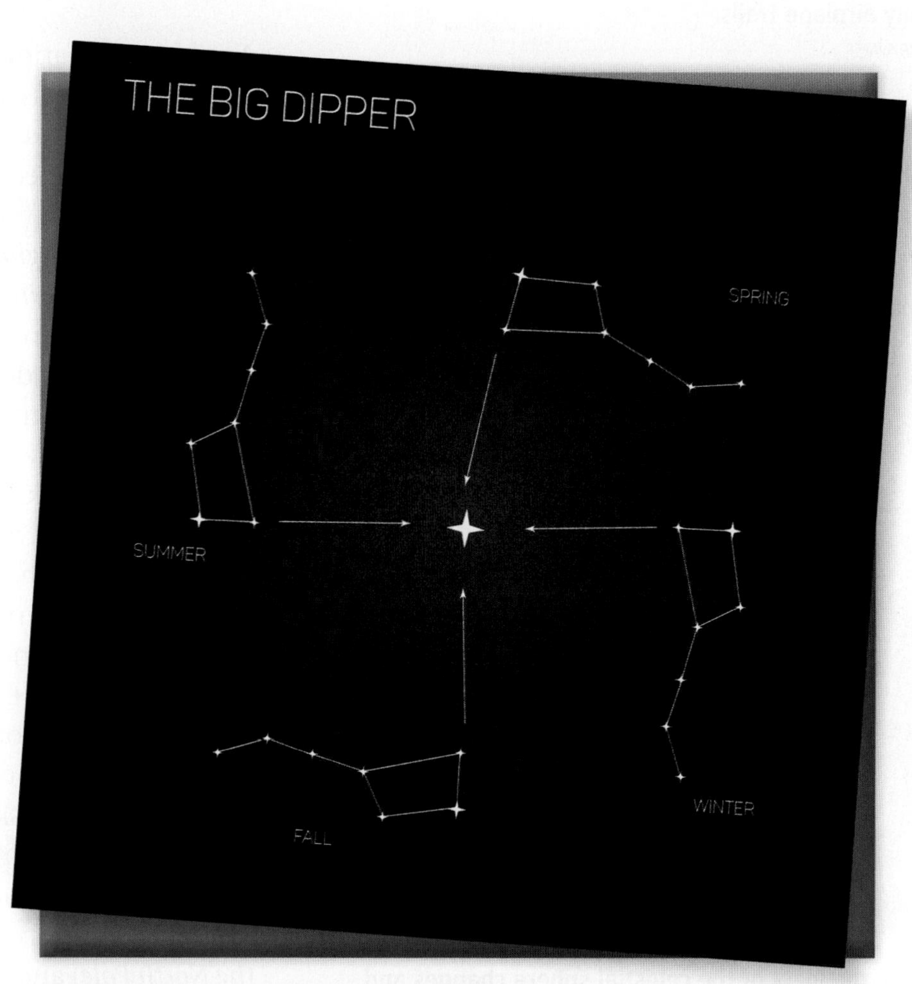

This image shows the location of the Big Dipper in the sky during different times of the year. The Big Dipper moves around the North Star.

Courtesy of KseniyaUN/Shutterstock

If you are at the North or South Pole the stars move in concentric circles in the sky. Concentric means *circles, arcs, or other shapes that share the same center*. The North and South Poles form the axis on which the Earth rotates, thus the stars appear in concentric circles at the poles. Time-lapse photos allow you to view the concentric circles made by the stars.

This time-exposure photo shows the concentric circles of stars over Kitt Peak National Observatory. The dotted lights on the horizon are actually airplane trails.

Courtesy of NOAO/AURA/NSF

Ancient civilizations believed that the earth was a sphere and was surrounded by a large sphere that rotated the Earth. The term celestial sphere refers to this *imaginary shell formed by the sky in which the center is the observer's position*. Think of nesting dolls where the Earth is the inner doll and fits inside the celestial sphere. We now know that the Earth rotates, which gives the appearance that the stars are actually rotating around the Earth. The concept of the celestial sphere is still used today by astronomers to describe and predict motions of the stars in the sky. They also refer to the north celestial pole (*the point on the celestial sphere directly above the North Pole*) and south celestial pole (*the point on the celestial sphere directly above the South Pole*).

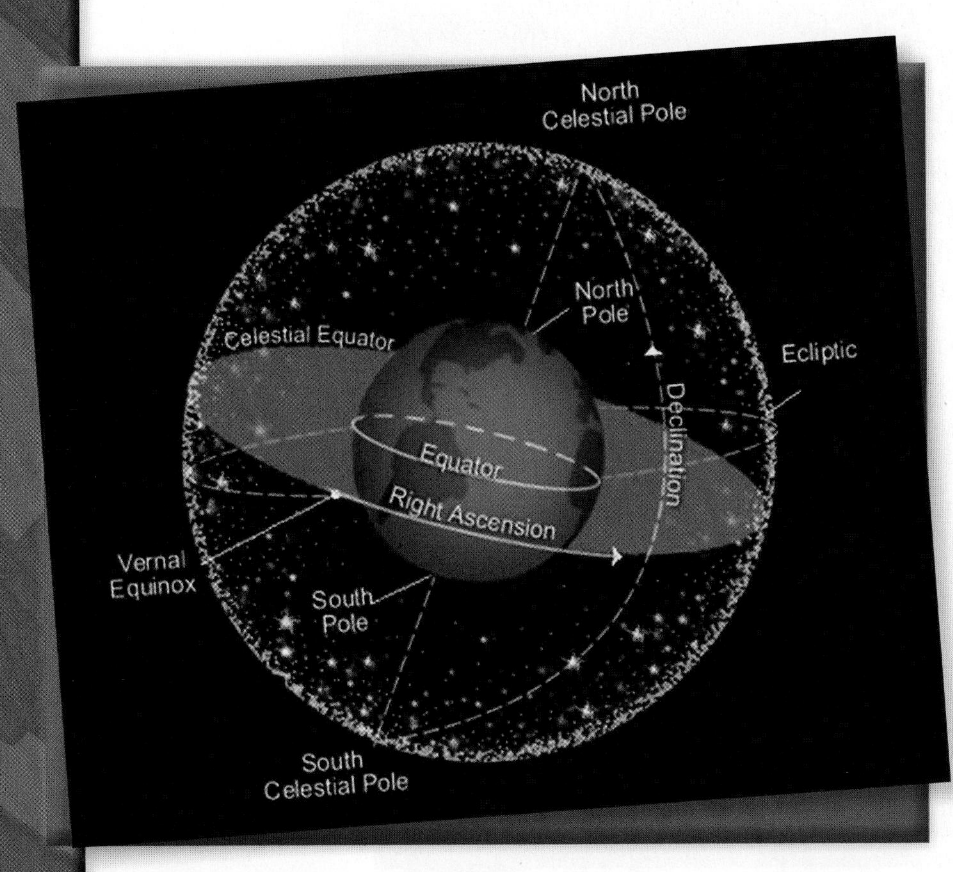

As the Earth rotates, the view of the celestial sphere changes and the objects in view appear to move.

Courtesy of the Lunar and Planetary Institute

Constellations

As defined earlier, a constellation is a group of stars that appear to create a pattern and were given a name. Ancient civilizations recognized patterns in the sky created by the stars and provided names for these patterns. The constellations were also used by ancients to identify directions and locations.

Many constellations were identified and named by ancient Middle Eastern, Greek, and Roman cultures. They identified constellations as characters that were important to their culture and lives. Today, stars are named for their location on the celestial sphere.

One of the largest and most recognizable constellations is Orion. It is a constellation that can be viewed around the world. According to Greek mythology, Orion, the son of Poseidon, was a gifted hunter. The most common story is that Hera (wife of Zeus) had a scorpion kill Orion. When looking at the night sky, look for Orion's Belt, the three stars in a row. Orion's constellation is holding a bow aimed at the constellation Taurus.

There are 88 constellations in the sky, and 50 were identified by the ancients. Constellations are still being identified and several "modern" constellations were identified in the 1500s, 1600s, and 1700s using telescopes. Pavo is a modern constellation that means Peacock. Pavo is made up of 10 stars and can be easily seen from the southern hemisphere. The constellation was named by Pieter Dirkszoon Keyser and Frederick de Houtman to fill the voids in the astronomy charts for the southern hemisphere.

Constellation Orion
Vector FX/Shutterstock

The name Pavo means peacock in Latin. In Greek mythology, the peacock was the sacred bird of Hera (wife of Zeus).
Elartico/Shutterstock

Did You Know?

Stars can actually move over time. This process takes tens of thousands of years or more to be noticeable to the naked eye. On average, a star will move approximately ½ a degree in the sky over 2,000 years. However, some stars move much faster. Barnard's star is the fastest star we know about and it moves approximately 5.5 degrees over 2,000 years.

The Motion of the Sun

Just like the stars, the sun seems to revolve around the Earth. As we now know, the sun has a fixed location, and the Earth revolves around it. It takes the Earth 365.25 days to completely revolve around the sun. Leap Day (February 29th) accounts for the extra day every four years.

Once a year the Earth rotates around the sun, and the sun circles the celestial sphere. The elliptic is *the circle formed by the Earth's orbit with the celestial sphere.* An ecliptic path of the sun does not circle Earth at the equator, it is tilted at an angle. Ecliptic is *the sun's apparent path during the year.* (Figure 1.1) If you recall the definition of solstice from earlier in the lesson, the equinoxes are the two locations where the ecliptic passes over the equator.

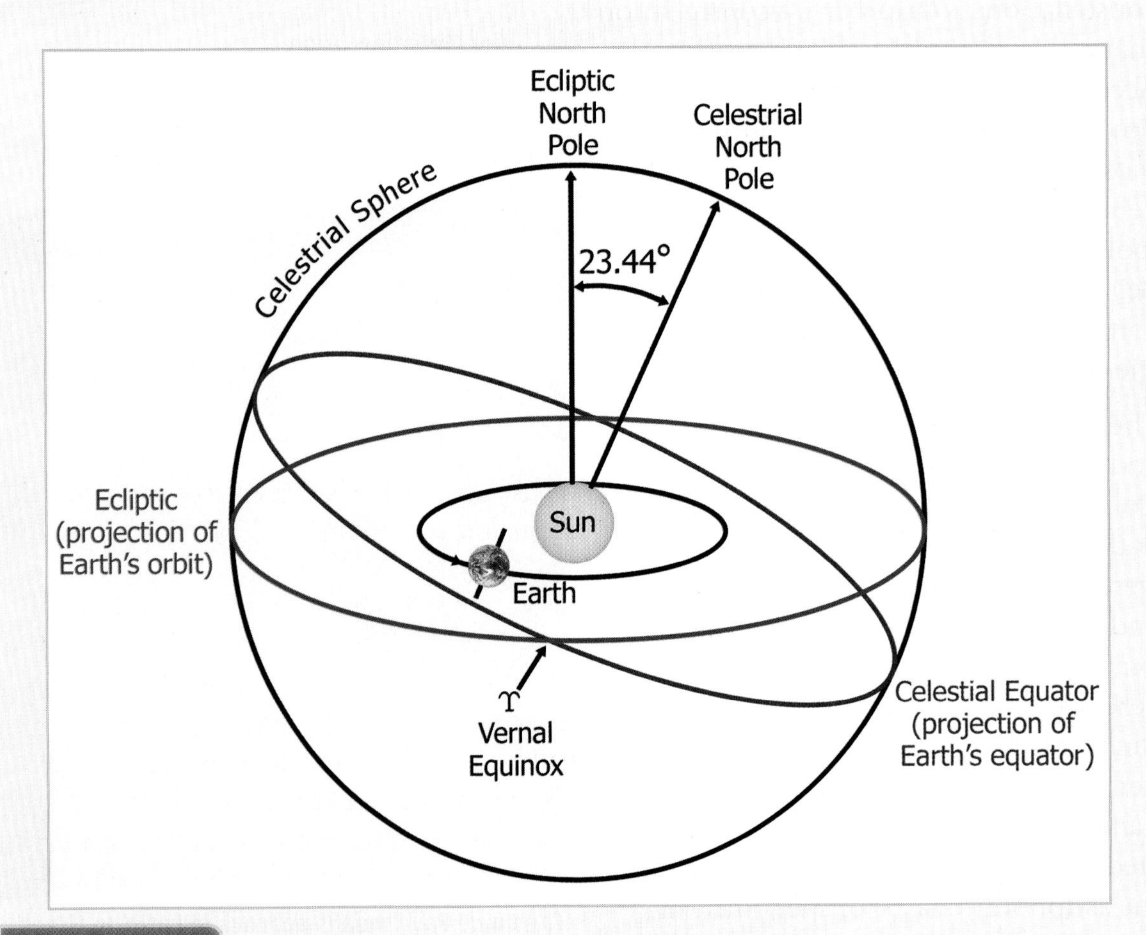

FIGURE 1.1

The ecliptic passes over the equator in two locations called the equinoxes.

The Right Stuff

Zodiacs

Along the ecliptic, the sun passes through a group of constellations referred to as the zodiac. Ancient civilizations used the zodiac constellations to determine the changing of the seasons. Based on the constellations seen in the sky, ancient civilizations could determine the calendar and seasons.

In fact, some cultures attribute the zodiac signs to a person's destiny or mindset. Many horoscopes, or forecasts of a person's future, are based on the zodiac sign that was in view during the person's birth month.

A simple geometric representation of the zodiac signs and constellations for horoscope with titles, line art illustration on the starry sky background.

Maria Gniloskurenko /Shutterstock

Did You Know?

A star is called a sun if it is the center of a planetary system. Our sun is the center of our planetary system so while we refer to it as the sun, it is also a star.

The Phases of the Moon

So far, we've discussed the stars and the sun and how ancient civilizations used them for navigation because they were set in their location. However, the Moon does actually rotate around the Earth. When the sun is on the opposite side of the Earth from your location, the Moon is illuminated by the sun. Based on the location of the Moon in its orbit, we see more or less of the Moon. This change in appearance of the Moon occurs in a cycle.

The full Moon occurs when the Moon is behind the Earth and the entire Moon is illuminated by the sun. The full Moon typically appears very large and bright on Earth. The smallest sliver of the Moon appears when the Moon is between Earth and the sun. As the Moon rotates around earth, we see more of the Moon and the sliver becomes larger. A half-Moon occurs when the Moon is at a right angle to between the sun and Earth.

This image shows how the phases of the Moon change based on the location of the Moon in its orbit around the earth.

Castleski/Shutterstock

The Right Stuff

Solar Eclipses

An eclipse *occurs when the Moon's rotation is very close to or on the ecliptic line*. There are two types of eclipses that may occur.

A solar eclipse *occurs when the Moon blocks the sun's path to Earth*. If you are in the right location, you can see a total solar eclipse, which means all of the light from the sun is blocked by the Moon; during this type of eclipse, it becomes dark on Earth during the day. However, from other locations, you will see only a partial solar eclipse, because the Moon is blocking only a portion of the sun. It is a fascinating experience to view a solar eclipse. On average, a total solar eclipse occurs every 18 months. Partial solar eclipses can occur up to four times in a year. The trick is being in the right location to view the eclipse.

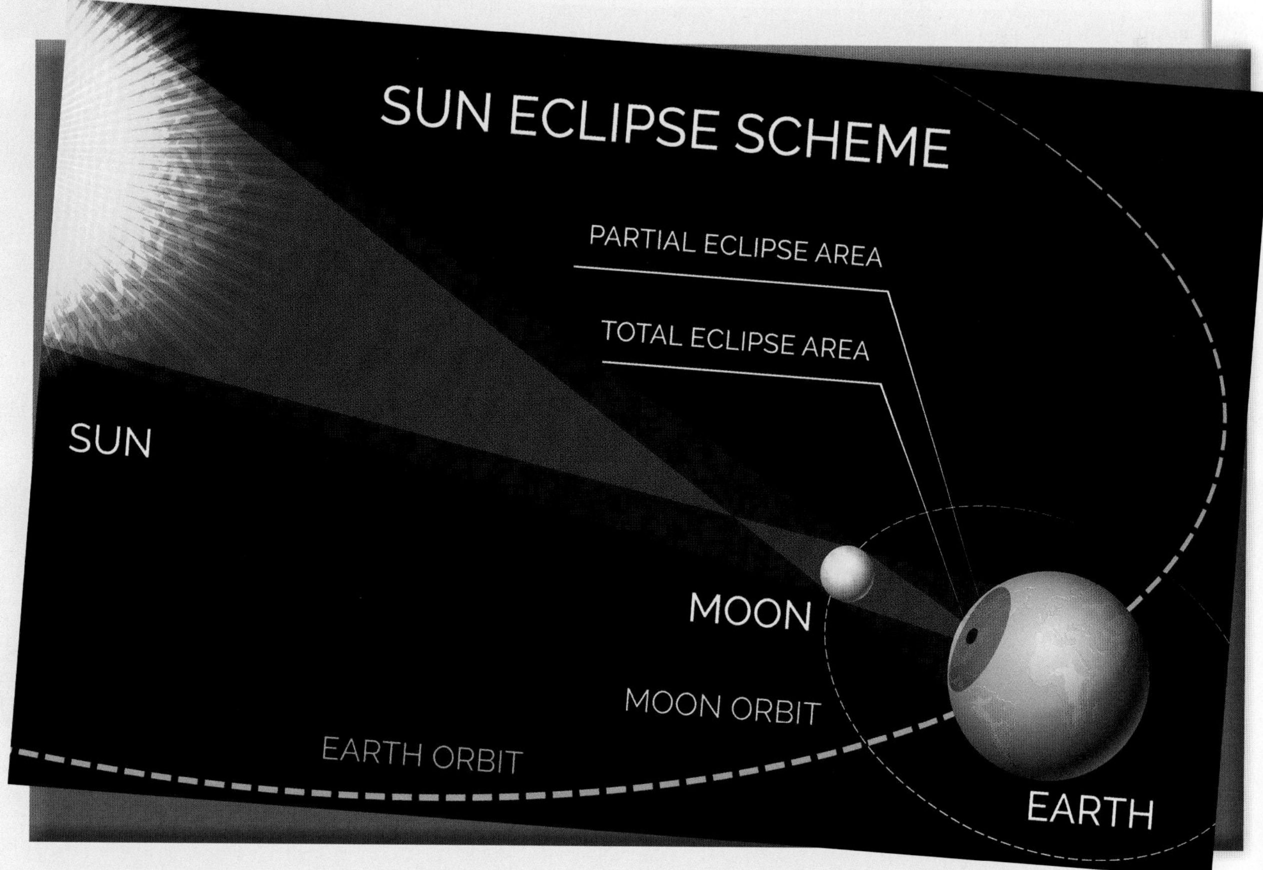

Solar eclipse occurs when the Moon travels between the Earth and the sun. On August 21, 2017 much of North America was treated to a total solar eclipse.
vectortatu /Shutterstock

The Right Stuff

Lunar Eclipses

A lunar eclipse *occurs when the Moon passes directly behind Earth and into its shadow.* When this occurs the Earth's shadow actually blocks the sun's light from illuminating the face of the Moon. A lunar eclipse can only occur during a full Moon, when it passes through Earth's shadow. A lot of pieces must be in place for a lunar eclipse to occur.. For this reason, lunar eclipses only occur about twice a year.

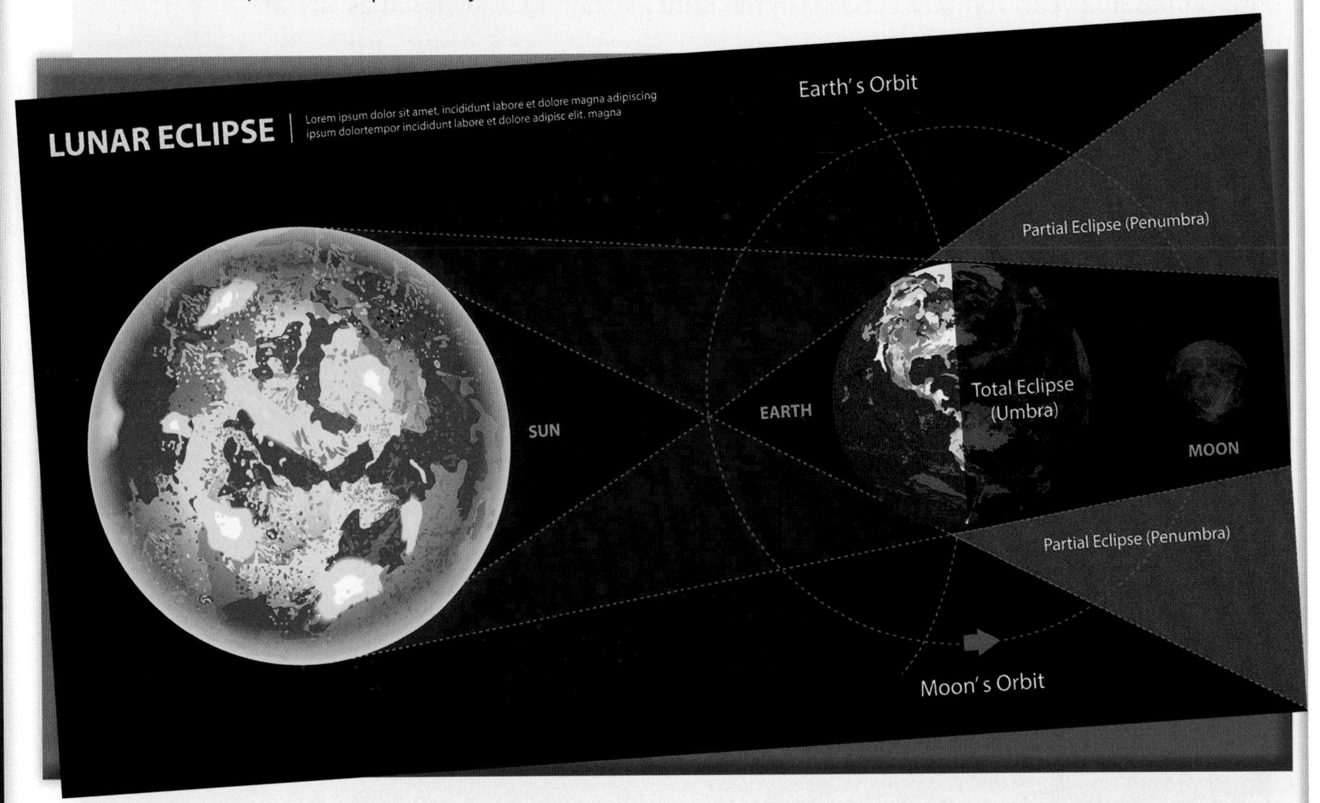

This diagram shows the alignment of the Earth, sun, and Moon that must occur for a total lunar eclipse.

Natee Jitthammachai /Shutterstock

Planetary Motion

Planet comes from the Greek word meaning "wanderer." A planet is *a celestial body moving in an elliptical orbit around a star*. There are eight planets in our solar system, and five of these can be seen with the naked eye: Mercury, Venus, Mars, Jupiter, and Saturn. Depending on the planets' distance from the sun, they may orbit the sun faster. For example, Mercury orbits the sun every 88 days, whereas it takes Jupiter 12 years to make a full orbit around the sun.

Early astronomers were confused by the motion of the planets. It appeared as though they bounced around the sky and changed course frequently. In reality, this phenomenon is attributed to the elliptical or oval shape of the orbit and the speed at which a planet orbits the sun.

Today, it is easy to spot a planet in the sky at night! There are apps, such as Sky Map, that you can load to your phone to identify objects in the night sky. If you point your phone's camera at the night sky, Sky Map will identify the constellations, stars, and planets that you are viewing.

In this image, you can see the Moon, Venus, and Mars in the night sky.
Irina Kuzmina/Shutterstock

The Greek Earth-Centered Model

Ancient civilizations observed the night sky and began developing scientific models from the data they gathered. Some of these astronomers viewed Earth as the center of the universe, while others believed that the sun was the center of the universe.

The Babylonians recorded and cataloged their observations of the skies for centuries. Using the data they gathered, they could predict the time between full Moons and eclipses. They also used the data to create a nearly accurate calendar.

The Egyptians used astronomy predictions to plant agriculture and identify seasons. The Great Pyramid is aligned with the cardinal points, and many temples are aligned along the axis of the rising midwinter sun, signifying to Egyptians that they should begin to prepare for planting in the spring.

The Chinese kept meticulous records of their observations. Initially, the data was used to record time accurately. As with the Mayans, Chinese astronomy was more astrology. Unlike other cultures charting the stars at this period, astrologers were separate from astronomers, and their job was to interpret occurrences and omens portrayed in the sky. The Chinese developed an extensive system of the zodiac designed to help guide the life of people on Earth. Zodiac is *an imaginary belt in the sky that includes the paths of the planets and is divided into twelve constellations*.

Perhaps the most influential study of astronomy came from the Greeks. The Greeks studied astronomy so that they could understand how the universe worked. The work of the ancient Greeks laid the foundation for Western astronomy. The keen minds of the Greek philosophers were the first to attempt to find a universal theory. They tried to uncover the hidden laws of creation long before Einstein, Hawking, and the great theoretical physicists, who are the distant descendants of the Greek astronomers, and certainly speak the same language of mathematics.

Ancient Greek Astronomers

One of the first Greek astronomers was Thales, who lived around 600 BC. Thales sought to understand the universe. He studied the Sun and stars as balls of fire rather than believing they were gods. Some believe that Thales successfully predicted an eclipse in 585 BC.

Anaximander (uh-nak-suh-man-der) was a Greek philosopher who created a cosmological model. Cosmological is *a branch of philosophy dealing with the origin and general structure of the universe*. He proposed a theory explaining how the universe worked and revealed what he felt were the hidden processes guiding the movement of the planetary bodies.

Pythagoras (Pith-AG-oh-ras) was a Greek philosopher and mathematician who put forth the theory that the earth was spherical. He proposed this idea because he noticed that ships disappear below the horizon when they sail away, implying that the surface of the earth is curved. Pythagoreans were those that held the same beliefs as Pythagoras. The Pythagoreans also proposed the theory that the universe was controlled by a central "fire." The central "fire" was believed to be invisible to the human eye. Pythagoreans did not believe the sun was the central "fire." Pythagoreans are credited as the forerunners, 2,000 years ahead of Nicolause Copernicus' publications, who viewed the Earth as a planet rather than the center of the universe.

This image portrays a sketch of Pythagoras, the famous Greek philosopher and mathematician.

Drawhunter/Shutterstock

Aristotle's Earth-Centered Views

The Greek philosopher Aristotle agreed with the Pythagoreans that the Earth and Moon were spherical, but he put the Earth as the center of the universe rather than the central "fire." Aristotle believed that the planets and stars must move in a circular pattern. This belief stemmed from the fact that Aristotle could not see any signs of the Earth moving. Aristotle argued that if you threw a ball straight up into the air, it should land behind the thrower, given no wind, if the Earth was moving. In reality, the movement of the Earth cannot be felt when standing on its surface.

Whereas, if a person moves, changes can be seen on Earth. For example, when you drive down the road you can see the position of the trees change. Parallax is *the apparent displacement of an observed object due to a change in the position of the observer.* We now know that parallax occurs in astronomy as well. The stars are so far away from the Earth, that the shift in position appears quite small. This is called stellar parallax and was discovered in 1838. We will cover stellar parallax in a later lesson.

Ptolemy's Model

While Aristotle greatly influenced ancient Greek scientists, it was the work of Claudius Ptolemaeus, or Ptolemy (tol-uh-mee), a Greco-Roman mathematician, astronomer, astrologer, and geographer born in Alexander, Egypt in 100 A.D., that was most widely accepted. For 1,400 years Ptolemy's model of the universe was the conventional model used by the scientific community.

Ptolemaic Model

The Ptolemaic Model, or Geocentric Model, was based on the view that Earth was a stationary object and the center of the universe. All other heavenly bodies rotated around Earth. His theories were based on perfect circles. The Earth, Moon, sun, and heavenly bodies were all perfect circles that orbited in perfect circles (Figure 1.2).

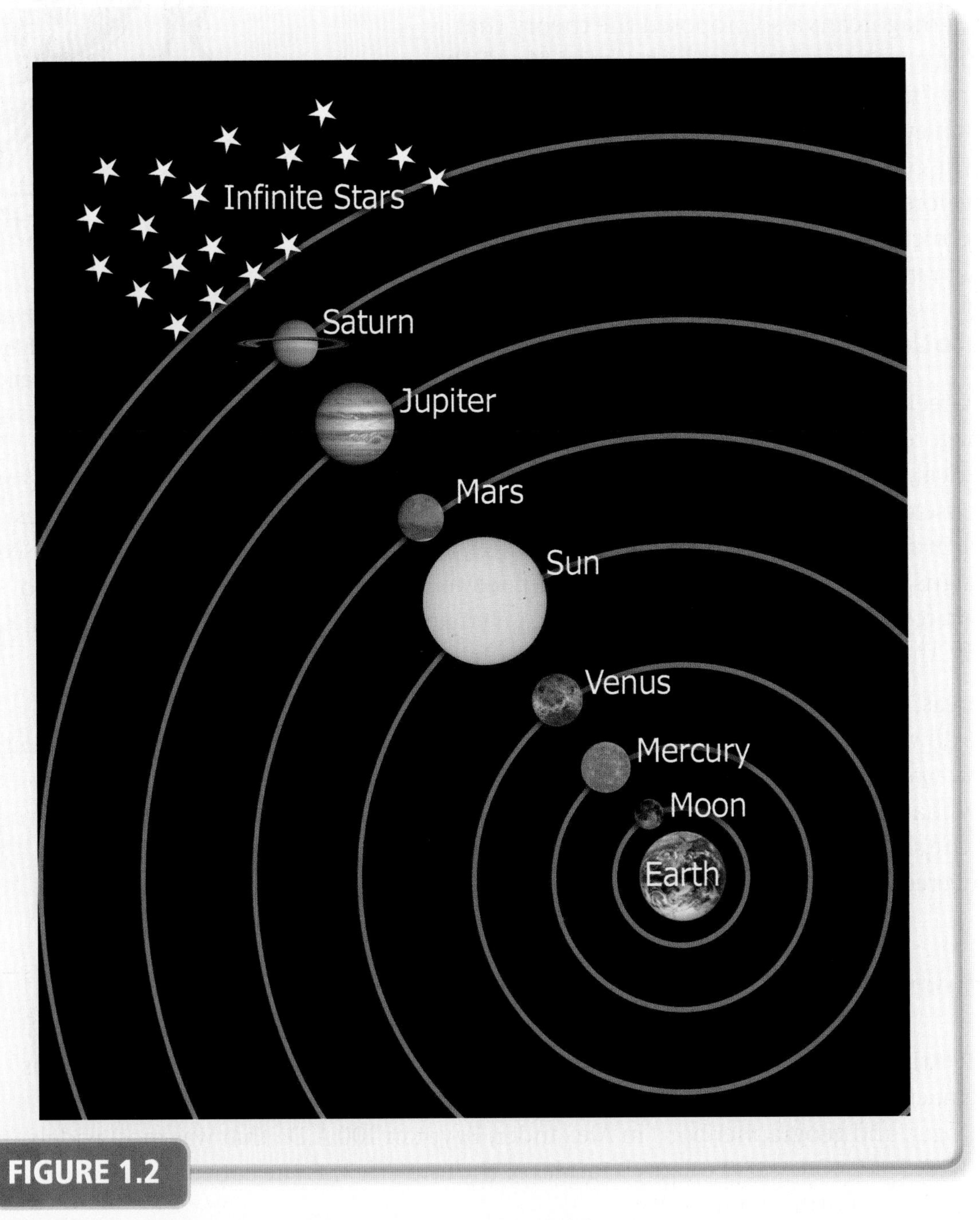

FIGURE 1.2

Ptolemy believed that all celestial objects orbited the Earth.

The heavenly bodies were seen as perfect entities, unlike the Earth. The Moon changed positions amongst the stars at night. It was believed to have imperfections; because the Moon was between the imperfect Earth and the perfect heavens, it was bound to have flaws.

The Ptolemaic model became more complicated as you looked farther than the Moon. Ptolemy believed the heavenly bodies followed circular motions because they were attached to unseen revolving solid spheres. An epicycle is *a small circle in which the center of the circle moves around a larger circle*. Ptolemy used epicycles to account for the motions of the planets and heavenly objects (Figure 1.3) Ptolemy believed that all planets revolved around Earth in a circular orbit and all planets also revolved around a smaller epicycle so essentially, they were in two orbits. Ptolemy compiled his theories and knowledge of the ancient Greeks in his book, Almagest.

This image portrays a sketch of Ptolemy, the Greco-Roman philosopher, astronomer, and mathematician.

Everett Historical/Shutterstock

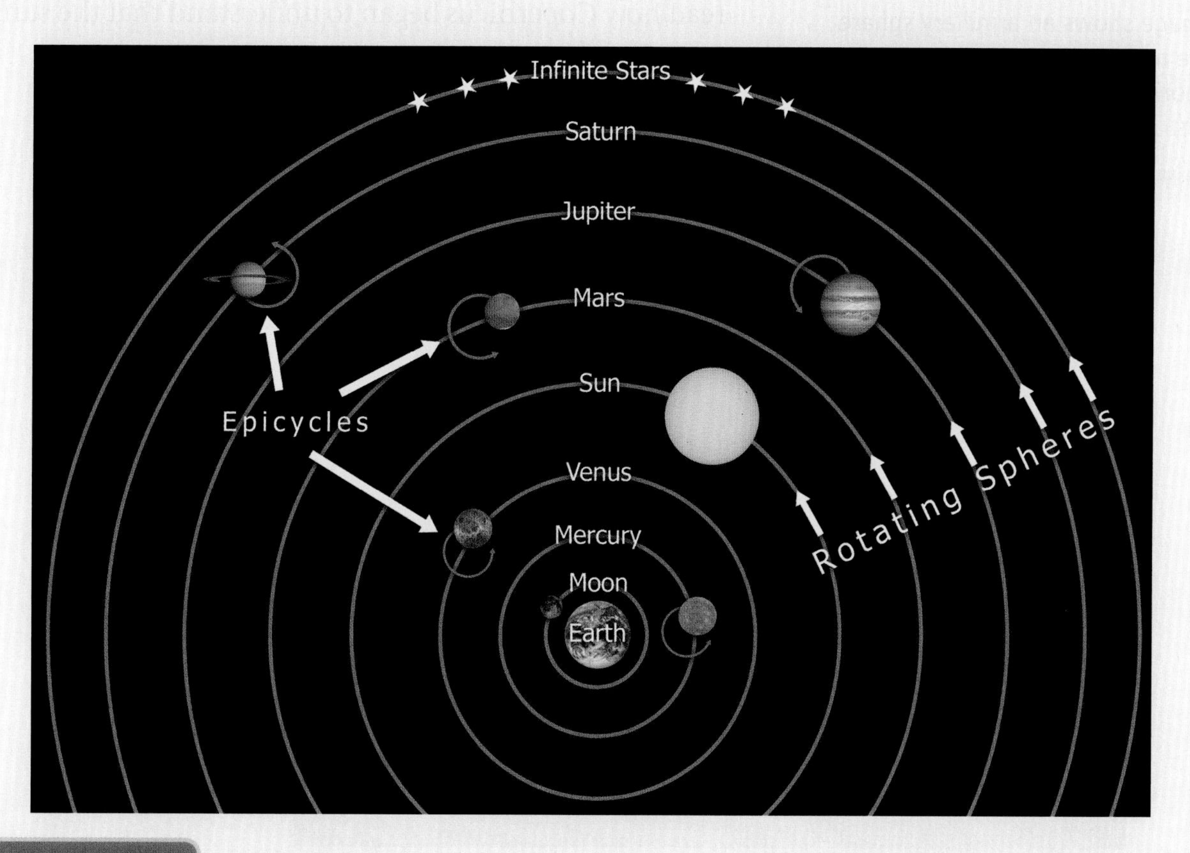

FIGURE 1.3

Epicycles were used to account for the movement of heavenly bodies.

Did You Know?

To help explain his theories to his students, Ptolemy used an armillary sphere to demonstrate the three-dimensional motions.

This image shows an armillary sphere like the one Ptolemy used to explain his theories.
Elenarts/Shutterstock

One of the issues with Ptolemy's theory was the movement of Mars. Depending on the time of the year, Mars may look like it is moving east or south. While epicycles were used by Ptolemy to explain the retrograde motion or opposite motion of Mars, these very motions can be used to explain why the Earth cannot be the center of the universe.

Ptolemy's model was not accurate, but his theories are still being used today. Have you ever visited a planetarium? Modern planetariums build equipment using Ptolemy's model to give the appearance of the sky as if you are viewing it from a stationary Earth. The equipment in the planetarium provides uniform movement in perfect circles of the heavenly bodies. While one motor moves the projector in a large circle, the other creates an epicycle to give you an accurate view of the night sky.

It would be almost 1,400 years until Nicolaus Copernicus disputed the Ptolemaic Model. In the next lesson, we will read how Copernicus began to understand that the sun was truly the center of our solar system.

The Bangkok Planetarium uses Ptolemy's theories to create a realistic view of the night sky to observers.
KOKTARO/Shutterstock

Lesson 1 Review

Using complete sentences, answer the following questions on a sheet of paper.

1. What is the large sphere that ancient civilizations believed orbited the Earth called?
2. How do constellations get their names?
3. What is one of the largest and most recognizable constellations?
4. When would a star be called a sun?
5. What is the difference between a solar eclipse and a lunar eclipse?
6. Why did Pythagoras believe that the Earth was a sphere?
7. Why did Aristotle believe that planets and stars must move in a circular pattern?
8. What was Ptolemy's model?
9. How are Ptolemy's theories used today?

APPLYING YOUR LEARNING

10. Many early astronomers thought that Earth was the center of the universe. Describe how astronomers may have come to this conclusion.

LESSON 2

Astronomy and the Renaissance

Quick Write

After reading the opening passage about Aristarchus's ideas, consider at least one way that the course of history could have changed if his work had been accepted sooner.

Learn About

- Copernicus and the Sun-centered model
- Kepler's laws of planetary motion

IN November 2011, a NASA satellite captured spectacular images of Aristarchus (Aris-tar-cuss), a gigantic crater on the moon. This lunar impact crater – deeper than the Grand Canyon – is named after the ancient Greek astronomer and mathematician, Aristarchus of Samos (310 BC–230 BC). Why was such an honor awarded to Aristarchus? Although he is not as well-known as, say, Copernicus, Galileo, or Newton, Aristarchus was the first to say the Sun, and not the Earth, was the center of our universe. Worthy of a crater named in his honor? Most definitely.

We know little about Aristarchus. Only one of his written works still exists, titled, *On the Sizes and Distances of the Sun and Moon*. In his work, he places the Sun in the middle of the solar system and calculates the sizes of the Moon and Sun. He also tries to figure out how far they are from the Earth.

Aristarchus was one of the first astronomers to observe the Moon during a lunar eclipse. He wanted to calculate the angle and size of the Earth. He realized that the Sun was so much more massive than the Earth. His deduction? Tiny objects (like the Earth) should orbit around large ones and not the other way around.

We also credit Aristarchus with first proposing that the Earth and the planets orbit the Sun. He further suspected that the stars we see in the night sky are distant suns, part of a massive universe far beyond the accepted size.

Peers rejected these theories as being far too revolutionary. Further, his theories contradicted religious teachings based on the Earth-centered model. Scientists did not revisit Aristarchus' theories until the sixteenth century.

Imagine how different history may have been if we had accepted the work of Aristarchus sooner!

Vocabulary

- heliocentric
- Copernican Revolution
- Scientific Revolution
- geocentric
- astronomical unit (AU)
- predictive power
- ellipse
- perihelion
- aphelion

Aristarchus Crater (left) is one of many lunar impact craters named after a famous astronomer.
Manuel Huss / Shutterstock

Copernicus and the Sun-Centered Model

Nicolaus Copernicus was a Polish astronomer who studied the idea of a Sun-centered model of the universe in the early sixteenth century. There were several reasons why Copernicus chose a Sun-centered model over Ptolemy's Earth-centered model.

Ptolemy's model was used to predict the future position of a planet for a given night, years into the future. Over time, however, Ptolemy's model would change the planet's position as much as two degrees! Scientists had to adjust each planet's position on their charts. Copernicus realized these changes were a signal: Ptolemy's model had major flaws.

The idea that Earth was the center of the universe had become a part of religious teachings. Copernicus was a devout Catholic, so this Earth-centered idea presented a religious challenge for him. The heliocentric model, *where the Earth and other planets orbit the Sun*, would provide good data to the scientific community. It would also establish an accurate calendar that would help Roman Catholics observe holy days at the right time.

The Right Stuff

Copernicus: Launching a Scientific Revolution

Throughout Copernicus's life, and even after his death, Ptolemy's model was the most accepted. However, publication of his most famous work, *On the Revolutions of Heavenly Spheres*, initiated the Copernican Revolution. The Copernican Revolution presented *a shift from Ptolemy's geocentric model to the heliocentric model with the Sun at the center of the Solar System*. It also encouraged a new approach to scientific thought. His work is commonly cited as the start of the Scientific Revolution.

Nikolaus Copernicus (1473–1543)
Nicku/Shutterstock

The Scientific Revolution *refers to changes in thoughts and beliefs in modern science that occurred from 1550-1700*. The Scientific Revolution dared society to look at the world in a new and different light.

The Roman Catholic Church opposed many of the discoveries that sparked the Scientific Revolution. Many believe Copernicus had a good relationship with the church. Besides being a famous astronomer, Copernicus was respected as a priest and an active member in the church. In fact, he dedicated his most famous works to Pope Paul III.

The use of epicycles, or circular orbits, to predict the positions of planets was not always correct. Ptolemy's model explained these changes, but without precision. For example, it was not possible for the geocentric model, *where the Earth is the center of the universe*, to explain the changes in the brightness of the planets or their motions.

Did You Know?

The Scientific Revolution did not happen overnight. Instead, it was a process of discovery. From ancient Greece to the early Renaissance, scientists built on the work of their predecessors to make scientific advances of the modern world possible.

Copernicus's Heliocentric System

Copernicus's heliocentric view of the universe shows the Earth's movement around the Sun and our four seasons.
Designua/Shutterstock

Copernicus is well known for coming up with the concept of a heliocentric solar system, in which the Sun, rather than the Earth, is the center of the solar system - a most significant finding of the times! A lesser-known fact is that we credit Copernicus with establishing the four seasons. That is right! We can attribute the designation of spring, summer, fall, and winter to Copernicus's theory.

In addition, Copernicus thought that the Moon revolved around the Earth. So, the Earth circles the Sun and the Moon circles the Earth.

Copernicus's description of the solar system was more detailed than the scientists before him. His model also provided a more accurate formula for calculating the planets' positions and their distances from the Sun. For example, Copernicus predicted that Venus is 0.7 times (and Mars 1.5 times) farther from the Sun than the Earth is. Compared to today's values, Copernicus's calculations are very close to accurate. Without a scale, he could not calculate the exact distances though. At the time, Copernicus did not have an accurate value of the astronomical unit (AU), which is *the average distance between the Earth and the Sun*.

In 1672, Italian astronomer Gian Domenico Cassini used parallax – the effect of an object appearing different when viewed from different positions – to measure the actual distances. Today, astronomers use spacecrafts and radio signals, to measure the planets' distances from the Sun.

Evaluating the Copernican Model

Although Copernicus accurately placed the Sun in the center of our system, he kept many of Ptolemy's elements in his theories. This caused a number of flaws in his predictions. Like Ptolemy, Copernicus assumed that the planets move in circular orbits and at constant speeds. He also used epicycles as a means to address these two flaws. Based on the basic data available at the time, the Copernican model also provided flawed predictions for planetary motion.

Copernicus missed the mark in predicting accurate planetary motion, but his model satisfied two other important criteria for scientific models. His model passed the test of simplicity and predictive power. The Ptolemaic model required different rules to explain why the observed movements of Mercury and Venus differ from other planets, but the heliocentric model naturally (and simply) accounted for the motions of Mercury and Venus inside the Earth's orbit.

Predictive power, the second criterion for scientific models, *refers to the ability of a given theory to allow predictions to be made*. For example, Copernicus thought, correctly, that stellar parallax, which is defined as the shift in position of a nearby star when seen from two different places, would exist. By observing this shift from two different locations, a measurement of distance could be made. This theory allowed a prediction to be made and tested.

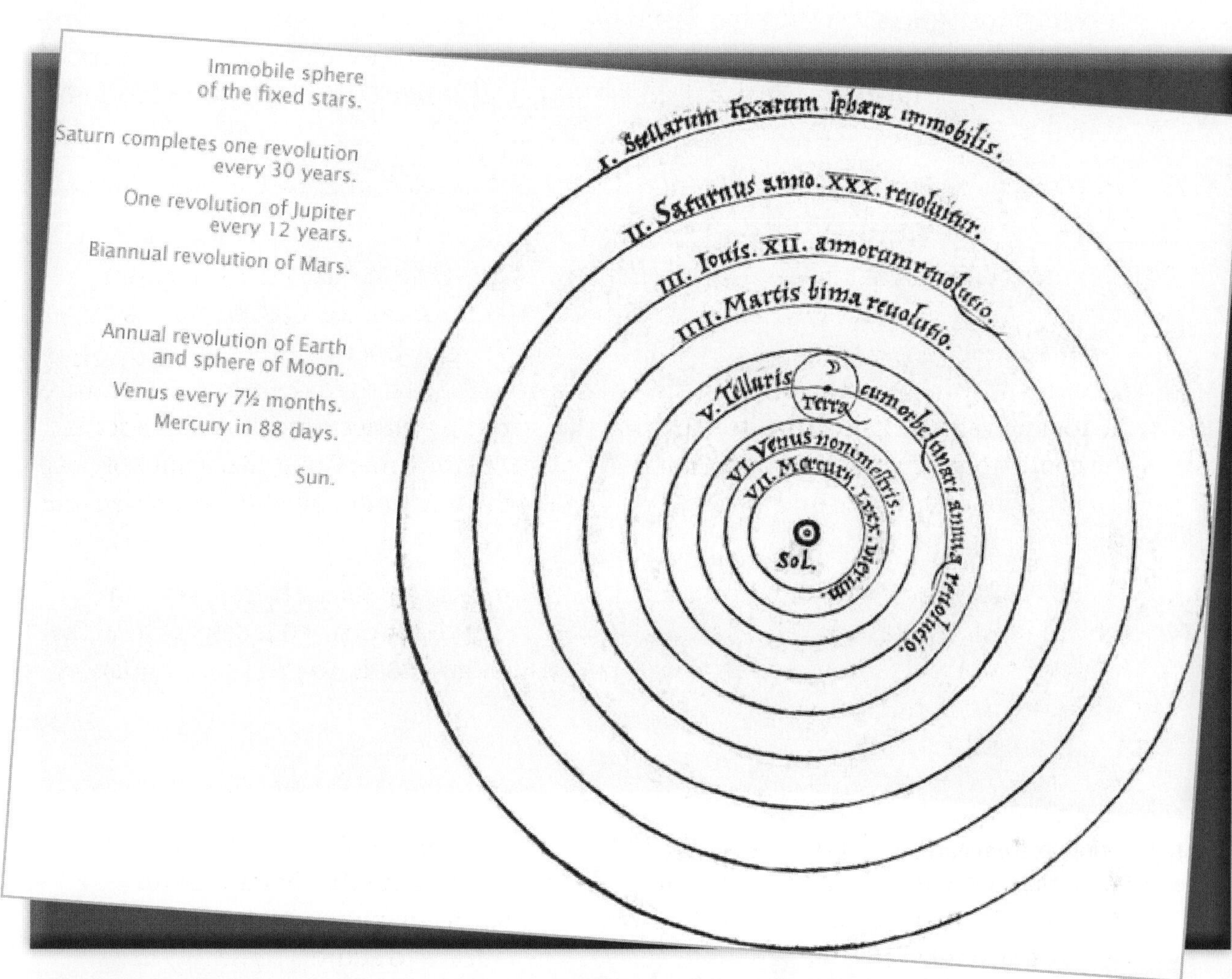

Copernicus' heliocentric view of the universe in which the Earth, along with the other planets, rotated around the Sun.

Courtesy of NASA.gov.

Did You Know?

Stellar parallax for one of the closest stars is one second of arc, which is the equivalent to the thickness of a piece of paper viewed from across a large room. The detection of such a tiny angle requires the use of a large telescope, which is why it could not be detected until the invention of the telescope. Stellar parallax arc is greatly exaggerated in most textbooks to make it more visible.

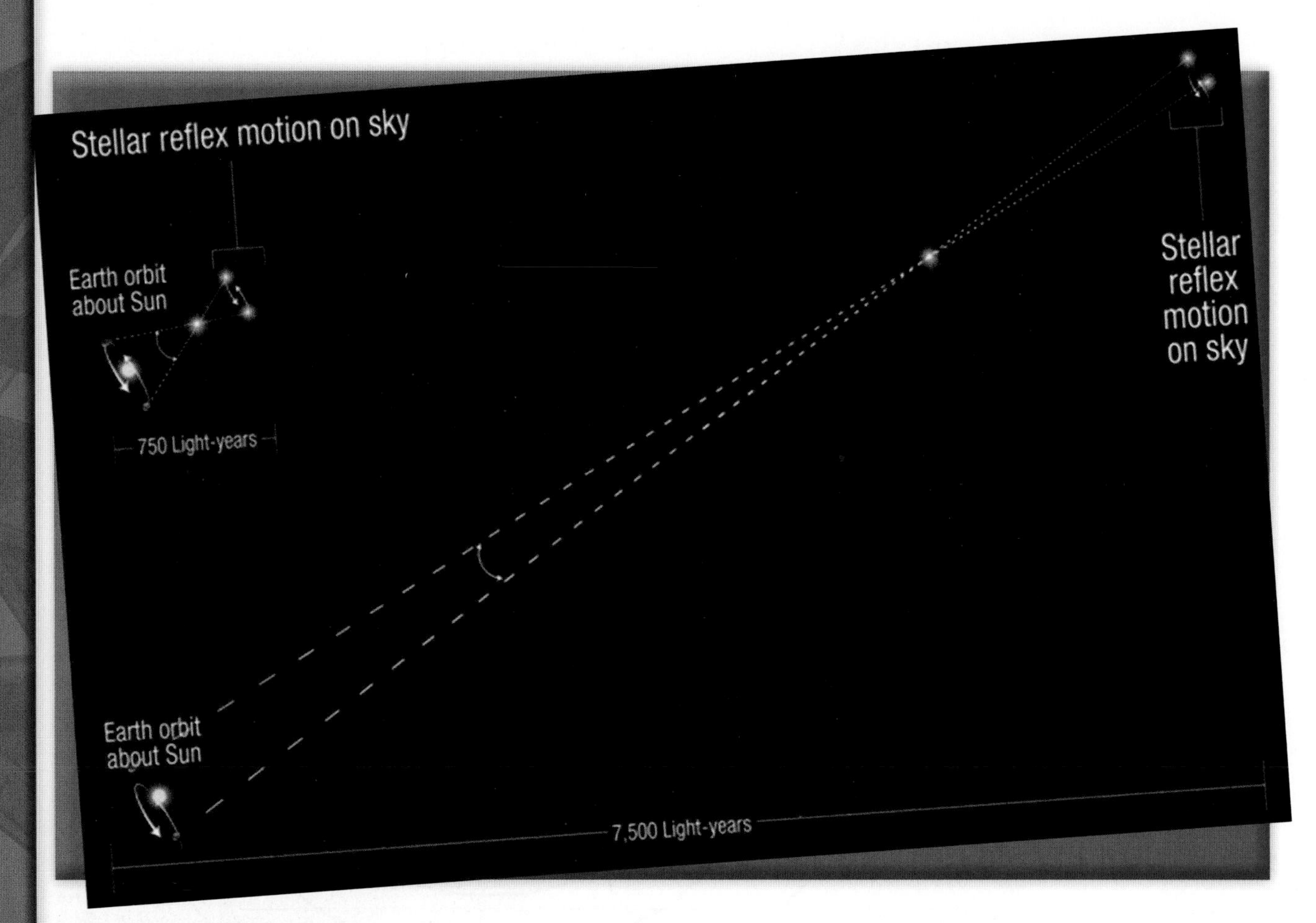

Stellar parallax from earth

Courtesy of NASA/ESA, a. Field/STScI

Tycho Brahe was the most meticulous astronomical observer of his time. Tycho studied hundreds of stars and completed the first modern stellar atlas.

Glada Vänner / NASA

Tycho Brahe's Observations

Although Copernicus's model passed the test for predictive power and simplicity, his theory for understanding the planets' motion was flawed. Tycho Brahe, a Danish nobleman, astronomer, and writer was determined to address the situation. Tycho spent decades observing and recording great amounts of data on planet positions. He made the most accurate observations of his time, and no one before him had attempted to make so many observations.

Tycho, however, believed the Earth was in the center of the Sun's orbit. Like Copernicus, he also thought other planets orbited the Sun. He based his beliefs on the lack of observed stellar parallax. Remember, stellar parallax occurs when a celestial object is so far away that its shift in position appears small.

As with the scientists before him, Tycho's work was not wasted. His data would go on to lay the foundation for the important work of his assistant – Johannes Kepler.

Kepler's Laws of Planetary Motion

Johannes Kepler, a German–born mathematician and astronomer, worked toward a model of planetary motion that would fit the data observed by Tycho. Specifically, Kepler worked hard to make sense of his predecessors' observations in finding an orbit for Mars.

The Right Stuff

Johannes Kepler

Born into a poor family, Kepler became a key figure in the seventeenth century Scientific Revolution. He used the observations of other scientists during and before his lifetime to find an orbit for Mars. Following the Copernican model and working alongside Tycho Brahe, Kepler developed his laws of planetary motion, which earned him the position of Imperial Mathematician to the Holy Roman Emperor. Kepler's laws of planetary motion are still relevant to this day.

Johannes Kepler (1571-1630)
iryna1/Shutterstock

Kepler's First Law: The Significance of the Ellipse

What are Kepler's three laws of planetary motion? Why are they important? Let's consider each one in detail. With Tycho's data at hand, Kepler worked hard to make his observations of Mars' positions, match up with a circular orbit. Over time, Kepler found that an ellipse, not a circle, would best fit the data. Kepler's first law therefore replaced circular planetary motion with elliptical motions.

What is an ellipse? *An ellipse is a geometrical shape where each point on the ellipse is at the same total distance from two fixed points, or foci.* ("Foci" is the *plural of focus*).

An ellipse can be drawn by pushing two pins (the foci) into a board, separated by a random distance. Connect them tightly with a loop of string. Place a pencil within the loop and move it around the pins, keeping the loop tight. The string acts as a guide to draw the shape of an ellipse. Each point on the ellipse is at the same total distance from the two foci. (See Figure 2.1)

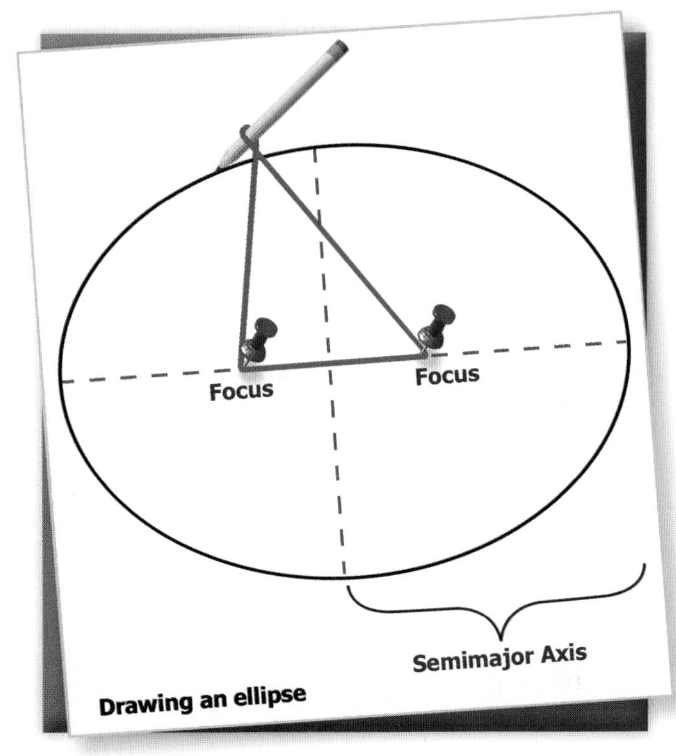

Figure 2.1

Kepler's Second Law: The Planets' Changing Speeds

Let's next look at Kepler's second law of planetary motion. While observing Mars's positions, Kepler noted changes in the speed of the planet as it moved closer to the Sun.

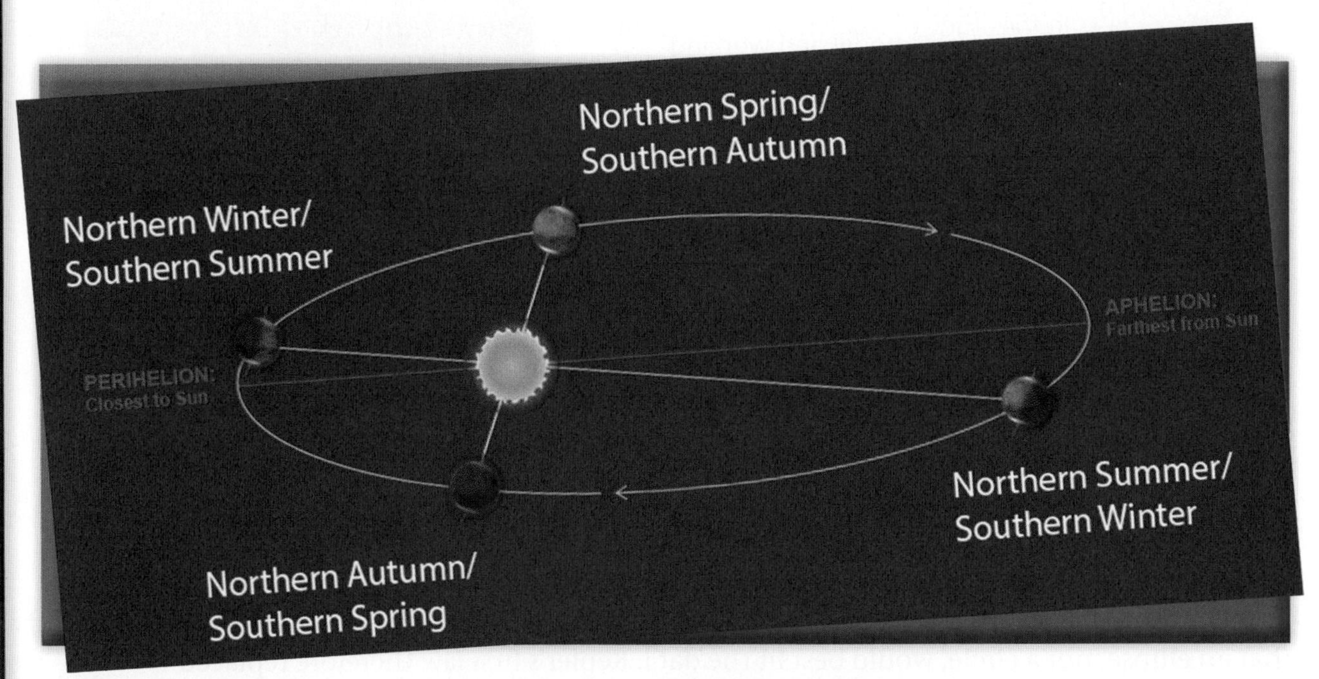

During its orbit around the Sun, Mars sweeps out a triangular area that is both wide and short. Near perihelion, the triangular shape is short and thick. Near aphelion, the triangular shape is long and narrow. Despite where the planet is in orbit though, the two triangular areas are equal in area.

Courtesy of NASAgov.

These observations led to Kepler's second law, stating that planets move fastest at perihelion and slowest at aphelion. Perihelion refers to *the planet's closest approach to the Sun*. Aphelion refers to *the planet's farthest distance from the Sun*.

Kepler's Third Law: The Implied Force of Gravity

What do we have left to uncover? It is the relationship between the time it takes a planet to orbit the Sun and its distance from the Sun. Kepler would be the first to suggest that a force (gravity) kept planets near the Sun. Kepler would also be the first to provide an equation for calculating the relationship between the time it takes a planet to orbit and its distance from the Sun.

Using Brahe's data, Mars has an orbital period of 1.881 years. Because we know the period, we can calculate the distance of Mars from the Sun. And given a distance, we can calculate an orbital period.

Similar to the astronomers before him, Kepler provided more data for a heliocentric model of the solar system, replacing circular orbits with ellipses and observing the changing speeds of the planets. He laid a foundation for ongoing scientific study that would continue the Scientific Revolution and challenge the views of the world at the time.

Did You Know?

Besides discovering three major laws of planetary motion, Kepler made major advancements in the study of modern day optics. He applied laws of light refraction to study the path of light through the eye. He successfully showed that an image is formed on the retina and that it is inverted. This discovery made it possible to explain other visual events that would further studies in astronomy and the treatment of disorders and diseases of the eye.

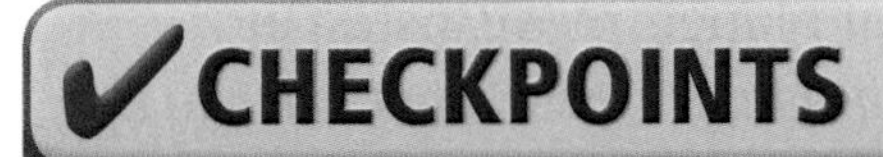

Lesson 2 Review

Using complete sentences, answer the following questions on a sheet of paper.

1. Why did Copernicus pursue a model different than Ptolemy's?
2. What was the main difference between the Ptolemy model and the Copernican model?
3. What was the Scientific Revolution?
4. What were the major flaws in the Copernican model?
5. What did Kepler's first law of motion prove regarding planetary orbits?
6. According to Kepler's second law of motion when do planets move fastest?
7. According to Kepler's third law of motion what keeps planets near the sun?

APPLYING YOUR LEARNING

8. Consider how scientific theories evolve over time and how scientists, like Kepler, used the observations of Copernicus and Tycho, to form new laws and theories. Think about a field of study or a career that you would like to pursue; describe innovations or discoveries that have already been made and how you would make improvements.

LESSON 3

The Enlightenment and Modern Astronomy

Quick Write

If scientists find a habitable planet in space, what do you think the impact on society would be? Would you want to be part of an expeditionary trip to a new planet?

Learn About

- Galileo and the telescope
- Newton's laws of motion and gravity
- Einstein and relativity

Work by early astronomers continues to influence society and modern astronomers, who build on the knowledge of early astronomers.

For example, in 2016, scientists made an exciting and monumental discovery. Just 4.2 light-years away from earth—extremely close when looking into space—a planet was found orbiting the star Proxima Centauri in the habitable zone. A light-year is how astronomers measure distance in space; it's how far a beam of light travels in one year: a distance of six trillion miles. Life could have once lived on, or may possibly still live on, this planet. The planet is dubbed Proxima b and is about 1.27 times the size of Earth.

So what next? How do scientists further explore this planet? The Breakthrough Prize Foundation—whose board members include Facebook founder Mark Zuckerberg, entrepreneur Yuri Milner, and, before his death, physicist Stephen Hawking— plans to send a microchip-sized spacecraft to target this new planet and its star. The plan is a hugely expensive operation and an engineering feat. When the spacecraft launches, it will take 20-25 years for the spacecraft to reach Proxima b. A lot of work is still required to make this expedition possible.

But just because the planet is in the habitable zone, doesn't mean that it is actually habitable. Scientists are using Earth as a model to determine what it might be like on the newly found planet. Scientists are studying the orbit of Proxima b and simulating that orbit, using Earth as their model. Early research shows that the atmosphere may not be ideal and may allow exposure to stellar radiation on the planet's surface. Only time and research will determine if Proxima b is habitable, but it's an exciting endeavor.

Vocabulary

- inertia
- acceleration
- gravitational force
- mass
- principle of equivalence
- space-time

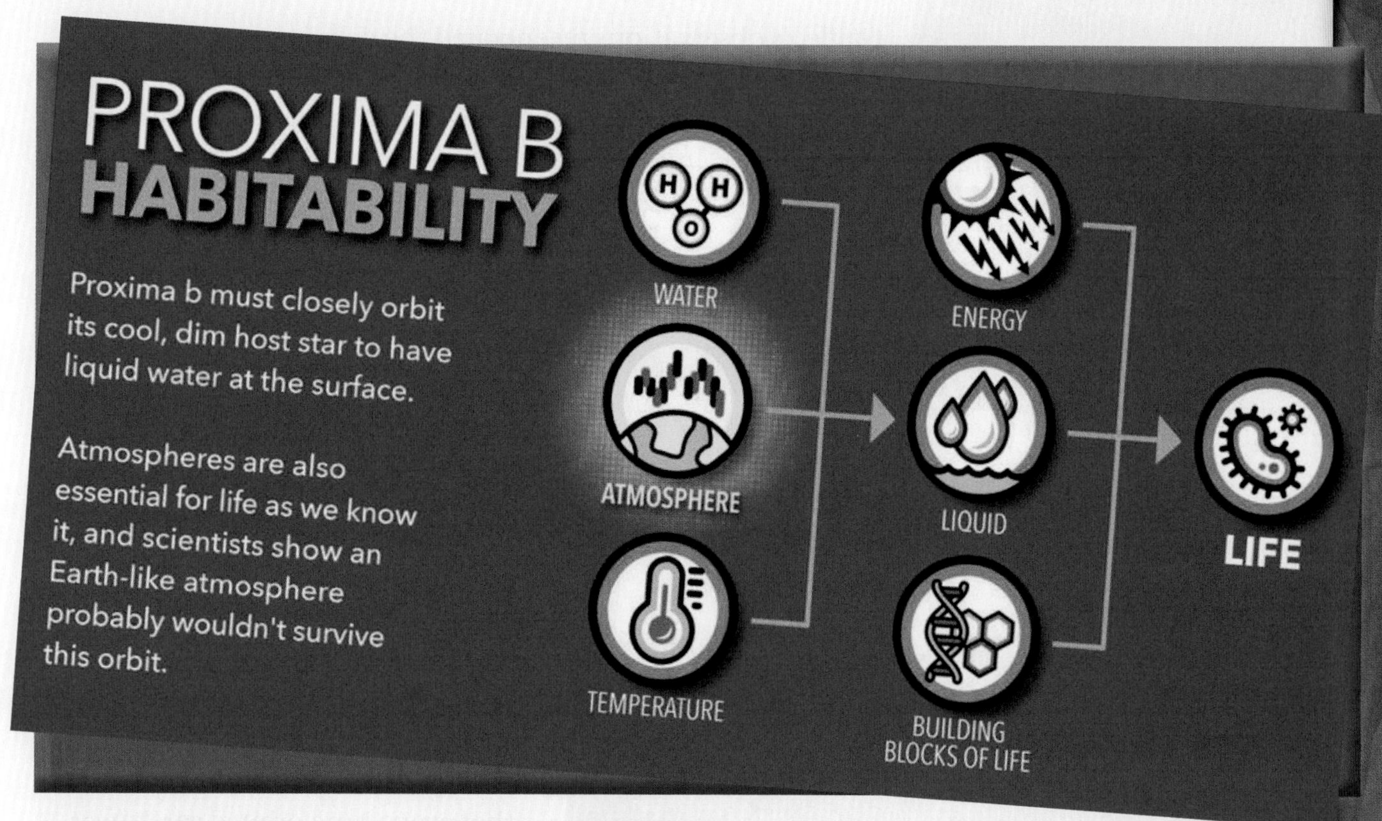

This image depicts the resources needed for habitability.

Courtesy nasa.gov.

Galileo and the Telescope

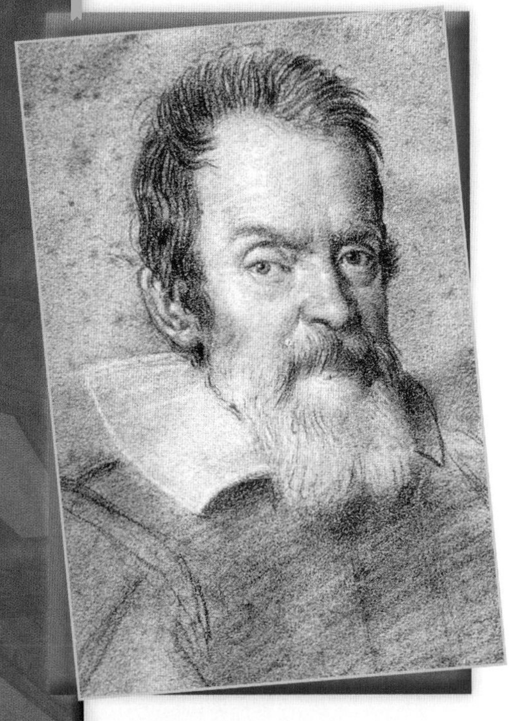

Galileo (1563-1642), portrait by Ottario Leoni ca. 1624

Everett Historical/Shutterstock

Galileo is often referred to as the father of modern astronomy because his observations revolutionized our views of the universe. Galileo di Vincenzo Bonaiuti de' Galilei was an Italian mathematician and physicist born in 1564. Galileo was fascinated by the newly invented "spyglass." He studied the device and built his own telescope in 1609, which could magnify objects 20 times greater than the spyglass. One of Galileo's biggest achievements was this improvement of the telescope.

So, what did Galileo see in his telescope when he viewed the stars?

He noticed that the Moon was not smooth, as previously thought; it had mountains and valleys, just like Earth. He saw many more stars than could be seen with the naked eye. And he observed the moons of Jupiter and Venus.

Galileo's observations regenerated the debate between the heliocentric and geocentric theories. The Ptolemaic model was integrated into Christian theology, and Galileo's observations of a heliocentric universe did not sit well with the Catholic Church.

Galileo's Observations

Jupiter is the largest planet in our solar system, so Galileo began observing Jupiter. He made a fascinating discovery as he studied Jupiter. The planet had four "stars" around it. As he continued to observe the stars, Galileo noticed that the "stars" were orbiting Jupiter and were actually moons. The four moons—Io, Europa, Ganymede, and Callisto—are now known as the Galilean moons. Remember, the Ptolemaic model, which assumed that everything orbited earth, was widely approved of at this time. Galileo's observations provided strong evidence for the heliocentric model.

This image depicts an artist's rendering of Jupiter and the Galilean moons.

Kirschner/Shutterstock

After studying Jupiter, Galileo turned his focus to Venus, which is the brightest celestial object in the sky, (other than the Sun and Moon). Venus is often referred to as "the evening star," even though it actually is not a star; it's a planet visible to the naked eye.

When Galileo viewed Venus through his telescope, he observed phases similar to the phases of our moon. As Venus orbits the sun, the sun's light reflects off of Venus, giving us the same phases we see in the moon. As Venus wanes from the full phase it appears to get larger to the naked eye because it is getting closer to Earth. The crescent view of Venus will actually look larger than the full view of Venus from Earth because Venus is closer during the crescent phase. The phases displayed by Venus could only be explained if Venus was orbiting the sun, rather than Earth. This observation provided more evidence for the heliocentric model. If Ptolemy's theory were accurate, then Venus would only appear on Earth as a crescent.

Winter Moon, Venus, and Silhouette of Lone Willow Tree, Reflected on Quiet Slough, Central California.

Terrance Emerson/Shutterstock

Did You Know?

Jupiter has 53 moons! The most well-known are Io, Europa, and Callisto. The largest moon in the solar system, Ganymede, also belongs to Jupiter.

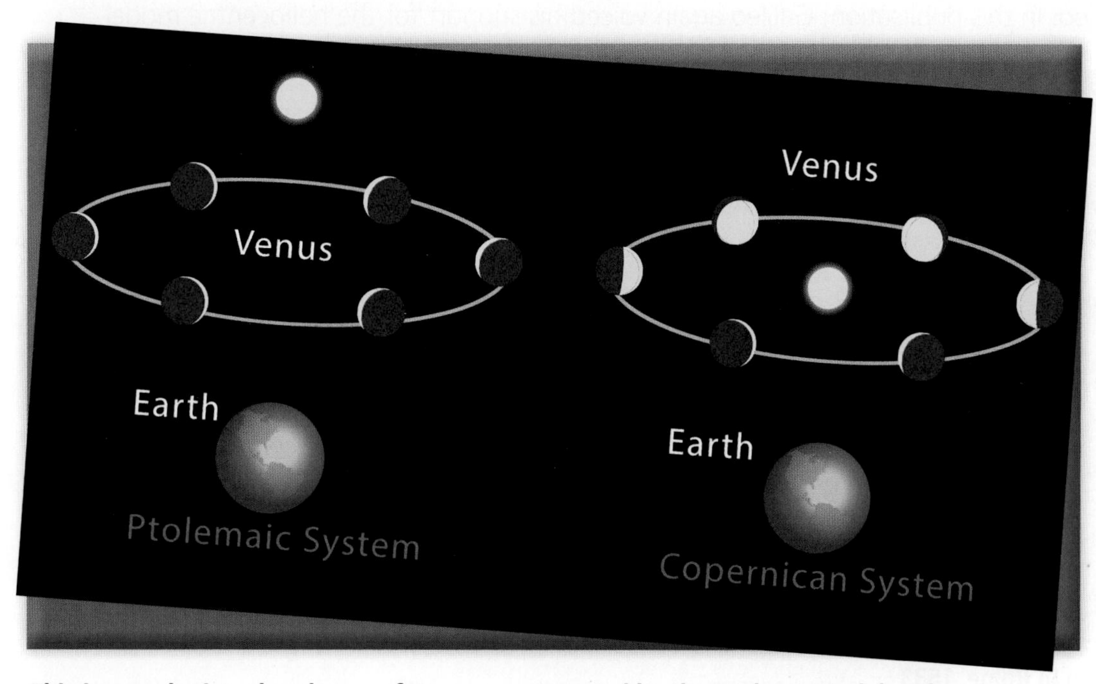

This image depicts the phases of Venus as presented by the Ptolemy Model and Copernican Model.

The Right Stuff

Galileo

Galileo Galilei was born in Pisa, Italy, on February 15, 1564. He was the oldest son of seven children. Vincenzo Galilei, Galileo's father, was an accomplished musician and taught his son how to play the lute. Although Galileo was a skilled lute player, he was torn between becoming a Catholic priest or a doctor. His father encouraged him to become a doctor, so Galileo began medical courses at the University of Pisa when he was 17 years old.

However, things soon changed for Galileo when he attended a mathematics lecture. Galileo was captivated by mathematics and saw it as a method to explain the universe. He then decided to study mathematics and physics because they interested him much more than medicine. Surprisingly, Galileo never completed his degree.

Galileo published his first book about hydrostatic balance at the age of 22. He then went on to teach art in Florence, but math was his passion, and he was soon awarded the Chair of Mathematics at the University of Pisa. There, he made his discoveries on Jupiter and Venus. As you've probably deduced, Galileo became a believer in the heliocentric model that Copernican proposed.

Galileo kept quiet about his beliefs for a long time, but in 1613 he published Letters on Sunspots that voiced his support for the heliocentric model. His book created quite the controversy. Remember, Galileo had considered becoming a Catholic priest and was involved in the Catholic church, yet the church had adopted the Ptolemy model and integrated it into their teachings.

In 1616, the Roman Catholic church had condemned Copernicus's book and banned members from supporting its writings. Many other religious leaders followed suit, such as Martin Luther and John Calvin.

In 1632, the Church in Florence approved the publication of Dialogue Concerning the Two Chief World Systems by Galileo. In this publication, Galileo again voiced his support for the heliocentric model.

The Trial of Galileo in Rome, 1633

Everett Historical/Shutterstock

In 1633, Galileo was summoned to Rome on charges of heresy based on his latest publication. Galileo was sentenced to life in prison; however, due to his age, the sentence was lessened to house arrest. He spent eight years on house arrest, until his death in 1642. However, house arrest did not stop Galileo from continuing his research and publishing his findings.

It wasn't until 1835 that the Catholic Church approved everything written by Galileo. In 1992 Pope John Paul II formally acknowledged the mistake of condemning Galileo.

Newton's Laws of Motion and Gravity

Sir Isaac Newton was born a year after Galileo's death, 1643. By the time Newton went to college, the scientific revolution was underway. Society was eager for new ideas. Newton built upon the work of Galileo, Kepler, and others to create the first unified model of how the universe works.

Did You Know?

Isaac Newton was responsible for inventing calculus and providing clarity in the field of optics.

Newton's research led to his laws of motion and gravity. Newton's research carries an enormous influence on society and the exploration of astronomy and space. His contributions to humanity cannot fully be weighed as they are too great. His research also helped set the stage for Einstein, who further explored the concepts of gravity, light, and matter.

Waypoints

Newton knew his accomplishments would not be possible without the work of early astronomers. He is quoted as saying, "If I have seen further, it is by standing on the shoulders of giants."

Newton's Three Laws of Motion

In AS200, we talked a bit about how aircraft actually fly. More than 300 years ago, Sir Isaac Newton discovered the principles of motion that make flight possible, while he was working in mathematics and physics. In 1687, Newton published his findings in Philosophiae Naturalis Principia Mathematica.

Sir Isaac Newton (1643-1727) examining the nature of light with the aid of a prism.

Everett Historical/Shutterstock

Newton's First Law of Motion

Waypoints

"Every object persists in its state of rest or uniform motion in a straight line, unless it is compelled to change that state by forces impressed on it."
– Sir Isaac Newton

Newton's first law of motion states that every object will remain at rest or in uniform motion in a straight line unless compelled to change by the action of an external force. You may have heard this referred to as inertia. Inertia *is the tendency of an object to resist a change in motion*. Newton's first law is often referred to as the Law of Inertia.

Newton's first law is an observation of cause and effect. If there is zero velocity or an absence of force, then an object will stay at rest; an object with motion will stay in motion at a constant speed. A simple example can help explain the first law of motion. Let's say you are told to walk down the street going straight. If you are walking in a specific direction, you will always go that direction unless you are compelled to change. So, unless you tell your body to stop or turn, you will just keep walking in that direction. The same can be said if you are standing still. If you are standing still and nothing happens to you, you will continue to stand still. This is a very basic example of Newton's observations.

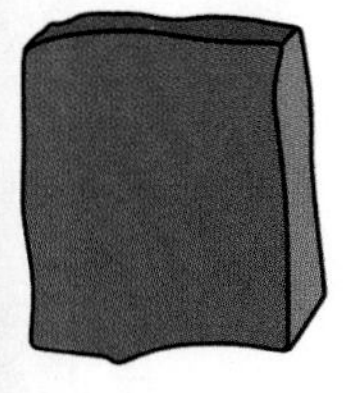

This image provides examples of Newton's first law of motion.

udaix/Shutterstock

Newton's Second Law of Motion

Newton's second law explains what happens when an external force is introduced. Newton found that acceleration (*the act of increasing speed or velocity*) of an object is directly related to the magnitude of force that is applied to the object. Newton developed a formula to calculate the amount of force required to produce a specific acceleration. If force is already known, the formula can be used to determine the acceleration rate. We can look at the formula in two ways:

Force = mass x acceleration ($F = ma$)

OR

Acceleration = Force / mass ($a = F/m$)

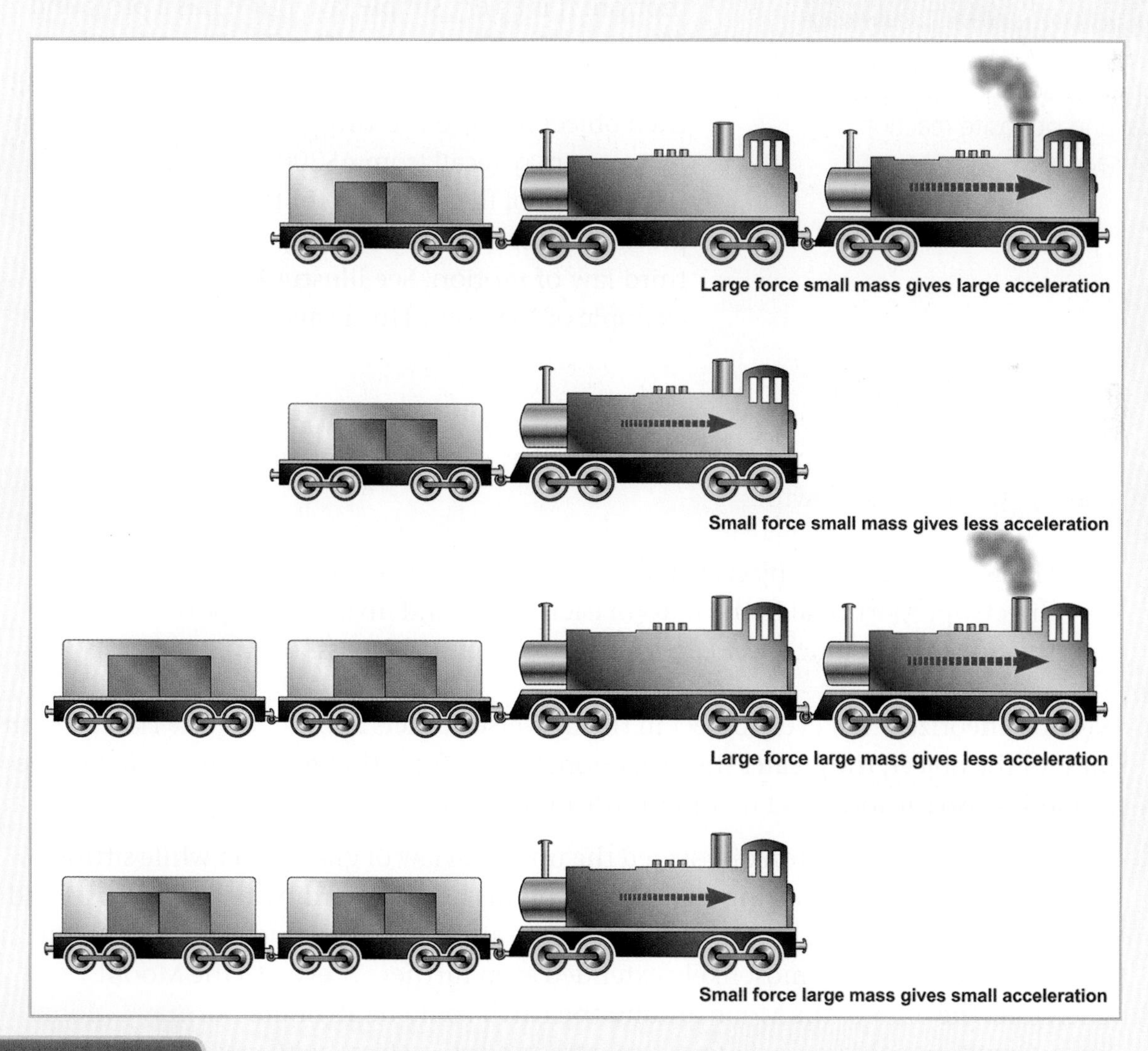

FIGURE 2.1

This image provides examples of Newton's second law of motion.

Fouad A. Saad/Shutterstock

"Force is equal to the change in momentum (mV) per change in time. For a constant mass, force equals mass times acceleration."
– Sir Isaac Newton

Look at the trains in Figure 2.1. In this figure, we can see the difference in acceleration, force, and mass. In the first train, we have a large force and mass, so we get a large acceleration. In the second train, we have a small force and small mass, so we have less acceleration. In the third train, we have a large force and mass, which gives us less acceleration than the first because of the larger mass. In the fourth train we have a small force with a large mass, so our acceleration is also less.

"For every action, there is an equal and opposite reaction."
– Sir Isaac Newton

Newton's Third Law of Motion

You have probably heard of Newton's third law of motion. It is a very simple law, but it has a profound effect on scientific inquiry. The third law of motion basically says that if object A exerts force on object B, then object B is also exerting an equal force on object A. As you may recall from AS200, this law explains the concept of flight. The lift of the wing and the production of thrust by the engine demonstrate the third law of motion. See illustration on page 43 for example of Newton's Third Law.

The Law of Gravity

In his publication, Newton also described his law of universal gravitation or the law of gravity. According to Newton:

> "Between every two objects there is an attractive force, the magnitude of which is directly proportional to the mass of each object and inversely proportional to the square of the distance between the objects' centers of mass."

Newton theorized that every object in the universe attracts another object. The larger the mass of the object, the greater the attraction. So, the force that makes objects fall to the ground is also the force that maintains the orbit of planets.

There is a story that Newton discovered the universal law of gravitation while sitting under an apple tree and having an apple fall on his head. According to Newton's second law of motion, the force, or "gravity," pulled the apple to the ground. Newton figured that this force of gravity most likely extended even further . . . even to the Moon! Therefore, the orbit of the Moon around the Earth could be attributed to gravitational force. Gravitational force *is the force that attracts any two objects with mass*. This same principle can be applied to humans. Earth exerts a force (weight) that keeps you from floating off into outer space.

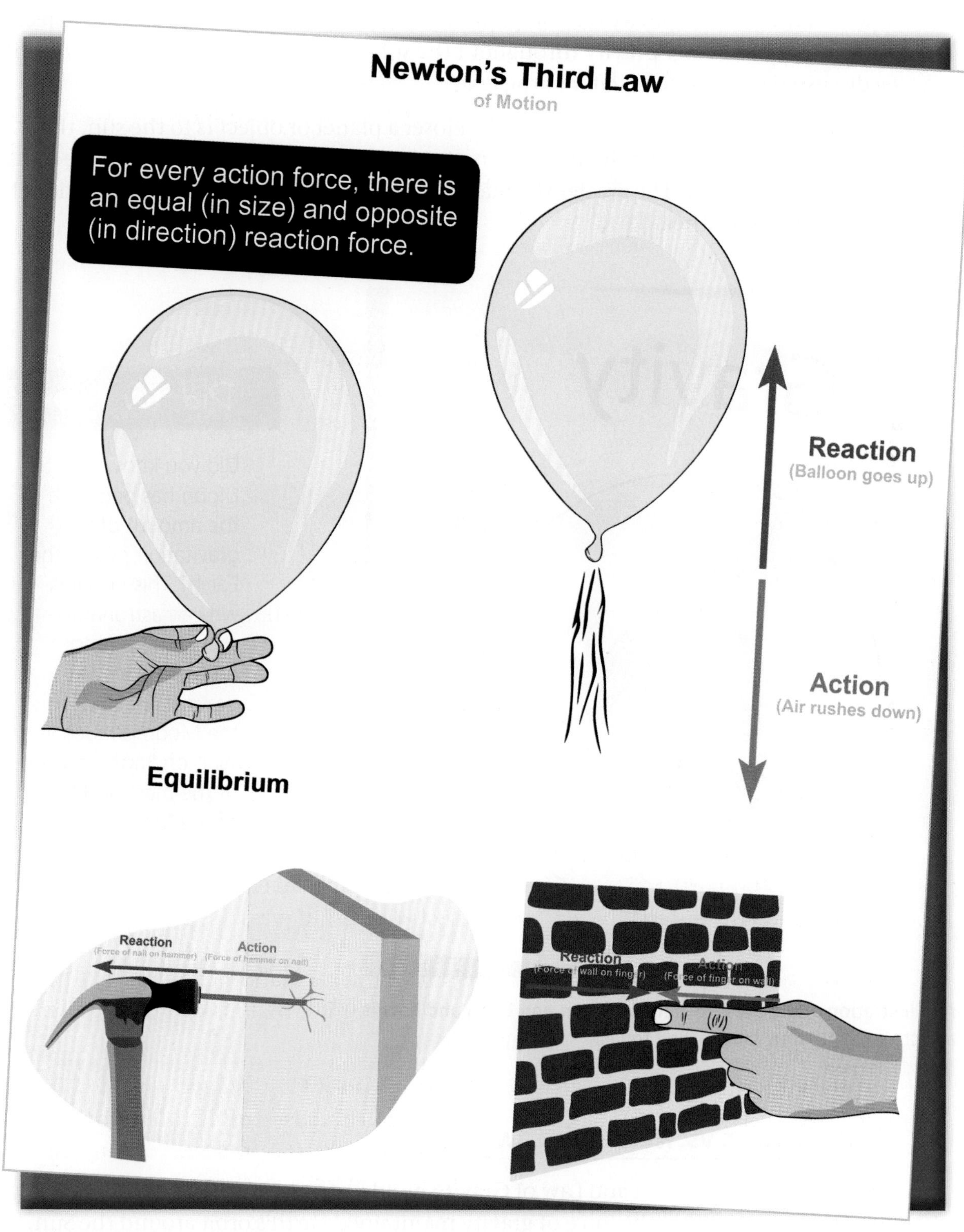

This image provides an example of Newton's third law of motion.
udaix/Shutterstock

Newton's theory of gravity explains many occurrences in space that were unable to be explained. For example, Newton's theory explains that planets travel at different speeds based on the force of gravity on that planet. And gravity is used to explain why objects speed up when falling. Gravity throughout the solar system affects space travel and will be discussed in more detail later in the text.

Newton's theories also explained that the closer a planet or object is to the sun, the greater the gravitational pull. In addition, the gravitational force depends on the mass (*weight*) of the object. The greater the mass of the objects, the greater the gravitational pull toward each other.

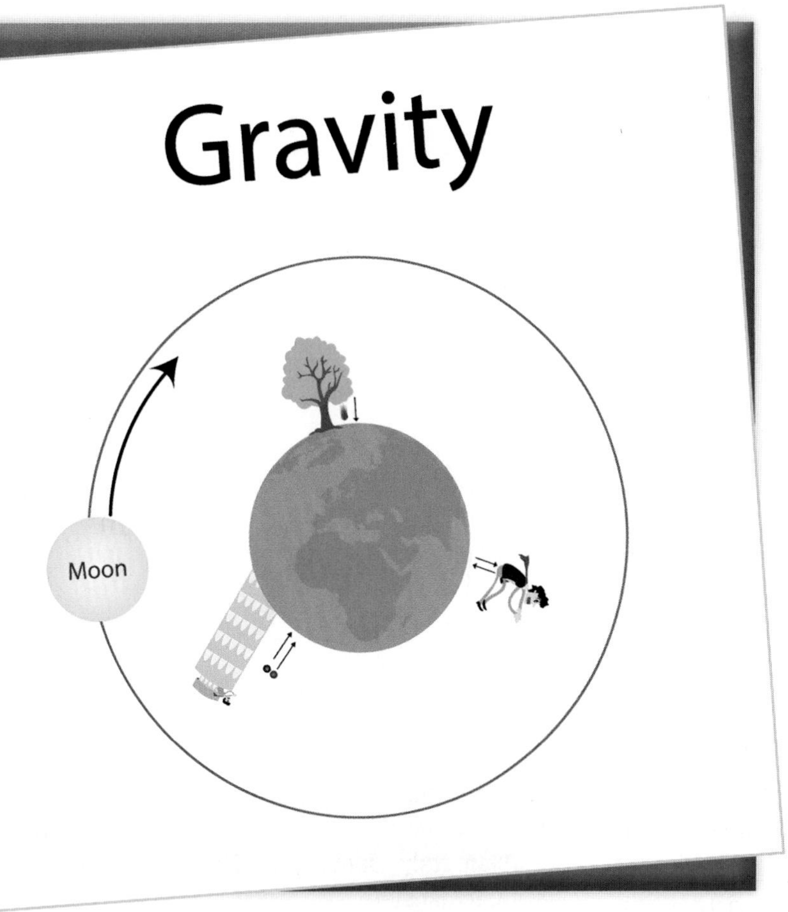

An illustration of gravity between the Moon and Earth and how it affects an apple tree, human, and tower.
Nasky/Shutterstock

Did You Know?

Did you know that the Moon has one-sixth the amount of gravitation pull as the Earth? This explains why an astronaut seems to "bounce" on the Moon. The gravitational pull on the Moon is much less than on Earth because of the mass and size of the Moon.

Kepler's Laws vs. Newton's Laws

Newton's Laws of Motion and Law of Gravity build upon the work of Kepler. Kepler was the first to suggest that the force of gravity maintained Earth's orbit around the Sun, but his findings were based on observation. Kepler was able to describe the motions without an explanation of why they moved a certain way. Newton was able to prove these findings and expand on them using principles of nature.

Einstein and Relativity

Albert Einstein published his special theory of relativity in 1905. According to Einstein, as objects approach the speed of light, common sense is unreliable. Einstein found that the speed of light is always the same. This observation is hard to understand since the prediction cannot be observed. The consistency of the speed of light is now considered a law of nature. Scientists have proven this theory many times and have shown that Newton's laws of motion are inaccurate as speed increases.

Einstein also made predictions on what happens when objects approach the speed of light. Einstein concluded that mass can become energy, which is represented mathematically by the equation E=mc2. The first explosion of a nuclear bomb confirmed Einstein's equation in 1945. Another example is the conversion of nuclear power to electricity in a power plant.

> **Did You Know?**
>
> The speed is light of 186,000 miles per second. This is 10,000 times faster than the fastest rocket. If travelling at the speed of light, a rocket could go around the world 7 times per second.

Einstein and Gravity

In 1916, Einstein published his General Theory of Relativity. Einstein predicted that the light from the stars would be affected by gravity. Einstein also stated that gravity and acceleration are equivalent. The principle of equivalence is *the fundamental law of physics that states that gravitational and inertial forces are of a similar nature and often indistinguishable.* According to Newton, this means that if a person was free falling in a windowless building and the building was also moving, then the person would not be aware of the buildings movement. Using Einstein's theory, the person would not know if the free fall was caused by the fall or the movement of the building.

Milan, Italy - January 13, 2017: Scientist Albert Einstein
spatuletail/Shutterstock

Einstein coined the term space-time, which refers to *a combination of space and time.* Einstein viewed the universe as sitting on an imaginary rubber sheet. When an object that is heavy is sitting on the rubber sheet, the object creates a dip that can affect other objects. If we look at the sun, the sun is a large object and causes a dip in the space-time continuum. This dip created by the sun can affect other objects, such as a comet's path or a star's light. Space-time is a difficult concept to visualize and the rubber sheet was the method that Einstein used to visualize the affect of gravity on other objects.

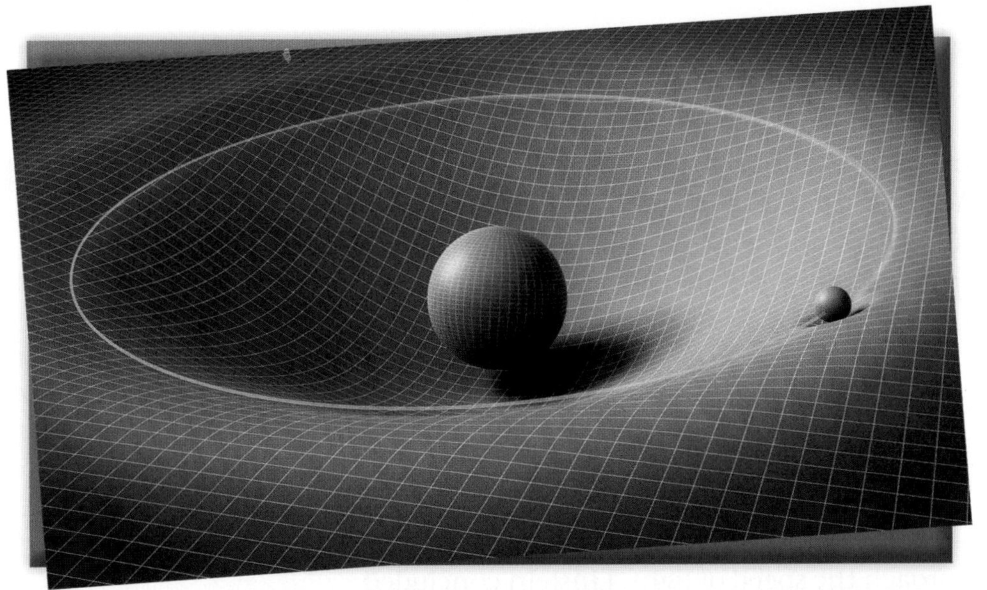

This image depicts the "rubber sheet" and the effect of an object's mass on gravity.

posteriori/Shutterstock

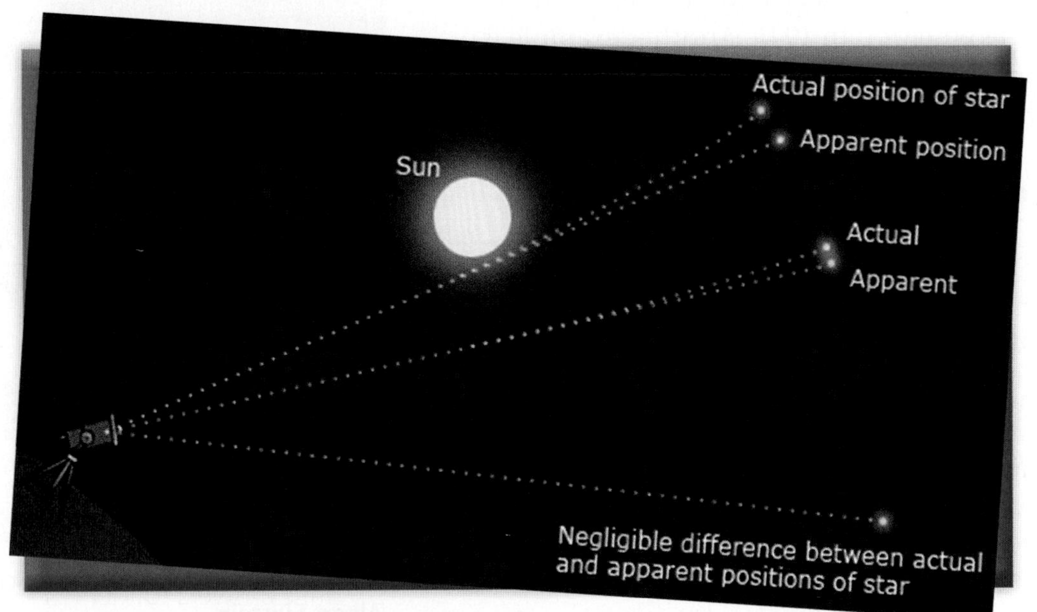

According to Einstein, the light from stars that passed closest to the sun would show the greatest degree of bending.

Courtesy of Wikipedia Commons

Einstein's General Theory of Relativity was proven by British astronomer Sir Arthur Eddington on May 29, 1919. Eddington decided to lead an experiment to test Einstein's theories during a solar eclipse. To prove Einstein's theory, scientists would need to observe the size and position of the stars closest to the sun. This proved impossible during normal circumstances, but during a total solar eclipse scientists could study these stars. Eddington set out to study the background stars around the sun during the 1919 solar eclipse and directly observe the effects of the warped space-time. The experiment was a success and proved Einstein's theory. It also superseded Newton's theory of inert space.

The Right Stuff

Stephen Hawking

Stephen Hawking was an English physicist who worked to combine Einstein's theory of gravity with quantum physics. He is an example of a modern-day scientist who built upon the work of those prior to him.

Hawking was born in in 1942. His father urged him to study medicine, but Hawking choose to study physics at Oxford University. He then continued his studies of the universe at the University of Cambridge.

In 1963, Hawking was diagnosed with a motor neuron disease and given less than two years to live. Hawking defied all odds, completing his PhD and going on to make many discoveries. As the disease spread, Hawking was confined to a wheelchair and, in 1985, lost the use of his voice. Using a speech-generated software program developed at Cambridge, Hawking was able to continue his research and speak on his findings.

Hawking's work with Einstein's theory of relativity led to the determination that Einstein's theory of space-time begins with the creation of the universe–the Big Bang–and therefore ends within black holes. Hawking researched black holes extensively and made many discoveries throughout his life.

Stephen Hawking passed away in March of 2018 at the age of 76. He lived 55 years after the diagnosis that gave him 2 years to live. He is considered one of the most brilliant theorists to ever live.

Stephen Hawking (1942-2018)
The World in HDR/Shutterstock

"All of my life, I have been fascinated by the big questions that face us, and have tried to find scientific answers to them. If, like me, you have looked at the stars, and tried to make sense of what you see, you too have started to wonder what makes the universe exist." - Stephen Hawking

The Right Stuff

Neil deGrasse Tyson's passion for astronomy can even be seen in his clothing. He's known for his colorful, astronomically themed vests.
Kathy Hutchins/Shutterstock

Neil deGrasse Tyson

We would be remiss to not discuss the astrophysicist Neil deGrasse Tyson. Tyson was born in 1958 in Manhattan and developed a passion for astronomy. Tyson's research focuses on stellar evolution, cosmology, galactic astronomy, and stellar formation, but he is perhaps best known for his jovial personality and passion for astronomy. Tyson has created a popularity for science that had been lacking in today's society. Tyson can regularly be seen hosting programs on PBS to educate others on astronomy and physics. Tyson has the ability to interpret very complicated theories and concepts so that the average person can understand the universe. His excitement for astronomy and physics is contagious, and he's become a popular advocate of astronomy.

Although most people still view the world using Newton's theories, scientists studying the universe have replaced Newton's theories with Einstein's theories. The understanding we have of the universe today would not be possible without Einstein. And, as scientists continue to explore the universe, a new theorist may develop theories that will replace Einstein's.

Lesson 3 Review

Using complete sentences, answer the following questions on a sheet of paper.

1. What was one of Galileo's biggest achievements?
2. What are the names of the Galilean moons?
3. What is Venus often called?
4. What is Newton's first law of motion?
5. What is Newton's formula for calculating force?
6. What does Newton's Law of Gravity theorize?
7. Who was the first to suggest that the force of gravity maintained Earth's orbit around the Sun?
8. How did Einstein view space-time?
9. Who was able to prove Einstein's theories in 1919?

APPLYING YOUR LEARNING

10. Using a modern example, provide a brief description for one of Einstein's theories in action.

CHAPTER 2

Panorama view of the universe with a space shot of the Milky Way galaxy and stars on a night sky background. The Milky Way is the galaxy that contains our Solar System.

Nuamfolio/Shutterstock

The Solar System

Chapter Outline

"Our solar system is actually a wild frontier, teeming with different, diverse places: planets and moons, millions of objects of ice and rock."

Carrie Nugent PH.D.
American Scientist and Asteroid Hunter

LESSON 1

The Earth and Moon

Quick Write

After reading the vignette about the composition of key ingredients that allow Earth to support life, which of the ingredients listed do you think is the most critical to life on Earth as we know it?

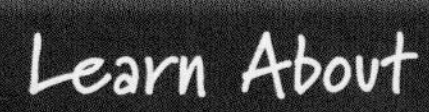

Learn About

- Earth – inside and out
- the Moon

Earth is a unique planet. It's currently the only one in our solar system that can sustain life as we know it. What makes Earth special? Let's examine the key ingredients that make life on Earth possible. First, we have the Moon; because of its stabilizing effect on our orbit, our climate is more stable and predictable. Secondly, the Earth's rotation is very stable; this regular rotation and frequency of day and night helps prevent extreme temperatures and promotes life.

Earth also has a strong, stable magnetic field. Without this magnetic field, we would be bombarded with cosmic rays and solar flares that would expose life on Earth to deadly radiation. Earth receives additional protection against radiation from the ozone layer. This layer of gas protects our atmosphere, which is made up of nitrogen, oxygen, argon, and carbon dioxide gases and provides the air we breathe. Earth's ozone layer not only absorbs solar radiation but also reduces temperature extremes between day and night.

Earth's Sun, which is called a yellow dwarf, is a rare type of star that's small and stable. Because the Sun is stable, it has a long life and is not expected to burn out for another five billion years. Because our Sun has such a long life and the Earth is around four billion years old, life on Earth was allowed to develop and flourish. Large stars tend to burn hotter and die off sooner, while smaller stars tend to expend large amounts of radiation.

Earth – Inside and Out

What are the components that make up Earth? What does Earth need to sustain living objects? Earth is in what scientists call the Goldilocks Zone. The Goldilocks Zone is *the habitable zone around a star in which planets can sustain life*. Earth is the only planet in the solar system with a large quantity of water on its surface and an atmosphere. In fact, 70% of Earth's surface is covered with water. Earth is the third planet from the Sun. An entire orbit around the Sun takes approximately 365 days.

The Earth itself is an extremely dense object. Density is *a measure of how much material (or mass) is packed into a given volume*. The density of an object can give us a clue as to its composition. Earth happens to be a large, rocky sphere hurtling through space. The spherical shape is due to the force of gravity on the Earth. The force of gravity crushes rock and pulls the Earth into its spherical shape. This process took millions of years. However, because of inertia, Earth is not a perfect sphere. Remember Newton's theories on inertia? Newton had suggested that the Earth's spin may cause the equator to bulge, and, indeed, it does. The result of inertia on the Earth is a bulge at the equator.

Composition of Earth

There are three main layers that make up the Earth. The core, mantle, and crust. The core can be thought of as an inner core and outer core. The inner core is a solid metal core that is comprised mostly of iron. The temperature of the inner core is hot enough to melt metal, but remains solid due the extreme pressure on the inner core. The outer core is a liquid layer made of iron and nickel. This liquid portion of the core creates the Earth's magnetic field.

Vocabulary

- Goldilocks Zone
- density
- plates
- margins
- seismic waves
- continental drift
- plate tectonics
- Mid-Atlantic Ridge
- ozone layer
- northern hemisphere
- southern hemisphere
- equinox
- magnetic field
- convection
- synchronous rotation
- perigee
- apogee
- tides
- Spring tides
- Neap tides
- precession

The Earth's mantle is partly liquid and makes up approximately 67% of Earth's mass. The crust is a rocky layer that is 5-30 miles deep. The crust is made up of various plates or *broken pieces*. As the plates move, the landscape on Earth can change. The *edges of the plates* are called margins, and when the margins crash together we end up with volcanic eruptions and earthquakes.

Through earthquakes, scientists can learn about the interior of the Earth. Seismic waves are *waves of energy that travel through the Earth during an earthquake*. The waves travel at different speeds, based on the material they are passing through. The speed of the waves helps scientists determine the type of material in the interior of the Earth.

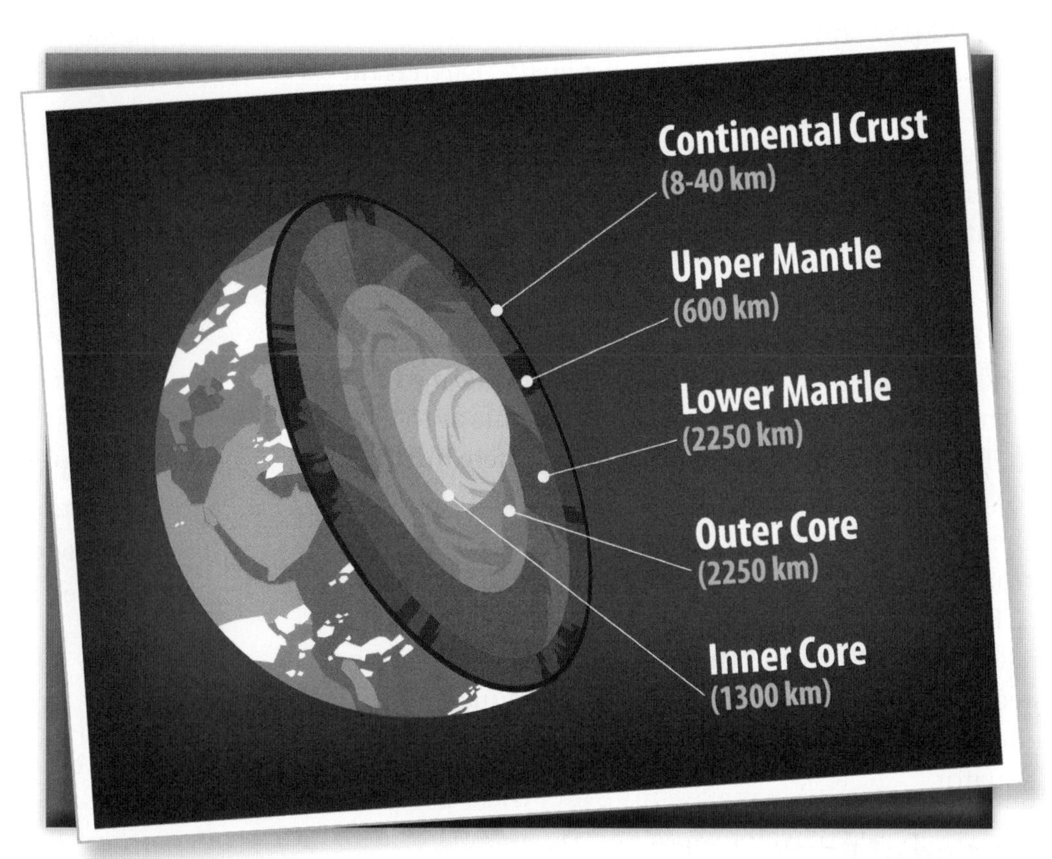

This diagram shows a section of Earth's crust and each layer.
Puslatronik/Shutterstock

In 1912, Alfred Wegener proposed the concept of continental drift, *the gradual movement of the Earth's surface*. He noticed that the coast of Africa and South America appeared as though they could join together like puzzle pieces. This is now known as the science of plate tectonics, *the theory used to explain the structure of the Earth's crust and its movement over time*. The shift of the Earth's surface is driven by hot magma rising and cool magma sinking inside Earth.

Before the continental drift, most of the continents were one large mass on Earth. After the continental drift, the continents started separating and the Earth took on its present form.
robin2/Shutterstock

Earth's Atmosphere

Earth is the only planet in our solar system with a breathable atmosphere for humans. Earth's atmosphere is made up of a mixture of gases. Oxygen is what we breathe and makes up 21% of the atmosphere. Nitrogen makes up 78% of the atmosphere; the last 1% is composed of water vapor, carbon dioxide, and ozone.

Earth's atmosphere is divided into layers:

- Troposphere – This layer extends 5-9 miles from the Earth's surface. This is the densest of all the layers and is where weather occurs.
- Stratosphere – This layer starts above the troposphere and extends about 31 miles high from the Earth's surface. The ozone layer, which *protects us from solar ultra-violet (UV) rays*, is included in the stratosphere.
- Mesosphere – This layer extends 53 miles from the Earth's surface. Typically, meteors that spiral toward the Earth are burned up in this layer.

Did You Know?

The Mid-Atlantic Ridge *is an area in the Atlantic Ocean where lava flows upward creating new mountains underwater*. The ridge supposedly caused the Americas and Euro-African land masses to split apart. In fact, the ridge continues to create new mountains underwater and pushes Europe and North America an inch farther apart each year.

- Thermosphere – This layer located above the mesosphere and below the exosphere and home to the international space station.
- Exosphere – This is the upper limit of the atmosphere.

The environment in each of these layers is vastly different. Figure 2.1 outlines the different in temperature and features of each layer of the atmosphere.

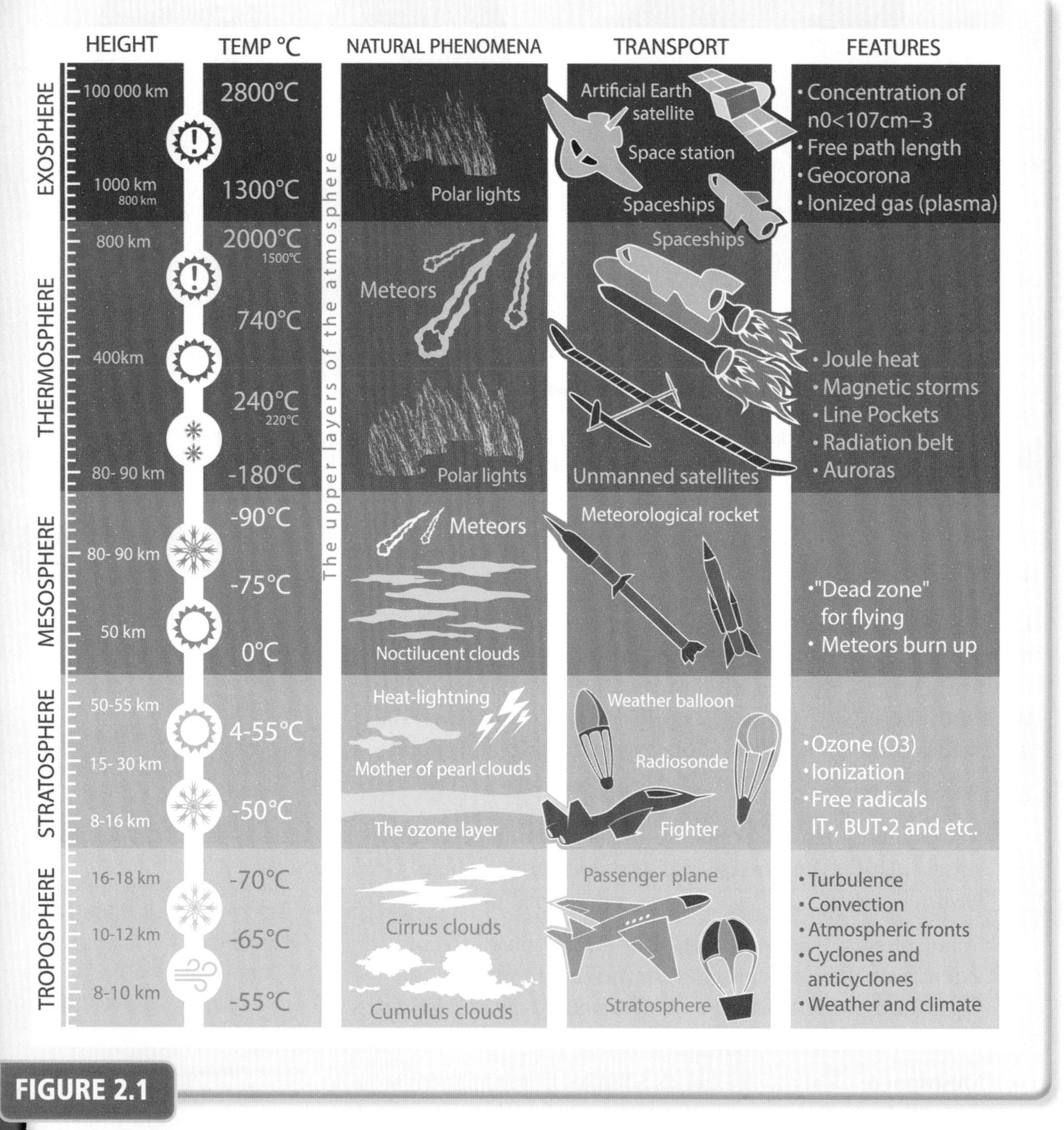

FIGURE 2.1

This diagram compares the structure of the Earth's atmosphere and the environment within each layer.
Panacea Doll/Shutterstock

The Seasons

Earth is divided into two hemispheres. The northern hemisphere is *the half of the Earth that lies north of the equator*. The southern hemisphere is *the half of the Earth that lies south of the equator*.

Earth rotates at a tilted angle on its axis. The tilt of the Earth determines the seasons in the northern and southern hemispheres. During the summer, the northern hemisphere is tilted closer to the Sun, which means the Sun is in the sky longer during the day. This causes the temperature to be warmer during the day, with less chance for the temperature to cool during the nights because the nights are shorter. The northern and southern hemispheres always experience the opposite seasons. For example, because of the tilt, when it is summer in the northern hemisphere, it is winter in the southern hemisphere. So, if you lived in Australia, you would be wearing shorts and enjoying summer in December!

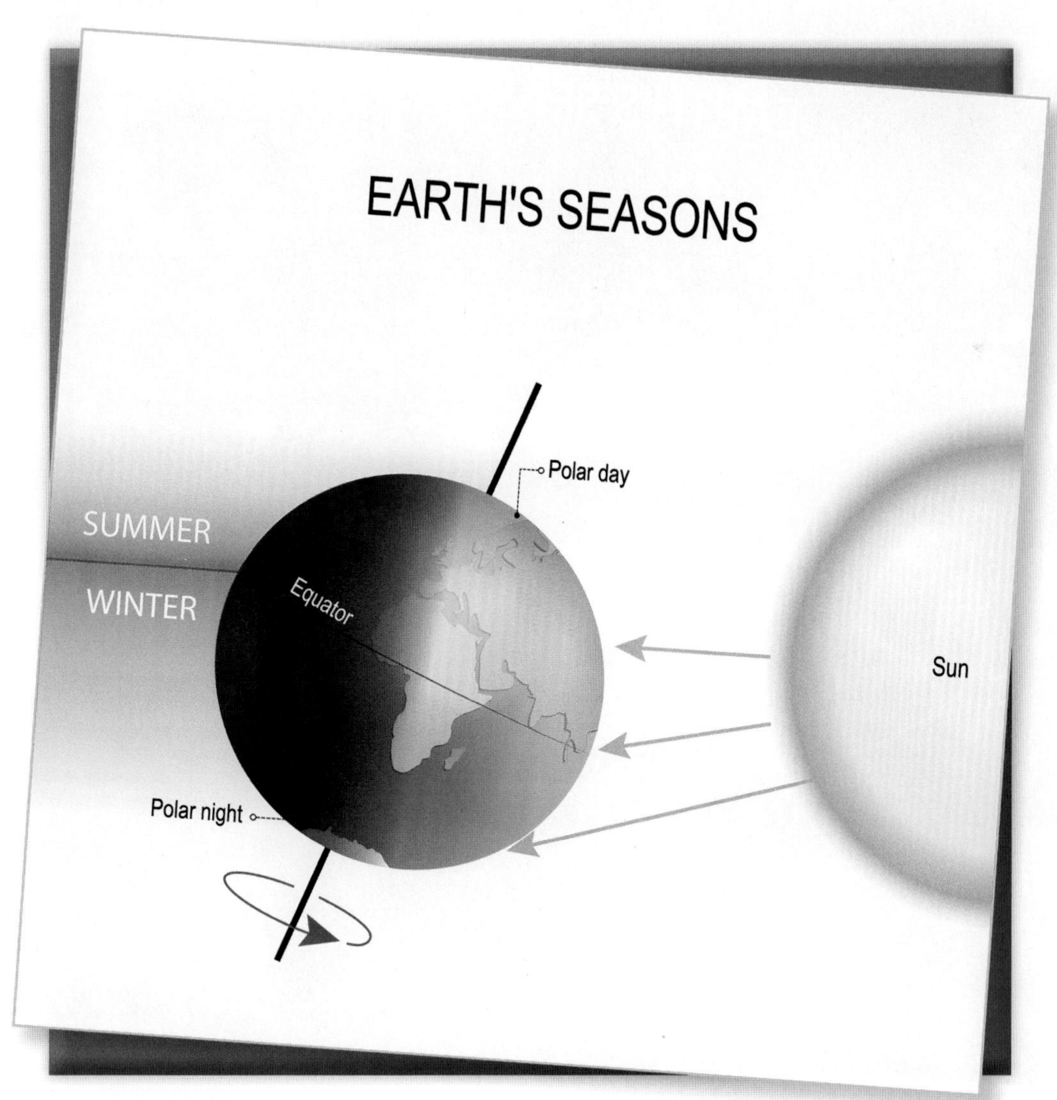

The tilt of the Earth's axis determines the season in the northern and southern hemispheres.

Designua/Shutterstock

The solstice occurs twice a year and is the time when the Sun is at its highest and lowest point in the sky. The summer solstice occurs on or about June 21st for the northern hemisphere; it is the longest day of the year because the Sun is at the highest altitude from the equator. The winter solstice occurs on or about December 21st in the northern hemisphere; it is the shortest day of the year. In the southern hemisphere, these two dates are reversed.

Another key concept to discuss is the occurrence of equinoxes. The equinoxes are *the points where the Sun crosses the celestial equator*. The equinoxes also help mark the changing of the seasons. On or about March 21st, we experience an equinox in the northern hemisphere; this date also represents the first day of spring. On or about September 22rd, another equinox occurs in the northern hemisphere, marking the first day of autumn. Again, these dates are reversed in the southern hemisphere.

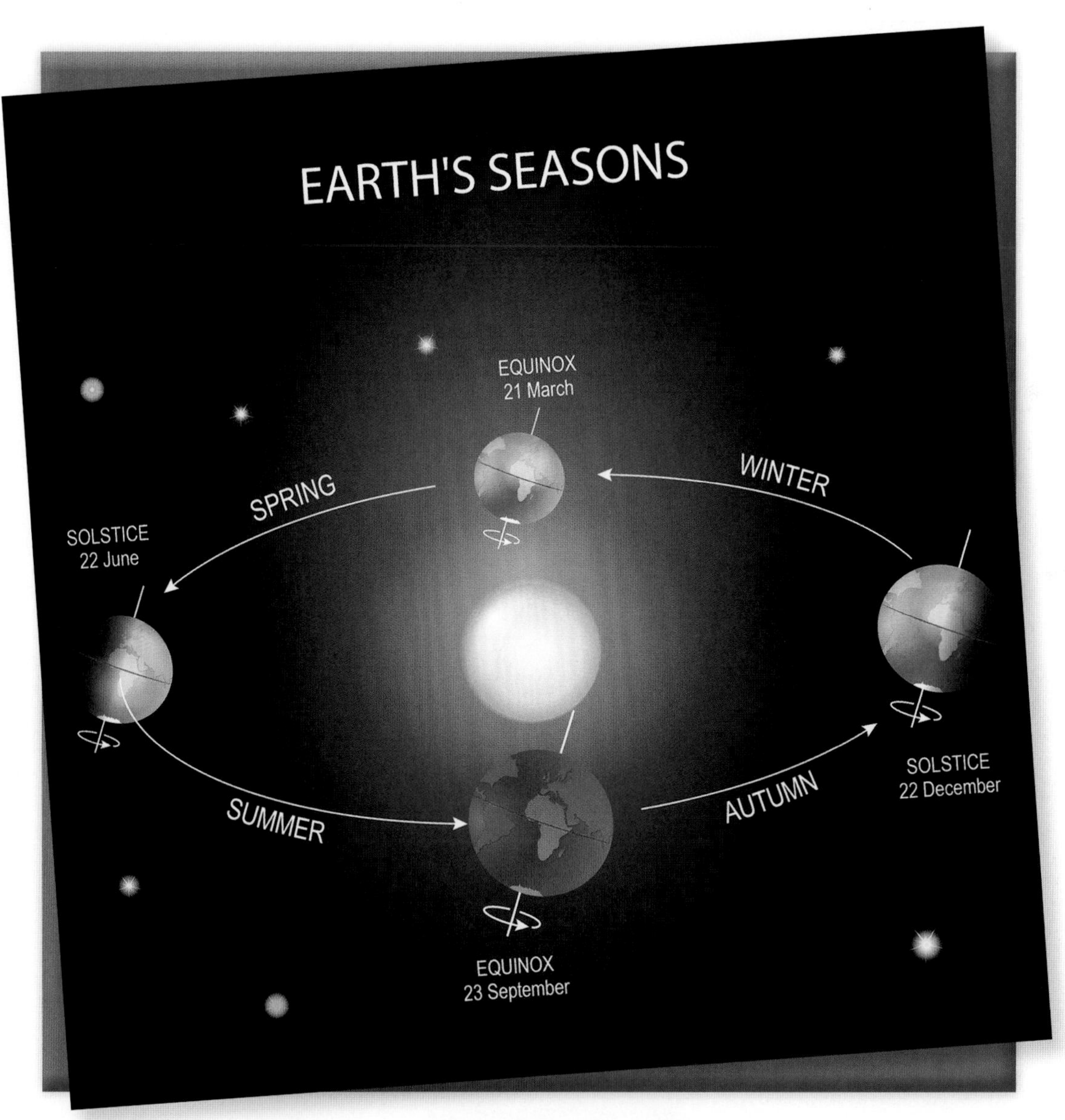

The effect of the solstices and equinoxes on Earth's seasons.
Designua/Shutterstock

Earth's Magnetic Field

Earth itself is a magnet and has a magnetic field in which *magnetic forces are communicated*. The compass is a tool that uses the Earth's magnetic field to determine which direction is north. Using the Earth as a magnet, the compass employs the magnetic pull of the North Pole to identify north to the reader.

In Figure 2.2, you can see the magnetic poles of the Earth close to the North and South poles. The magnetic poles of the Earth don't align directly with the geographic poles of the Earth. You'll also notice that the North Pole is actually the south magnetic pole and the South Pole is the north magnetic pole. The magnetic field lines represent the direction a compass would point when using the magnetic field. The more lines you see in a magnetic field drawing, the stronger the magnetic field.

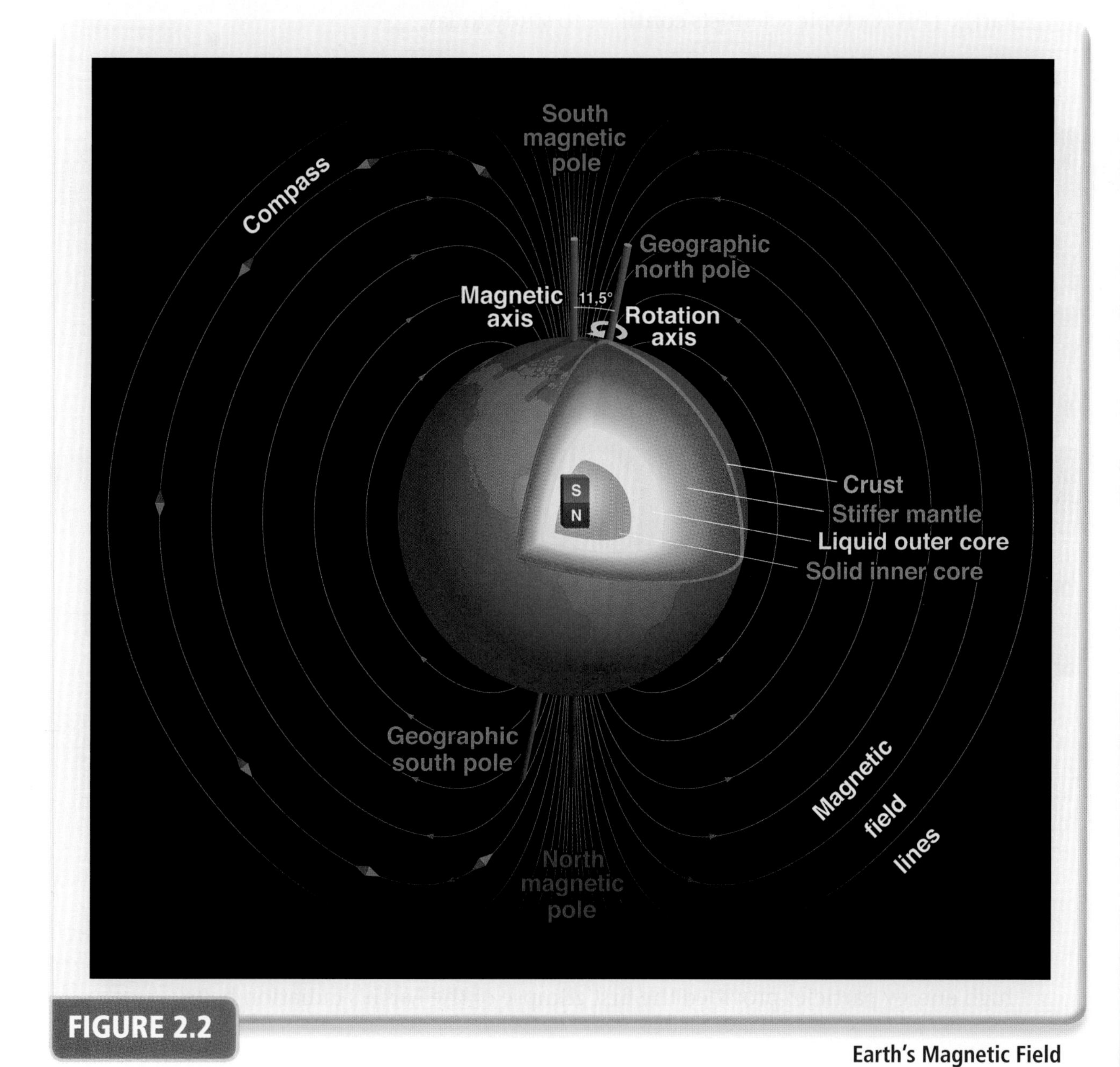

FIGURE 2.2

Earth's Magnetic Field

Peter Hermes Furian/Shutterstock

The magnetic field gets its origins from the electric currents flowing in the molten iron core of the Earth. Scientists aren't quite sure how these currents originated but suspect a combination of the rotational motion of the Earth and convection, *the generation of heat.*

Earth's magnetic field is a shield for us on Earth. It is constantly deflecting the subatomic particles that are being sent from the Sun. These particles could greatly damage the atmosphere and life on Earth; we have the magnetic field to thank for keeping us protected from these particles.

However, magnetic fields do not last forever. Earth's magnetic field is diminishing over time, and it is estimated that in 15,000 years it will disappear completely. Magnetic fields can also reverse. Over the past 170 million years, it is believed that the Earth's magnetic field has reversed over 300 times. The last reversal was 780,000 years ago. It is impossible to predict when the next reversal will occur, or what a reversal would mean for life on Earth. This is a topic scientists continue to study today.

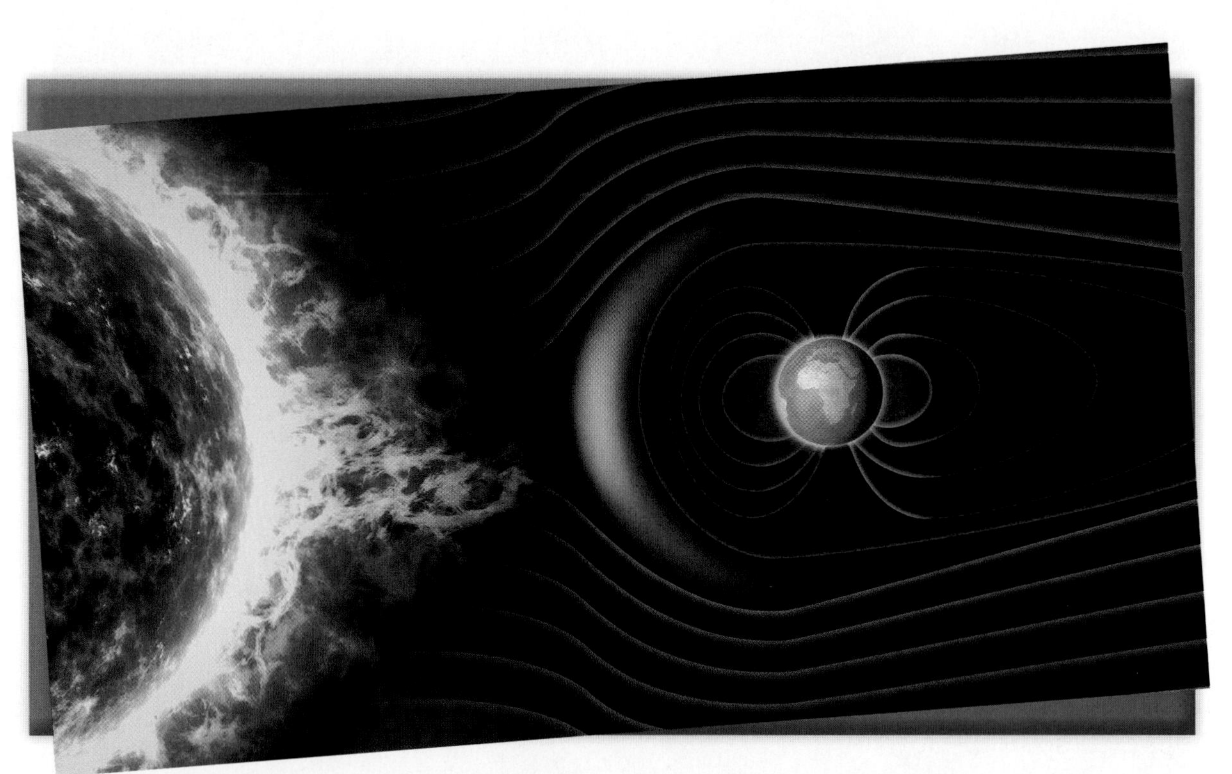

Earth's magnetic field, the Earth, the solar wind, the flow of particles deflected by the Earth's magnetic field. Elements of this image furnished by NASA.

Naeblys/Shutterstock

The Van Allen Belts

In 1958, the United States launched its first satellite into space, the Explorer 1. The satellite picked up radiation in an area that was considered void of all particles. These high energy particles provided the first glimpse of the Earth's radiation belts.

Two concentric rings of energy particles surround the Earth. They were named the Van Allen belts, after James Van Allen, a scientist who led the studies of the radiation data. The inner belt consists mostly of protons, and the outer belt consists of electrons. These particles are captured by Earth's magnetic field and spiral around the belt.

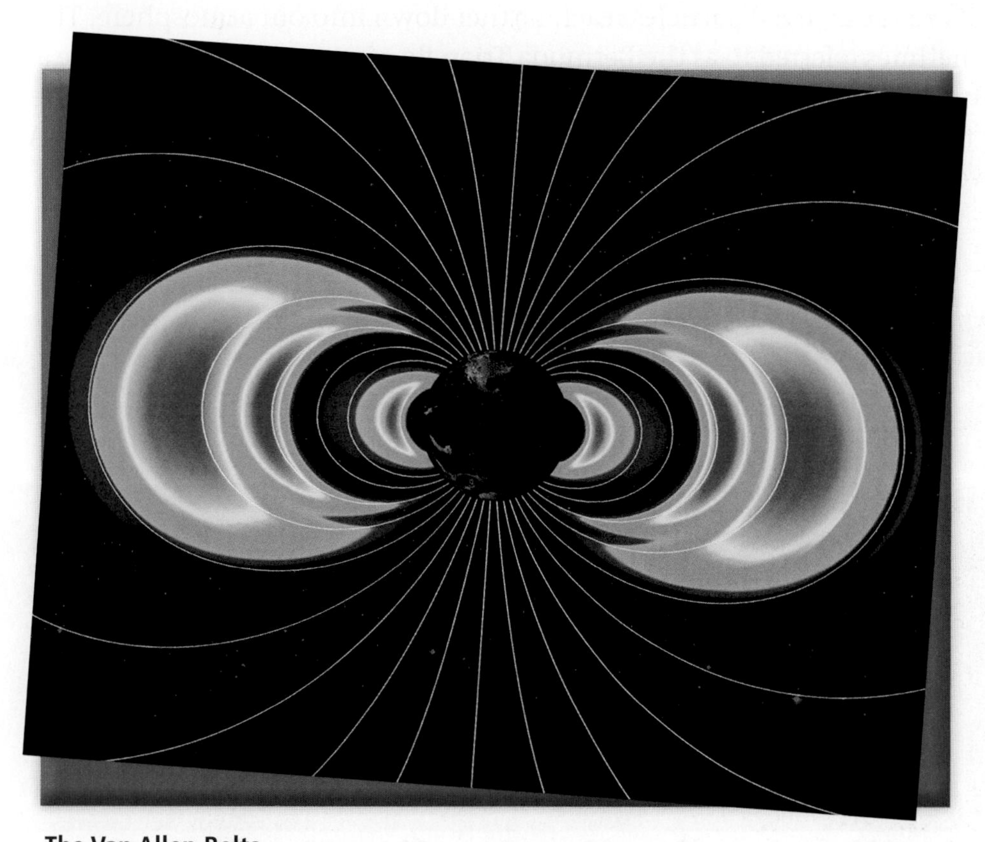

The Van Allen Belts

NASA's Goddard Space Flight Center/John Hopkins University, Applied Physics Laboratory

While scientists continue to study the Van Allen belts and their impact on Earth, we know the belts are responsible for one spectacular phenomenon: the Northern Lights. The Northern Lights are an aurora, which is a natural light display in the Earth's sky caused by charged particles colliding with atoms and molecules in our atmosphere. The fantastic light shows are typically seen in the Arctic region.

The Northern Lights in Norway

Stas Moroz/Shutterstock

The South Atlantic Anomaly

The South Atlantic Anomaly (SAA) is a region of dense radiation above the Atlantic Ocean off the coast of Brazil. Satellites and spacecraft that pass through this area are bombarded with high-energy protons. In this area, there is a dip in the Van Allen belts where charged particles reach farther down into our atmosphere. The area is sometimes referred to as the Bermuda Triangle of space, as it can produce anomalies in astronomical data and problems with electrical systems. Many satellites are shut down as they move through this area, and astronauts who travel through the SAA receive high levels of radiation.

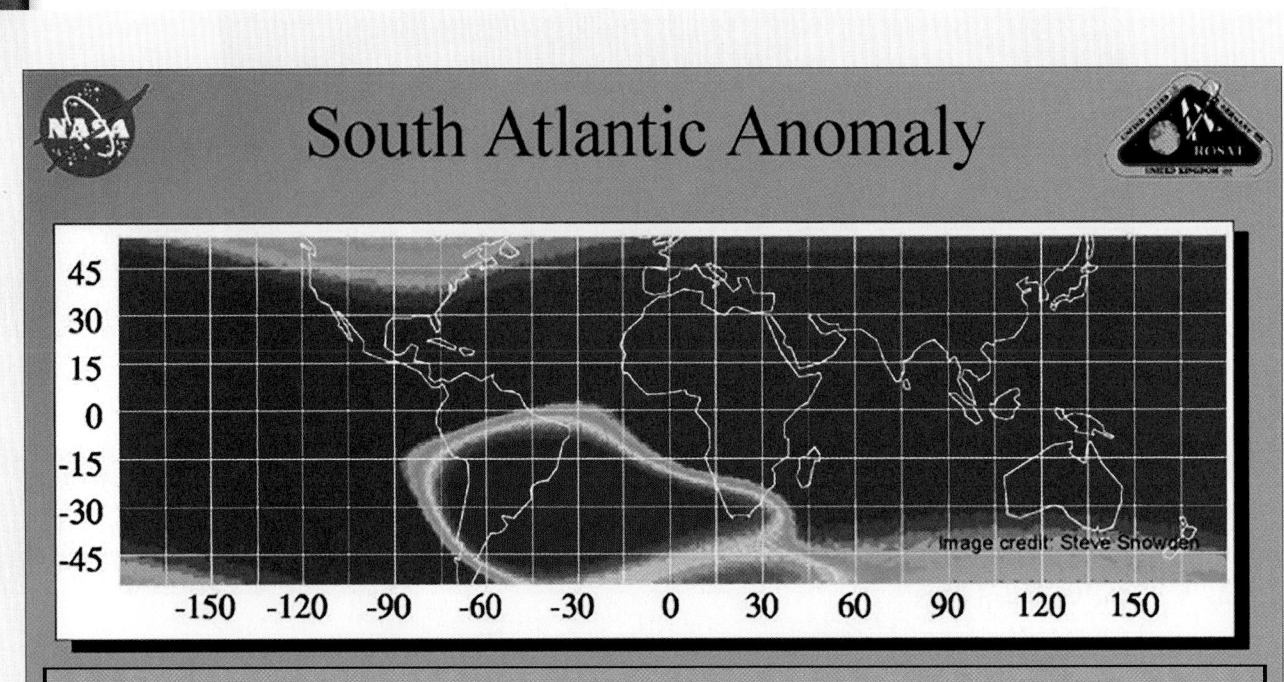

The South Atlantic Anomaly (SAA) is a dip in the Earth's magnetic field which allows cosmic rays and charged particles to reach lower into the atmosphere and interfere with communication with satellites, aircraft, and the Space Shuttle. The geologic origin is not yet known.

The enhanced particle flux in the SAA also strongly affects X-ray detectors, which are in essence particle detectors. The ROSAT PSPC had to be turned off during passage through the SAA to prevent severe damage. While the ROSAT HRI could be left on during the passage, it could collect no useful data. The light blue and green bands at the top and bottom of the image are due to an enhanced particle flux above Earth's auroral zones (particle belts).

FIGURE 2.3

South Atlantic Anomaly

NASA's Goddard Space Flight Center/Astrophysics Science Division (ASD)

The Moon

The Moon is only a quarter the size of Earth, but it is the fifth largest moon in the solar system. The gravity of Earth causes the Moon to spin. *The Moon spins very little and takes 27 days to complete one rotation, so we only see one side of the Moon*; this is called synchronous rotation. As a result, the Moon does not seem to be spinning but appears to observers from Earth to be keeping almost perfectly still. The Moon orbits the Earth once every 28 days. All the phases of the Moon can be seen within this 28-day period. The beginning of the cycle starts with the Moon growing until it reaches a Full Moon, and then waning until it is completely dark, or a New Moon.

The phases of the Moon.
Castleski/Shutterstock

How the Moon was Formed

The Moon was formed ~4.5 billion years ago, about 30–50 million years after the origin of the solar system, out of debris thrown into orbit by a massive collision between a smaller Earth and another planetoid, about the size of Mars. A planetoid is planet-size version of an asteroid. Initially, the Moon spun much faster. However, it is not perfectly spherical and bulges out slightly at its equator. These bulges along the Earth-Moon line caused a torque, slowing the Moon spin, much the same way a figure skater gradually opens to decelerate a spin. When the Moon's spin slowed enough to match its orbital rate, the bulge was in line with Earth, which is why we always see the same side of the Moon. In our solar system, almost all moons spin at the same rate as they orbit.

Earth would be a very different place if the Moon did not exist. Not only did the Earth slow down the Moon's rotation, but the Moon is slowing down the rotation rate of the Earth. Since the Moon's formation, the Earth's rotation has been slowing due to the friction of the tides caused by the Moon; in reaction to this exchange of energy, the Moon has been moving farther away from the Earth. In fact, at the time of the Moon's formation the Earth rotated much faster than it does today: a day on early Earth was only a few hours long. But the Moon, being small in relation to Earth, will take more than twice the age of the solar system to slow Earth's spin rate to the Moon's orbital rate.

The Distance from Earth to the Moon

How do astronomers actually measure the distance from the Earth to the Moon? There are two ways to measure the distance to the Moon: lunar eclipse and parallax.

Ancient Greeks used the lunar eclipse by timing how long it takes Earth's shadow to cross over the Moon. If you know the diameter of Earth (8,000 miles) and the speed of the orbit of the Moon (28 days), you can calculate the distance. Greek astronomers observed that the Earth's shadow was 2.5 times the size of the Moon and lasted about 3 hours. Aristarchus, a Greek astronomer, determined that the Moon was about 239,000 miles from Earth using this method.

The second method is using the parallax, the change of an objects position based on the viewpoint. If you view the Moon from one location and a friend views the Moon from 3,200 km (2,000 miles) away you can compare the images. You, your friend, and the Moon create a triangle. You know the distance between you and your friend and using simple geometry you can determine the distance to the Moon. Ptolemy was the first to use this method for determining the distance from the Earth to the Moon. He calculated the distances to be 27.3 Earth diameters (about 218,400 miles). His calculations were very close to the distance calculated today!

Typically, we calculate the average distance from the Earth to the Moon, because the Moon is not always the same distance from the Earth. At its farthest point, the Moon is 252,088 miles away. At its closest point, the Moon is 225,623 miles away. The average distance of the Moon to the Earth is about 239,000 miles. Today, scientists bounce lasers off the Moon's surface and use the speed at which the laser beam returns to accurately measure the distance from Earth to the Moon.

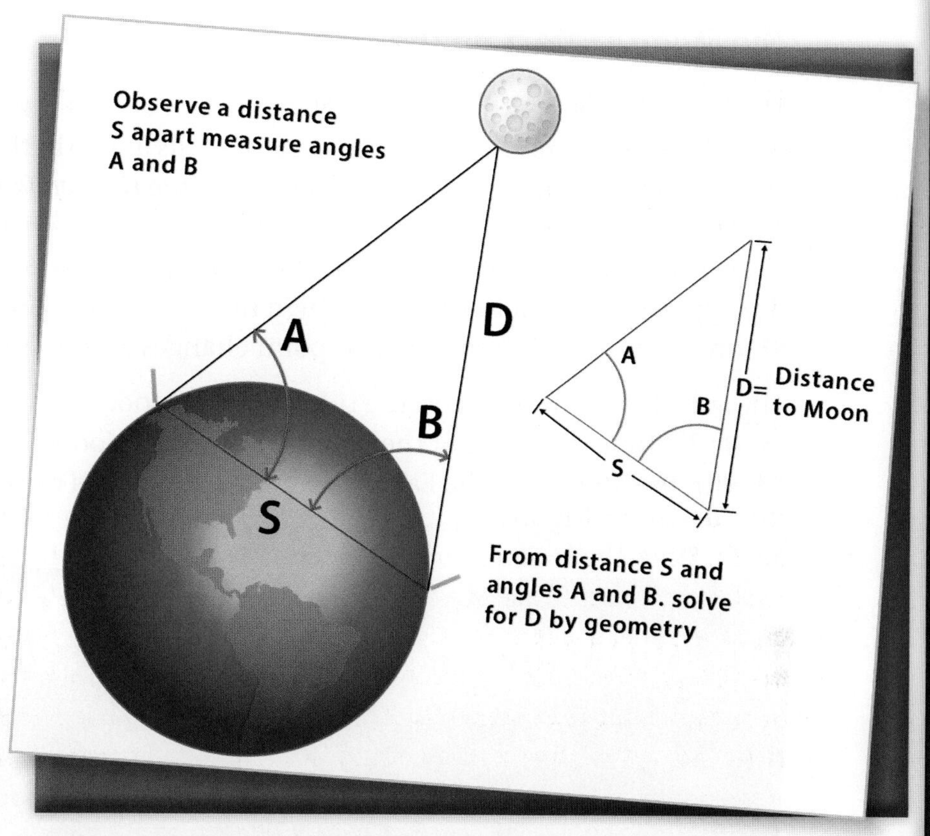

Using triangulation to calculate the distance from Earth to the Moon.

The Size of the Moon

Now that we know how far away the Moon is from the Earth, we can calculate the size of the Moon. Because it's impossible to determine the size of the Moon by just observing it, we rely on angular sizes and distances to determine the Moon's size. An object's angular size is determined by an object's size and its distance from the observer. Imagine the celestial sphere as a 360-degree circle. The angular size of the Moon, as seen from Earth, is half of a degree. Using this information, along with the distance to the Moon, we can calculate the size of the Moon using the small-angle formula. The Moon is approximately 2,160 miles in diameter using this calculation.

Although the size of the Moon is constant, the Moon may appear larger at certain times from Earth. This is due to the elliptical orbit of the Moon. When the Moon is at perigee, *closest to the Earth*, it appears much larger. When the Moon is at apogee, *farthest from the Earth*, it appears much smaller.

Did You Know?

The Moon (and Sun) appear larger in the sky when they are rising and setting. The Moon is not closer to the Earth during these times; it is an optical illusion. Take a piece of 3-ring binder paper and hold it at arm's length when the Moon is rising so that the Moon fits in the hole. Now wait until the Moon is fully in the sky and try the experiment again. The Moon will still fit in the hole.

The Moon's Effect on the Tides

The Moon has one-sixth the amount of gravity of Earth. But this is enough gravity to affect Earth's oceans. As the Moon's gravity pulls at Earth, the tides change. The tides are *the reaction of gravity that causes the sea levels to rise and fall*. Tidal changes occur when the Earth is nearest and farthest from the Moon. During high tide, the water surges up. During low tides, the water level drops. When one side of the ocean is at high tide, the opposite side is at low tide. The high tide is the point on Earth that is closest to the Moon. So, as the Earth rotates, this point changes and thus the tides change.

The Sun also has an effect on the tides, but the Moon is so much closer to Earth that it exerts 2.2 times the force of the Sun. Near the Full Moon and New Moon, the Sun lines up with the Moon and the Earth causing an intensified force on the tides.

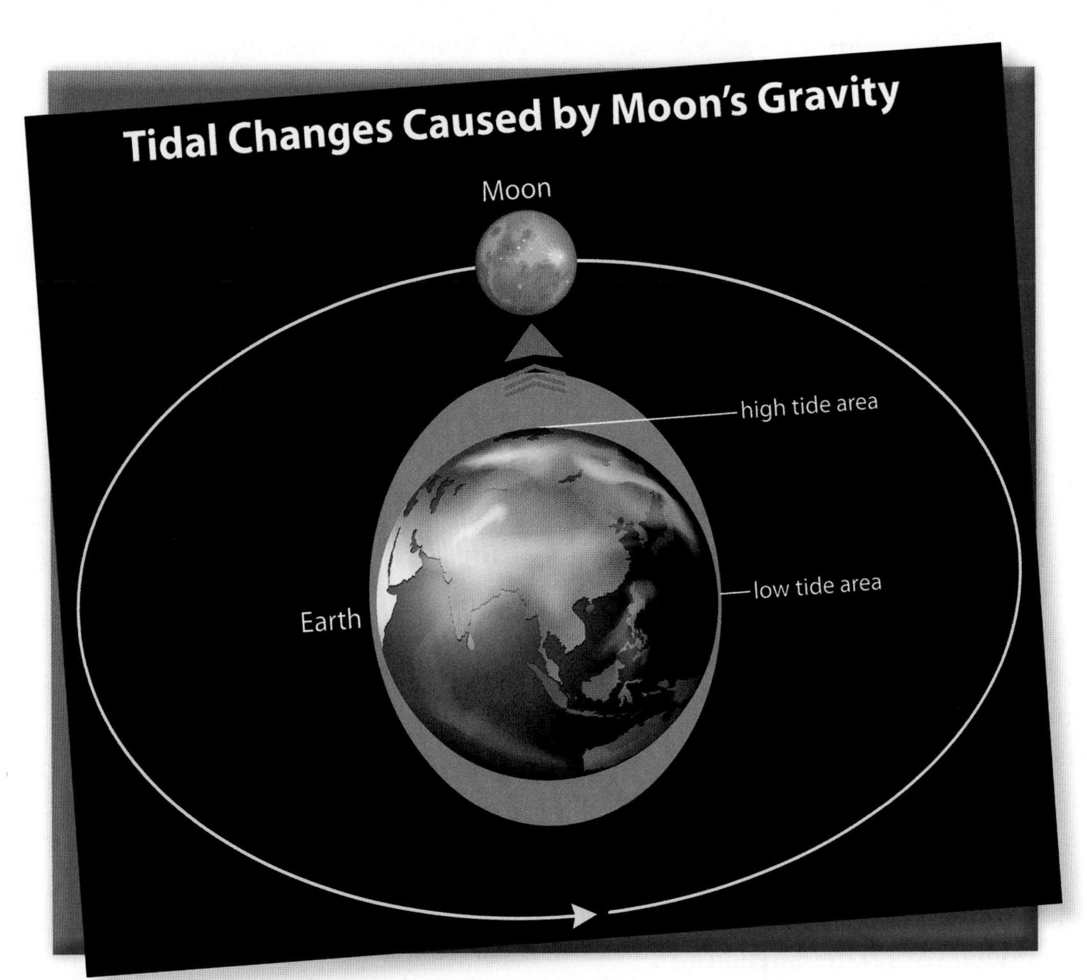

The tidal changes caused by the Moon's gravity.
BlueRingMedia/Shutterstock

Spring tides *occur when the Sun and Moon are aligned and cause the oceans to bulge more than usual, producing a higher tide than usual*. The spring tides occur twice each lunar month and are not associated with the seasons. The term "spring" is derived from the concept of the tide "springing forth." Neap tides *occur when the difference between the high tide and low tide is the smallest.*

Earth's Wobble

Earlier we discussed how Newton had suggested that the Earth's spin may cause the equator to bulge. The gravitational pull from the Sun and Moon isn't uniform across the Earth because of the bulge, which causes the Earth to wobble like a spinning top that is about to fall off its axis. The wobble is technically called precession and is defined as *the slow and gradual shift of the Earth's axis*. The force by the Sun and Moon affects the Earth's axis. A complete precession cycle takes approximately 26,000 years.

Earth's Wobble

This precession on Earth will eventually cause Earth's axis to move. So instead of the axis pointing to Polaris (the North Star), in 13,000 years Earth's axis will point somewhere near the star Vega.

Did You Know?

The gravitational pull of the Moon is slowing the Earth's rotation, which means our day lengthens by 2.3 milliseconds per century.

The Moon's Surface

The Moon's structure is similar to Earth in that it is composed of a core, rocky mantle, and crust. The inner core of the Moon is likely very small and consists mostly of iron. The rocky mantle is composed of dense rocks that have high levels of iron and magnesium. The mantle is about 825 miles thick and very dense. The crust is about 42 miles deep and is the top layer of the Moon's surface.

The composition of the Moon.
NASA/MSFC/Renee Weber

The surface of the Moon is very mountainous. In fact, the Moon has peaks that are as high as Mt. Everest. In addition, the Moon has many craters that were created when space debris, asteroids, comets, etc. crashed into it.

The surface of the Moon showing impact craters in the southern lunar highlands.
Procy/Shutterstock

Scientists used to think that the dark stretches of the Moon were oceans, so these were named Maria, which is Latin for "seas." These darkened spots are actually pools of hardened lava. Early in the history of the Moon, the interior was molten enough to produce lava from volcanoes or when a large asteroid hit the surface

Is there water on the Moon? Scientists have found traces of water on the lunar surface. They suspect that water may be more abundant on the slopes facing the lunar South Pole. However, the amount of water is comparable to a dry desert. Another study, in 2017, suggested that the Moon's interior could include water.

The Moon has a very thin atmosphere. The atmosphere on the Moon is not sufficient to sustain life, and the temperature varies widely between -274^{0}F to 257^{0}F. This thin atmosphere can explain why Neil Armstrong's footprint will remain undisturbed on the Moon for centuries.

Did You Know?

Gene Shoemaker was a geologist who helped prove that Arizona's Meteor Crater is an impact scar from a meteor. He also trained the Apollo astronauts and lobbied NASA to send a scientist to the Moon. Gene Shoemaker died in 1997, and in 1999 his ashes were taken to the Moon by the Lunar Prospector space probe. He is the only person to be buried on the Moon.

Lesson 1 Review

Using complete sentences, answer the following questions on a sheet of paper.

1. What is the Goldilocks Zone and why is Earth in this zone?
2. How can scientists learn about the interior of the Earth?
3. What makes up the Earth's atmosphere?
4. How are the seasons determined on Earth?
5. How does the magnetic field protect us on Earth?
6. What is the most recent theory on how the Moon was created?
7. What are two methods you can use to measure the distance to the Moon without lasers?
8. Why does the Moon appear larger at different times throughout the month?
9. How does the Moon affect the tides?
10. How long is a complete precession cycle for Earth?

APPLYING YOUR LEARNING

11. Suppose you lived 2,000 year ago. Provide a short description of observations you could make that would lead you to conclude that the Moon plays a role in the Earth's tides?

LESSON 2

The Sun and Its Domain

Quick Write

After reading the following vignette, consider all the possible uses of solar energy in everyday life. Describe at least one way that you could use solar energy.

Learn About

- the Sun's energy
- the Sun's core, atmosphere, and sunspots
- the solar system's structure

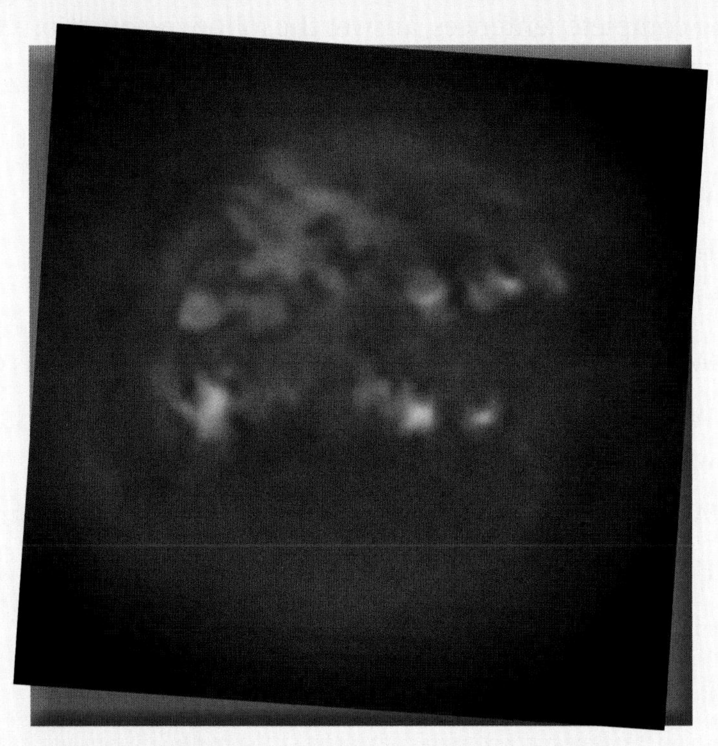

Energy from the Sun has powered life on Earth for millions of years.

Courtesy of NASA Space Place/NASA.gov

Throughout history, we have considered the Sun to be the "giver of life." Without the Sun, our land, water, and air would be frozen solid, and life as we know it would end. Lucky for us, the Sun is our partner and the center of our solar system. We use the Sun for light and heat, and since time began, we have been on a pursuit to harness its energy. The ancient Greeks, for example, built their homes to get the most sunlight during the cold winter months.

Today, solar energy is becoming a more common and sustainable electricity option for homeowners, business owners, and communities. Solar energy is *the solar radiation that reaches Earth*. We can collect this energy for heat and convert it into electricity.

One of the first attempts at using solar energy was in the 1830s. British astronomer John Herschel used a solar thermal collector box to cook food during an expedition to Africa. This device absorbs sunlight to collect heat. Some people use solar ovens today, to make backyard s'mores and homemade pizzas! Many homeowners use solar energy for heating water and generating electrical energy.

Solar thermal power plants also use solar energy to heat fluids to high temperatures. Some power plants have large solar photovoltaic (PV) arrays to convert sunlight into electricity for the areas they serve. *Photovoltaic refers to the production of energy from light.*

The pursuit of solar energy is sure to continue. Not only does it reduce electricity costs, it is renewable, abundant, sustainable, and cleaner for the environment. There are limitations on using solar energy, however. In winter (and at night), much less sunlight arrives at the Earth's surface. Similarly, cloudy days can pose a challenge to maintaining a constant energy supply from the Sun.

Vocabulary

- solar energy
- photovoltaic (PV)
- nuclear fusion
- gravitational contraction
- luminosity
- hydrostatic equilibrium
- conduction
- convection
- radiation
- photosphere
- chromosphere
- corona
- solar wind
- sunspots
- prominence
- solar flare
- theory
- asteroids
- meteoroids
- comets
- terrestrial planets
- Jovian planets
- dwarf planet

Photovoltaic panels.
wzlv/Shutterstock

The Sun's Energy

The Sun is our main source of light, heat and energy. It powers weather, climate, and life on Earth. Without it, we would cease to exist. Because it is so close to Earth, scientists study the Sun's physical and chemical properties to better understand most other stars in the sky.

Although our Sun is a star, it is not star shaped. Instead, it is a big, yellow ball of gas. This gas can be found in layers of mainly hydrogen and helium. In the Sun's core, *hydrogen atoms compress and fuse into helium*, a process called nuclear fusion. Every minute, the Sun converts enormous amounts of mass into energy. The Earth's surface receives that energy in just eight minutes!

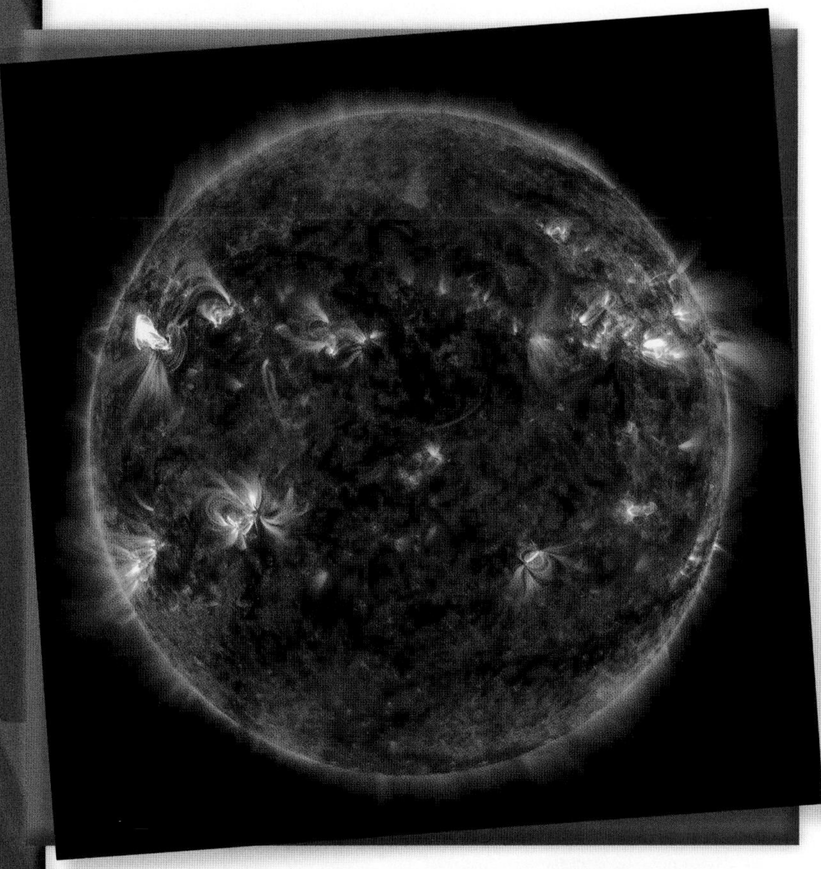

An image of active regions on the Sun from NASA's Solar Dynamics Observatory. The glowing hot gas traces out the twists and loops of the Sun's magnetic field lines.

Courtesy of NASA/SDO/AIA

The Source of the Sun's Energy

How does the Sun shine, year after year? How old is the Sun? Finding answers to these questions has challenged scientists and scholars for hundreds of years. Beginning in the mid-nineteenth century, a researcher and physics professor, Hermann von Helmholtz, suggested that the source of the Sun's energy was gravitational contraction, *or the shrinking and compression of gases, caused by gravity*. Another great scientist of the time, William Thomson (also known as 1st Baron Kelvin of Largs), also believed the main source of the Sun's energy was from gravitational contraction. With this belief, Lord Kelvin estimated the age of the Sun to be only 30 million years old.

Other scientists of the time, such as Charles Darwin, thought the Sun must be at least hundreds of millions of years old. For Darwin, only a Sun that old could explain the evolution of living things on Earth that depend on energy from the Sun.

Early in the twentieth century, scientific discoveries related to natural radioactivity suggested a new possibility: nuclear energy might be the source of the Sun's energy. Without evidence that the Sun contains many radioactive materials, scientists would continue to look further for answers.

About this time in history, Albert Einstein introduced his famous formula, E=mc2. Einstein's equation showed how mass and energy are interchangeable. Einstein proposed that a small amount of mass could, in theory, be converted into an enormous amount of energy. Einstein's formula would provide the direction needed to help scientists better understand the source of the Sun's energy.

Did You Know?

Scientists use spectroscopy to study the heat and chemicals of the stars by using a spectrum. A spectrograph breaks up light into a rainbow. Thin dark lines in the Sun's spectrum indicate different elements. Through spectroscopy, scientists know the Sun is made up of about 70% hydrogen gas.

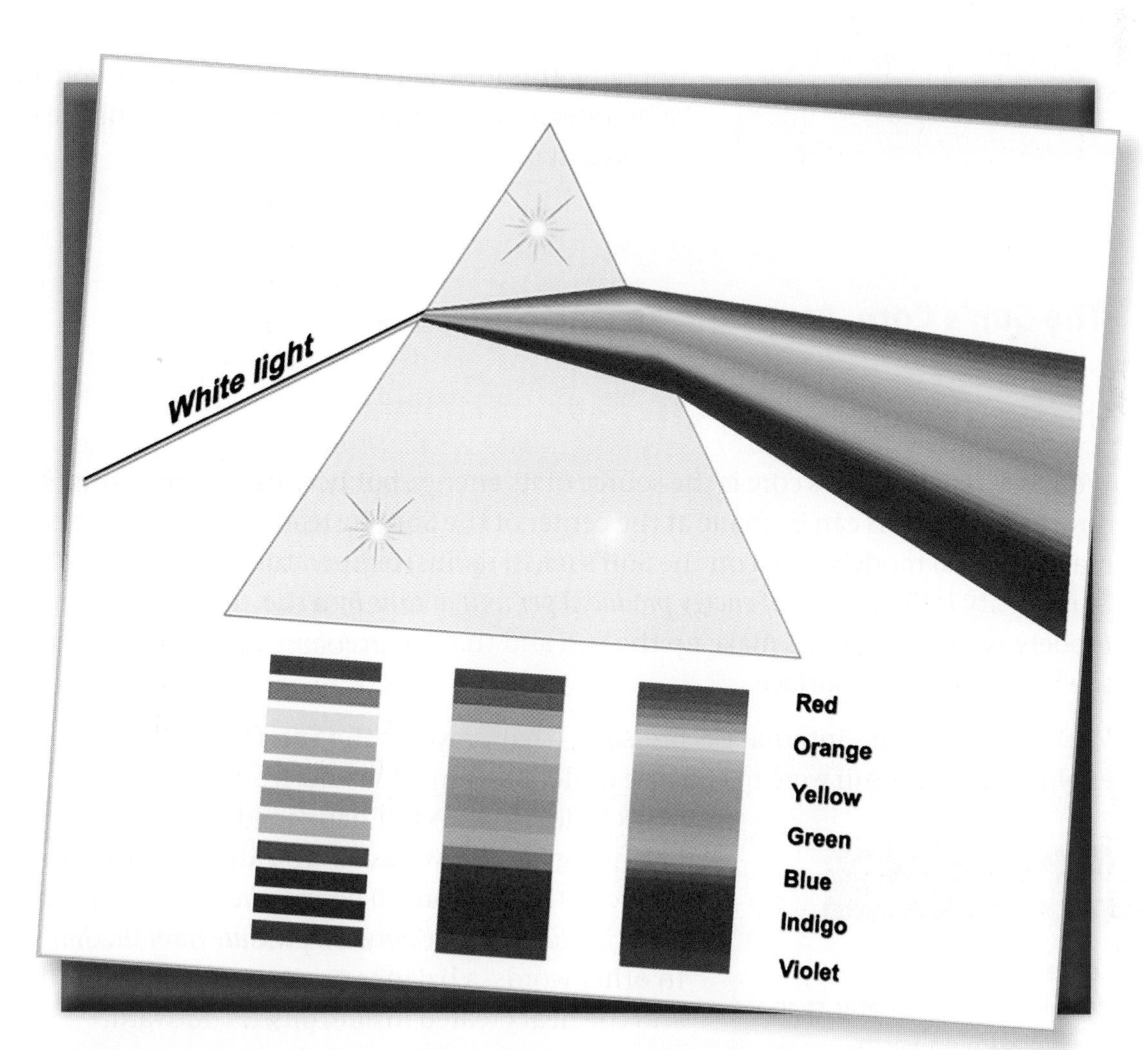

A spectroscope breaks down white light into a spectrum of colors. Each color on the spectrum represents a temperature, with red being the hottest and violet the coldest.
Fouad Asaad/Shutterstock

Nuclear Fusion

By 1920, scientists began to consider the high temperature inside the star as a source of the Sun's energy. With Einstein's formula at work, they measured the masses of many different atoms. These included hydrogen and helium. Astrophysicist Sir Arthur Eddington made a proposition in 1920. He proposed that the measurement of the mass difference between hydrogen and helium was significant. It meant that the Sun gives off heat and light by converting hydrogen atoms to helium. Earlier in the lesson, we defined this process as nuclear fusion.

To be exact, it takes four hydrogen atoms fusing together to form one helium atom. Each atom has a nucleus, which makes up most of an atom's mass. When the nuclei (plural of the word nucleus) combine, the result is a new nucleus with less total mass than the two that fused together. What happens to the lost mass? It is converted to energy. As shown in Einstein's formula (E=mc2), energy is equal to mass times the speed of light squared.

Applying this formula to nuclear fusion, the conversion of hydrogen to helium releases about 0.7% of the mass equivalent of the energy.

> **Did You Know?**
>
> Every minute, the Sun converts 240 million tons of mass into energy.

The Sun's Core, Atmosphere, and Sunspots

The Sun's Core

We know that the Sun's core is the source of its energy, but how do we know this? No direct observations can be made at the center of the Sun! Instead, scientists must use mathematical models based on the Sun's mass, radius, temperature, and luminosity. Luminosity *is the amount of energy produced per unit of time by a star*. These mathematical models assume that gases make up the Sun and that gas pressure is greater in the core of the Sun than on the surface.

With all of the Sun's internal gas pressure, what prevents the Sun from collapsing or exploding? If the Sun were to collapse under pressure, it would crunch down into a black hole. This has not happened, thanks to nature's balancing act, known as hydrostatic equilibrium. Hydrostatic equilibrium *occurs when compression due to gravity is balanced by outward pressure from the Sun's core*. In other words, a balance occurs as gravity pulls inward and the heat created inside, pushes outward.

> **Did You Know?**
>
> The outer pressure of the Sun is 340 billion times greater than the air pressure on Earth.

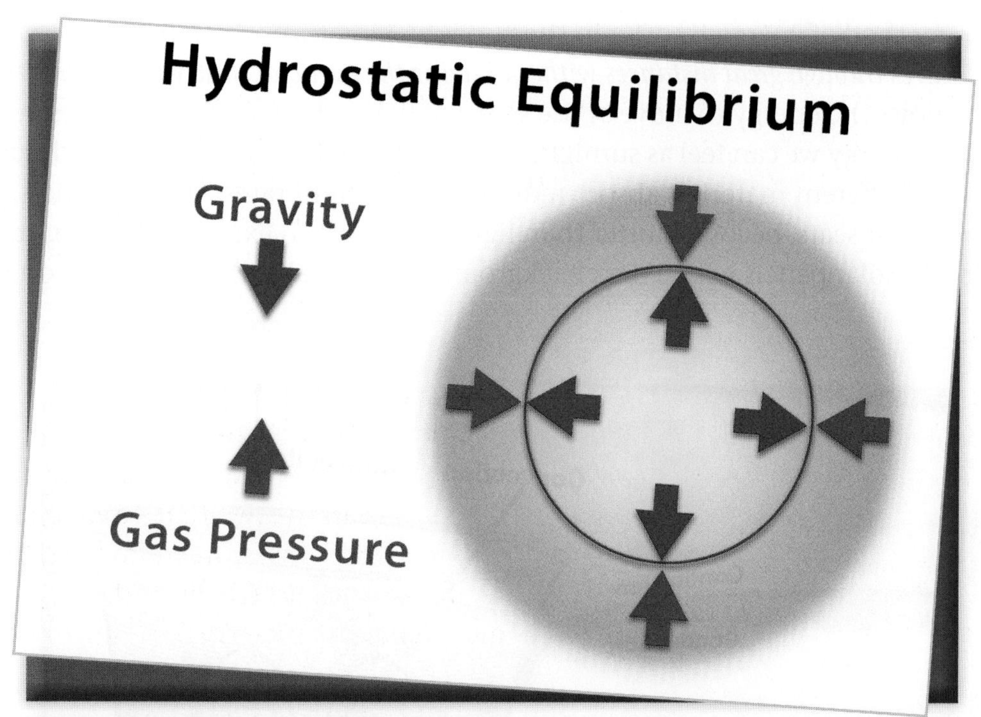

A star is somewhat like a balloon. The gas inside the balloon pushes outward and the rubber material supplies just enough inward compression to balance the internal gas pressure.

Modes of Solar Energy Transfer

The three basic methods of transferring energy on Earth, and also on the Sun, are conduction, convection, and radiation. Conduction *occurs in solids as energy transfers from atom to atom or molecule to molecule.* This energy transfer causes particles to vibrate faster without the particles moving out of their positions in the solid structure.

Different materials have different abilities to conduct energy, but conduction is usually high in metals. That is why we make most cooking utensils, such as pots and pans, out of metals but with nonmetallic handles. Conduction is not a major energy transfer method in the Sun, because the particles are too far apart.

Convection *occurs in liquids or gases when atoms move from one location to another.* Liquids and gases are made of atoms or molecules that are free to move about, instead of vibrating faster. They can move from one location to another to transfer their energy to another particle.

In a room with a heater, air molecules gain energy from the heat source. The molecules increase in speed, causing them to expand and rise until they encounter a barrier like the ceiling. Then, they transfer some of that energy to other molecules. As they lose energy, they slow down and contract, become denser, and move downward. This usually results in a circular pattern of movement that we can find in the Sun and here on Earth. Large birds we see flying effortlessly are using the updrafts of these convection currents to stay in the air.

The main method of energy transfer in the Sun is by radiation which is *the movement of energy as waves through great distances without having the particles closely packed together.* This also allows the energy to move at the speed of light which is 186,000 miles per second. The energy we can feel as sunlight shining through a window is a good example of radiation. Different materials absorb radiation at different rates. Dark materials, like black leather car seats, become hotter than lighter ones when exposed to sunlight. This is why we make solar panels with a dark background, so they absorb more radiation.

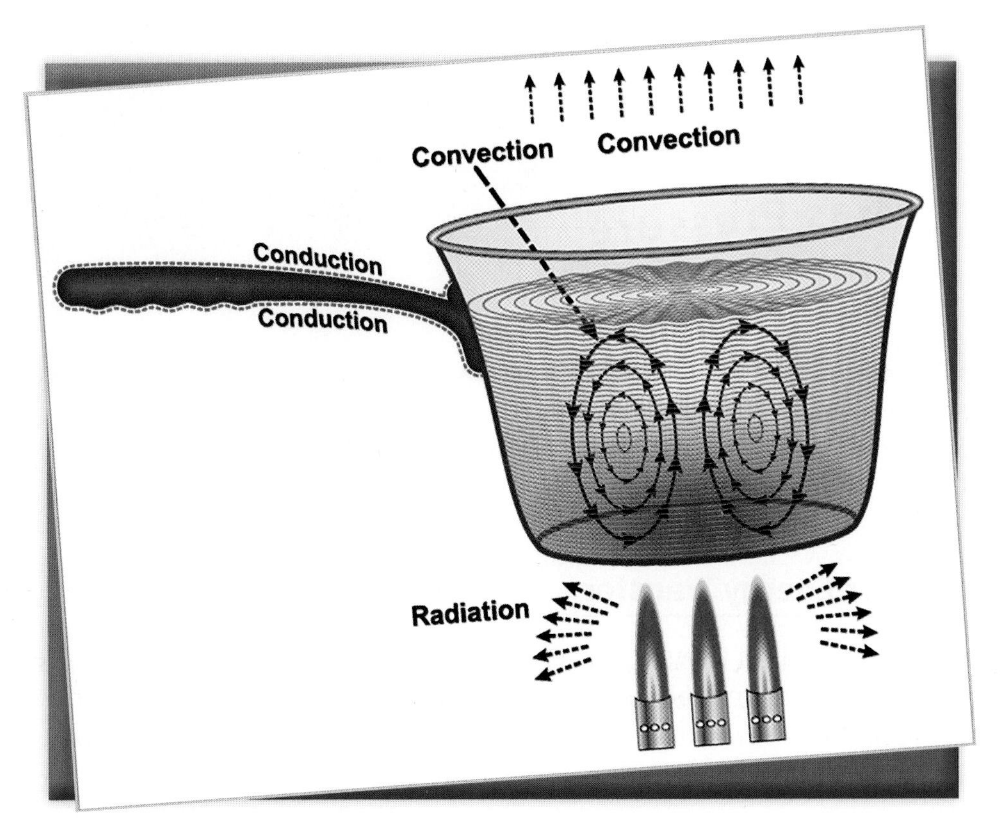

This diagram demonstrates the three methods of energy transfer, conduction, convection, and radiation.

Fouad Asaad/Shutterstock

The Layers of the Sun's Atmosphere

What comes to mind when you hear the word "atmosphere"? The mood of your favorite hangout? The air around you? The Earth's atmosphere? Like the Earth, the Sun has an atmosphere. Just as the Sun itself has layers, so does its atmosphere. The Sun's atmosphere has three layers. They are the photosphere, the chromosphere, and the corona. It is important to note that these layers do not have distinct boundaries as they appear in a diagram but gradually change from one layer to the next.

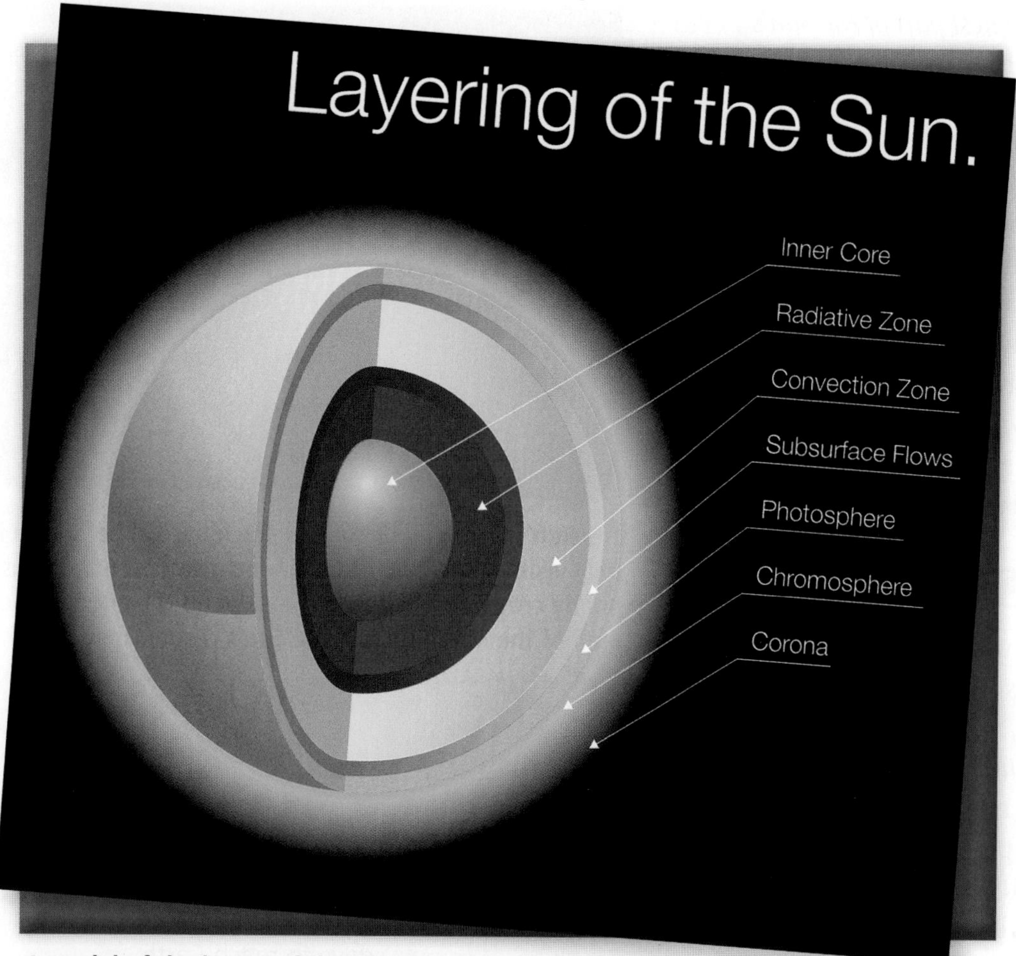

A model of the layers of the Sun. The inner layers are the core, radiative zone and convection zone. The outer layers are the photosphere, the chromosphere, and the corona.

John T Takai/Shutterstock

The deepest layer of the Sun's atmosphere is the photosphere. This is the layer we can see from Earth. It is about 300 miles thick. This layer is where the Sun's energy releases as the yellow-white light that we see. The brightness of the light varies as temperatures vary. Images of the photosphere show light and dark areas. These represent hotter and cooler regions as convection currents rise and fall. The average temperature in the photosphere is 10,000°F.

The layer just above the photosphere, and below the corona, is the *Sun's* chromosphere. This layer is between 250 miles and 1,300 miles above the photosphere. The temperature in the chromosphere also varies from 6,700°F to 20,000°F. The temperature rises as you move higher toward the top of the chromosphere. The light in this layer appears bright red during a solar eclipse.

The highest part of the Sun's atmosphere is the corona, or crown. This is the layer that extends out into space and is much hotter than the photosphere. It is about 1,300 miles from the photosphere. This layer has an average temperature of 3,500,000°F. Like the chromosphere, the corona is only visible during a solar eclipse and appears as white plumes of gas.

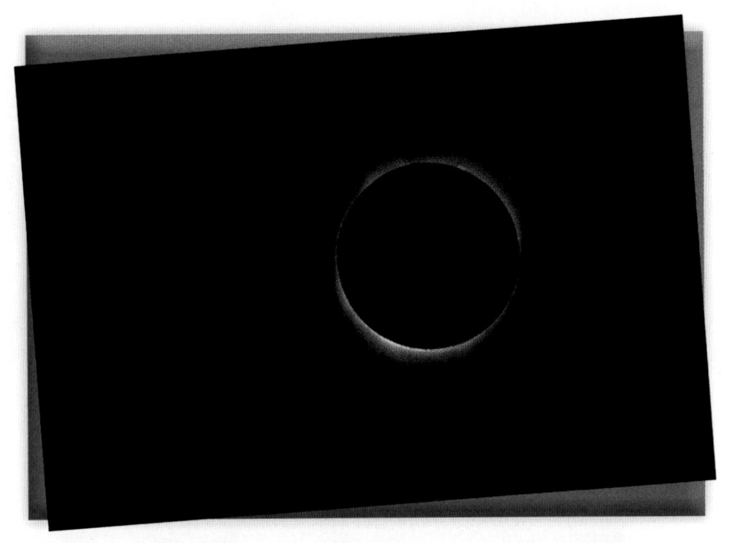

The Sun's visible-light corona, the inner part of which is only visible during a total solar eclipse, is seen here as a pearly crown. The reddish light around the edge of the limb of the Moon is from the Sun's chromosphere.

Courtesy of NASA.gov.

The Solar Wind

The outer part of the Sun's atmosphere creates a solar wind, which is *the emission of high-speed protons and electrons*. We call this a wind because it has the same effect on objects in its path as a strong wind blowing against a tree. It causes comets in space to form a "tail" that always points away from the Sun. These winds can also cause weather changes. They can affect radio signals and create the glowing lights called auroras around the North and South poles on Earth.

Did You Know?

The solar wind takes about four days to reach Earth.

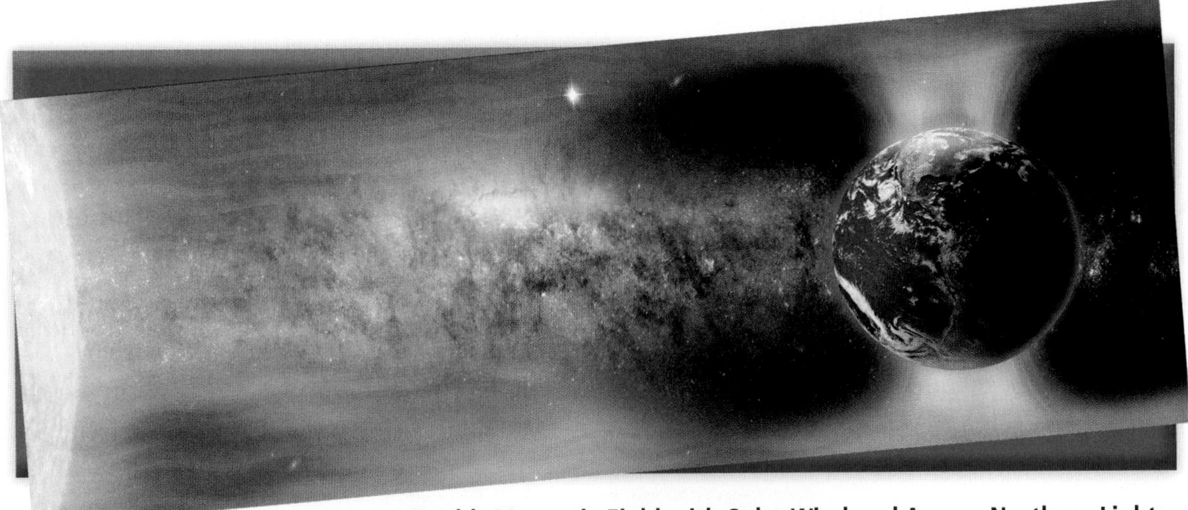

Earth's Magnetic Field with Solar Wind and Aurora Northern Lights. Elements of this image are furnished by NASA.

muratart/Shutterstock

The Impact of Sunspots and Solar Flares on Earth's Climate

Pictures of the Sun's photosphere show areas on the surface called sunspots. These *are dark regions that appear when hot gases cool during the convection process*. We have been able to see them for hundreds of years, but did not understand them until about 200 years ago. We now estimate sunspots to be about 2,000⁰ F cooler than the brighter areas of the photosphere. Although they are cooler, the magnetic fields in sunspots may be 1,000 times stronger than the bright, hot areas. These magnetic spots often appear in pairs with one being an N-pole and the other an S-pole, like a simple magnet here on Earth.

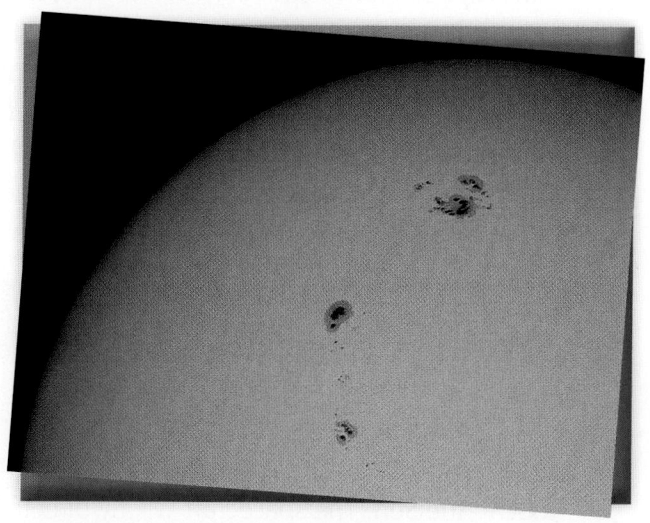

Sunspots on the Sun's photosphere
DeepSkyTX/Shutterstock

Sunspot activity occurs in cycles of about 11 years and ranges in number from almost none to hundreds. The life of a sunspot also varies from only minutes to a few months.

How are sunspots important to us on Earth? Studies between astronomy and geology seem to indicate that we have had ice ages on Earth at the same time as periods of low sunspot activity. Even during shorter periods, sunspots affect our climate.

The Sun's magnetic field can also create violent eruptions in the photosphere. Massive numbers of particles spew into the corona. Scientists view these areas of eruption as active regions of the Sun. One type of activity is a prominence, or *loop or arc of glowing gas extending from the photosphere into the corona*. We do not fully understand the cause of prominences. They usually last about a day and can be up to several hundred thousand miles long.

Sun's surface with solar flares, and splashes of solar prominences..
aeyaey/Shutterstock

Another powerful explosive activity on the photosphere is the solar flare. A solar flare is *an eruption on the Sun's surface*. As seen from Earth, a solar flare appears like a giant flame extending outward from the Sun. The solar flare gives off massive amounts of energy and radiation. Unlike the trapped gases that make up a prominence, the energy produced by a flare blasts out into space by the force of the explosion. Very large flares reach Earth in minutes and create the auroras in our polar regions. They may also affect radio signals on our planet.

The Right Stuff

The Parker Solar Probe

On August 12, 2018, NASA launched the Parker Solar Probe on a seven-year mission to "touch" the Sun. The spacecraft will fly directly through the Sun's atmosphere, as close as 3.8 million miles from its surface. The Parker Solar Probe will move around the Sun at speeds of up to 430,000 miles per hour. With the help of thermal technology that can protect the probe on its dangerous journey, the spacecraft's four instrument suites will study magnetic fields, plasma and energetic particles, and the solar wind.

Illustration of NASA's Parker Solar Probe approaching the Sun.

Courtesy of NASA/Johns Hopkins APL/Steve Gribben

The Solar System's Structure

"My Very Educated Mother Just Served Us Noodles." What does this phrase have to do with our solar system? For many, it is an easy way to remember the order of planets orbiting the Sun. Using the first letter of each word as a reminder of the first letter of each planet, in order, we have Mercury, Venus, Earth, Mars, Jupiter, Saturn, Uranus, and Neptune.

We once considered Pluto to be a planet. However, based on the 2006 International Astronomical Union definition, Pluto no longer qualifies. According to the definition, a planet must orbit a star, be big enough to have enough gravity to force it into a spherical shape, and be big enough that its gravity cleared away any other objects of a similar size near its orbit around the Sun.

Did You Know?

The Sun is nearly 1,000 times bigger than the rest of the planets put together. It has a diameter of almost 865,000 miles.

An illustration of the planets of our solar system.

Christos Georghiou/Shutterstock

We know the planets and their order of orbit around the Sun. This provides a solid start to visualizing our solar system. Of course, we have much more to learn about the planets. For example, why are the rocky inner planets so much smaller than the giant, gassy outer planets?

Jupiter, for example, is hundreds of times more massive than Earth. It is about five times the distance of Earth from the Sun. In comparison, Mercury is a little more than one-third the size of Earth. If Earth were the size of a tennis ball, Mercury would be about the size of a large marble.

What are the distances between the planets, using our tennis ball analogy? If you hold the tennis ball, Earth, you need to stand seven football fields away to be the same distance from the Sun. For the marble, Mercury, you would need to be 304 miles away to be the same distance from the Sun.

The Right Stuff

The Hubble Space Telescope

How do we know so much about the planets? Many stars and planets release or reflect light that we can see either with the naked eye or with telescopes. Just by looking at the light, scientists can determine the locations of stars and planets. The Hubble Space Telescope is the largest, most sensitive light-receiving telescope ever put into orbit. Hubble probes the universe with about 10x finer resolution and about 30x greater sensitivity to light than devices on Earth.

Other objects are either too far away to be seen or are behind something that blocks the light. We can detect their radio waves using radio telescopes. For other objects that do not emit radio waves, scientists rely on models and logic to detect planets around other stars.

The Hubble Space Telescope in orbit above the Earth. As light enters the main tube (at right), it strikes the main mirror (light blue disk at center), from which it can be directed to any of several instruments (behind the mirror). Elements of this image furnished by NASA.

Marcel Clemens/Shutterstock

How the Solar System Formed

Scientists believe that the solar system formed about five billion (5,000,000,000) years ago from gas and dust in space. To explain the origin and development of the solar system, we depend upon one or more theories. A theory is *an educated explanation of something but without real proof.* Although we cannot prove our solar system formation theories, they at least help to explain these facts:

1. The Sun is in the center of the solar system.
2. All of the planets revolve around the Sun, most in the same direction, and in nearly circular orbits.
3. The orbits of the planets are nearly parallel.
4. They all rotate in the same direction except for Venus and Uranus.
5. Most of the moons circle "their" planet in the same direction that the planet rotates and revolves in its orbit.
6. The orbits of the planets form a definite pattern from the Sun outward.
7. The denser planets are closer to the Sun.
8. The planets with solid surfaces, such as Earth, all have craters like those on our Moon.
9. The outer Jovian planets (such as Jupiter) have rings around them.
10. Within the orbits of the planets, there are asteroids (*small rocky objects orbiting the Sun mostly between Mars and Jupiter, although some have passed close to Earth*), meteoroids (*similar to asteroids but smaller*), and comets (*objects made of ice and dust and having a "tail" that points away from the Sun*) with their own motion patterns.
11. Other stars, besides the Sun, also have systems of planets.

Scientists divide theories about the development of the solar system into two categories. Evolution theories explain the creation of the solar system as evolving or forming from something else. René Descartes (1596-1650) was a famous philosopher, scientist, and mathematician. Descartes was responsible for developing the most popular theory of planetary motion in the seventeenth century. The famous scientist and philosopher stated that about 350 years ago the solar system may have formed from a spinning mass of material in the universe, like a tornado, which threw out pieces, such as the Sun and Earth. The problem with this theory is that it does not agree with scientific laws developed since Descartes' time.

Did You Know?

Descartes is mostly remembered for the saying, "I think, therefore I am.'

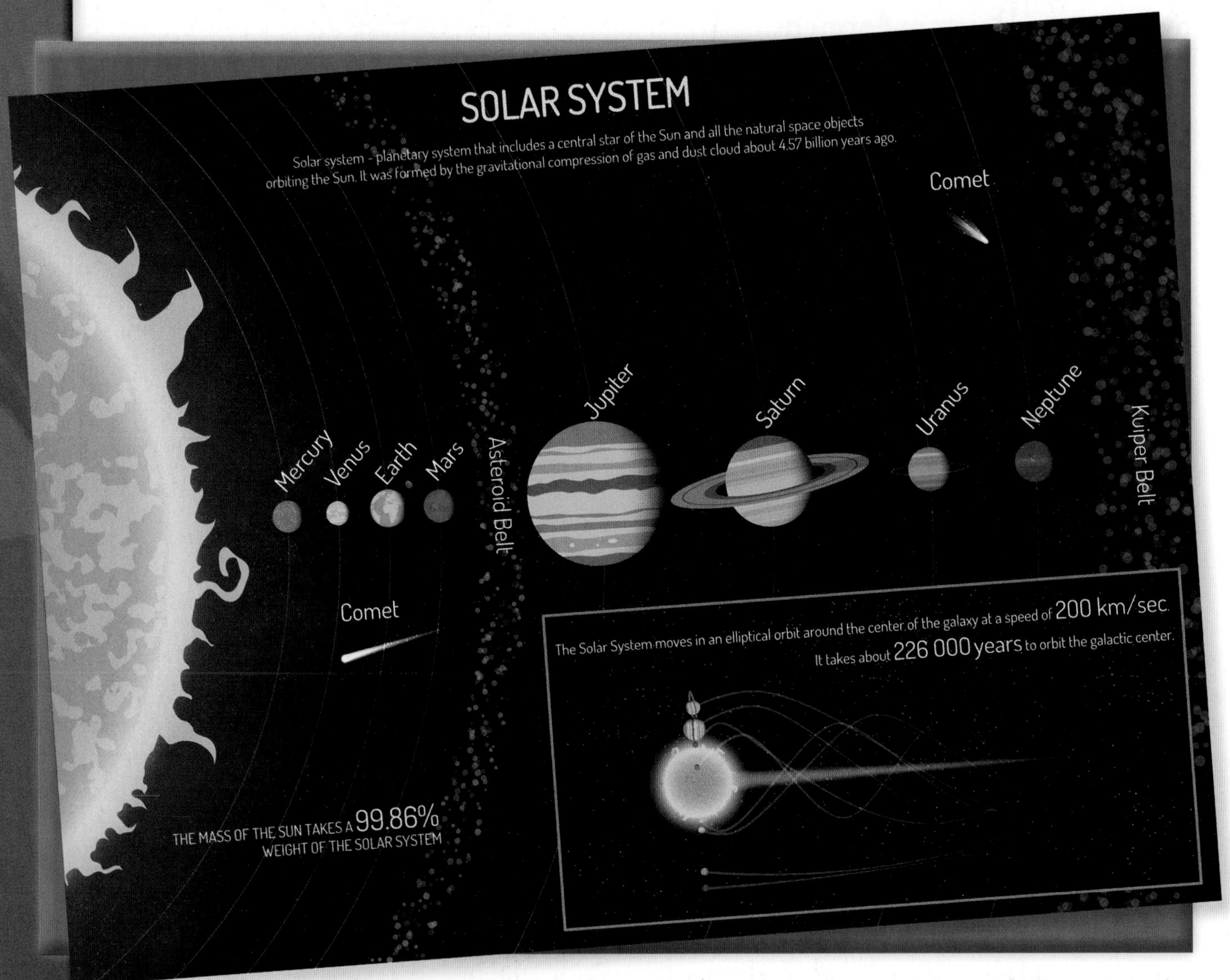

Any theory of the origin and evolution must explain the observed patterns in the solar system.
SkyPics Studio/Shutterstock

A second category of theories is catastrophe theories. Surprisingly, catastrophe theories do not mean something destructive like tornadoes took place. Instead, these theories suggest that the solar system was created during some "unusual" events. An early theory, about 250 years ago, suggested that a comet may have come close to the Sun and pulled material out to form the planets, like a car blowing leaves off a tree as it passes by at a high speed. This theory was later rejected as not being a real possibility, but it led to other theories that came later.

One such theory stated that the Sun may have been one of three stars that revolved around each other. (Astronomers have observed such formations in the universe in modern times.) Perhaps, the theory stated, this three-star system was not stable, and one star was thrown away from the other two, causing the formation of the Sun and its planets. There are two major problems with this theory. The material from the developing Sun would have been so hot that it would have scattered instead of coming together to

This image of Comet C/2001 Q4 (NEAT) was taken at the WIYN 0.9-meter telescope at Kitt Peak National Observatory.

Courtesy of T. A. Rector (University of Alaska Anchorage), Z. Levay and L.Frattare (Space Telescope Science Institute) and WIYN/NOAO/AURA/NSF

form the planets. Also, this catastrophe theory suggests that a star with planets around it would have been extremely rare, which has not turned out to be the case. Thousands of other systems with such an arrangement have been discovered to date.

Classifying Objects in the Solar System

We usually divide the planets into two groups. The terrestrial planets *are small and dense, with rocky surfaces and a great deal of metals in their core*. These include Mercury, Venus, Earth, and Mars. They are closer to the Sun, they have few (if any) moons, and they have no rings.

The outer planets, or Jovian planets, *are much larger and made up of gases*. These include Jupiter, Saturn, Uranus, and Neptune. They are lower in average density, and they all have ring systems and many moons. We often refer to them as the gas giants; they are made up mostly of hydrogen and helium.

Recall that Pluto was once considered to be a planet until scientists reclassified it. Today, Pluto is known as a dwarf planet. A dwarf planet *orbits the Sun, is large enough to be a spherical shape, but not large enough to have cleared away other objects of a similar size near its orbit*. Ceres, Haumea, Makemake, and Eris are four other dwarf planets.

The planets share many similarities and also significant differences. In the upcoming lessons we will explore more of the unique differences in their position, composition, and other features.

Did You Know?

Dwarf planet Ceres is the largest object in the asteroid belt between Mars and Jupiter and the only dwarf planet located in the inner solar system.

Lesson 2 Review

Using complete sentences, answer the following questions on a sheet of paper.

1. What is the main source of the Sun's energy?
2. What is hydrostatic equilibrium?
3. What is the main mode of solar energy transfer?
4. Compare and contrast the Sun's three atmospheric layers.
5. Why have studies shown that sunspots are important to Earth?
6. Describe one catastrophe theory of the solar system's formation.
7. What is Pluto's classification as a planet?

APPLYING YOUR LEARNING

8. Using the theory of planetary formation and the world we live in, describe how the Earth came to exist and how it relates to the rest of the solar system.

LESSON 3

The Solar System

Quick Write

Recall the criteria for effective scientific models–simplicity and predictive power –discussed in Chapter 1, Lesson 2. After reading the opening passage about the Planet X theory, jot down your thoughts on whether or not Planet X meets the criteria for a good scientific model.

Learn About

- terrestrial planets
- the outer planets and dwarf planets
- comets, asteroids, and kuiper belt objects

What lies beyond Pluto? According to Caltech scientists, there could be a planet! They believe "Planet X" to be about the size of Neptune, with a mass 10 times greater than that of Earth and an orbit 20 times farther from the Sun (on average) than Neptune. This means it could take between 10,000 and 20,000 years to make an unusual elliptical orbit around the Sun.

The existence of Planet X is only theoretical, and scientists do not have direct evidence yet. So why do scientists believe this planet possibly exists? Researchers studying the **Kuiper Belt**, *a ring of icy bodies outside of Neptune's orbit*, have observed that some of the dwarf planets and other small, icy objects tend to follow orbits that group together. One possible explanation for this behavior is a planet with controlling gravity. In 2015, Caltech astronomers Konstantin Batygin and Mike Brown announced their prediction–a planet with controlling gravity may be responsible for the unique orbits of smaller objects in the Kuiper Belt.

Researchers have not yet collected observable evidence. They continue to use mathematical modeling and computer simulations. Such tools help them further understand Planet X's orbit and the impact on its surroundings.

Batygin and Brown nicknamed the object "Planet Nine," but it has long been referred to as Planet X during scientific hunts for objects beyond Neptune. If Planet X is ever found, naming rights will be given to those who discover the planet.

Identifying the planet's location with powerful telescopes is the next logical step.

Vocabulary

- Kuiper Belt
- greenhouse effect
- red planet
- opposition
- chasms
- differential rotation
- oblate
- Galilean satellites
- Cassini Division
- sublimate
- asteroid belt
- trojan asteroids
- nucleus
- coma
- short-period comets
- long-period comets
- Oort Cloud
- meteoroid
- meteor
- meteorite

A distant view of Planet X looking back toward the Sun.
Courtesy of NASA/Caltech/R. Hurt (IPAC)

Terrestrial Planets

Recall the definition of terrestrial planets covered in a previous lesson. Terrestrial planets are small and dense, with rocky surfaces and a great deal of metals in their core. Terrestrial planets include Mercury, Venus, Earth, and Mars. They are closer to the Sun, they have few (if any) moons, and they have no rings.

We discussed Earth in a previous lesson. Our focus here will be Mercury, Venus, and Mars. Let's start with Mercury, since it is closest to the Sun.

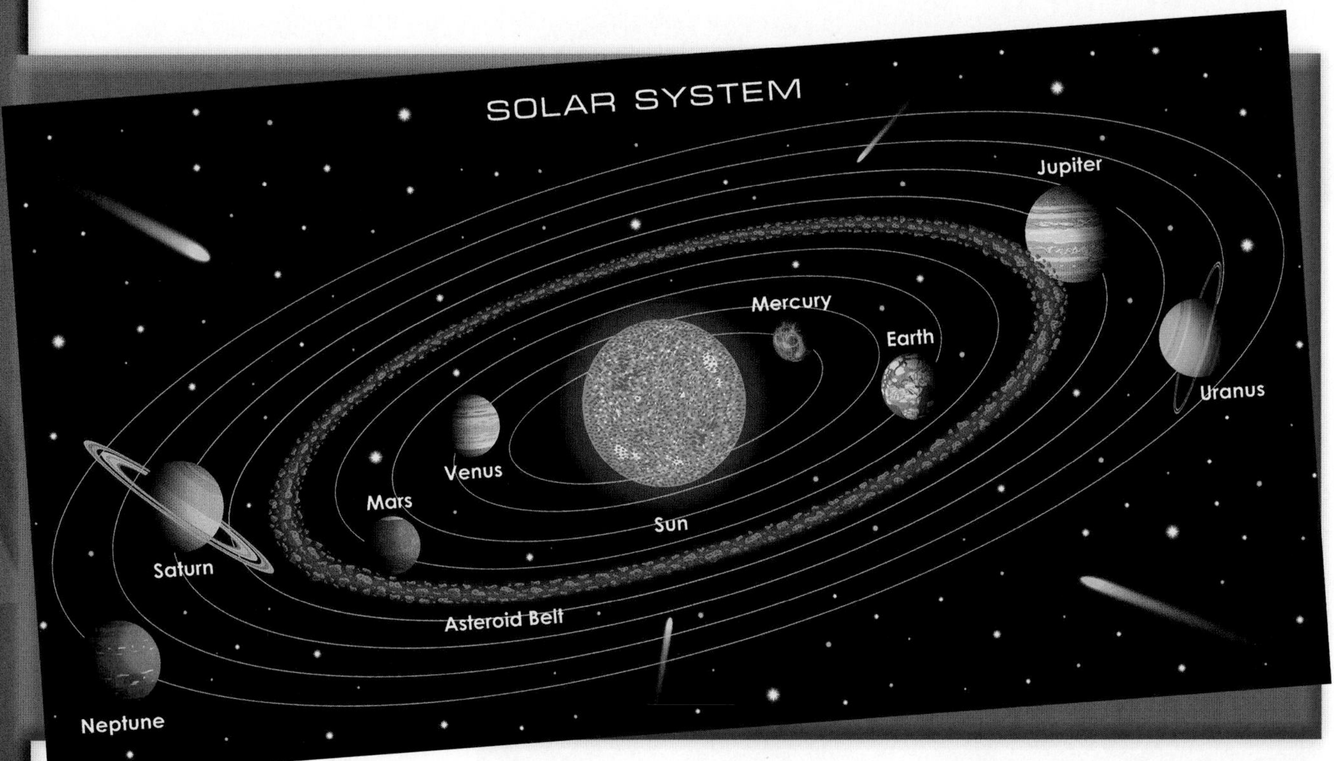

The solar system.
D1min/Shutterstock

Mercury

A picture of Mercury taken by the space probe, Messenger, in 2008. The crater walls are not as steep as those on the Moon, the craters are not as deep, and the ejected material landed closer to the impact site because of Mercury's greater surface gravity.
NASA, JHU APL, CIW

Mercury is a little more than one-third the size of Earth. If Earth were scaled down to the size of a tennis ball, Mercury would be about the size of a large marble. Using that same scale, Mercury would be about 304 miles away from the Sun.

From Earth, Mercury usually appears as a yellowish bright star. At night, we can see it in the western sky, setting about an hour after the sun. In the morning, it appears in the eastern sky, rising about an hour before the Sun.

Even though Mercury is closest to the Sun, it is not the hottest planet in our solar system. But it is the fastest planet, orbiting around the Sun every 88 Earth days. In fact, Mercury gets its name from one of the fastest ancient Roman gods!

Mercury's surface is much like Earth's moon, marked by billions of years of asteroid impacts. Large asteroids crashed into the planet long ago.

They created very large impact craters, including the Caloris Basin. We can also find extinct volcanoes, huge plains, and vast cliffs on the surface of Mercury.

At 3,033 miles in diameter, Mercury is the smallest of the eight planets. Determined by gravitational pull, its mass is only 0.553 of Earth's mass, this means that if you weighed 150 pounds (lbs.) on Earth, you would only weigh 57 lbs. on Mercury. That's because Mercury's gravitational pull is only about .378 of Earth's. Like the Earth, iron and nickel form Mercury's core, making it high in density. Researchers believe a rocky mantle surrounds the core, topped by a solid crust scarred by asteroid impacts.

Two missions have visited Mercury. The first, Mariner 10, was in 1974-1975. The second, MESSENGER (MErcury Surface, Space ENvironment, GEochemistry and Ranging), flew past Mercury three times before going into orbit around Mercury in 2011. These missions gave us clear pictures of the planet's surface. They also detected a weak magnetic field on Mercury.

Mercury's egg-shaped orbit takes the planet as close as 29 million miles and as far as 43 million miles from the Sun. Remember that Mercury orbits the Sun every 88 Earth days. It also rotates every 59 Earth days, rotating three times in two orbits. That means there are 176 Earth days between one Mercury sunrise and the next!

Every 11 years, early risers have an opportunity to see five naked-eye planets in early morning skies during late January through February.

Courtesy of NASA/JPL-Caltech

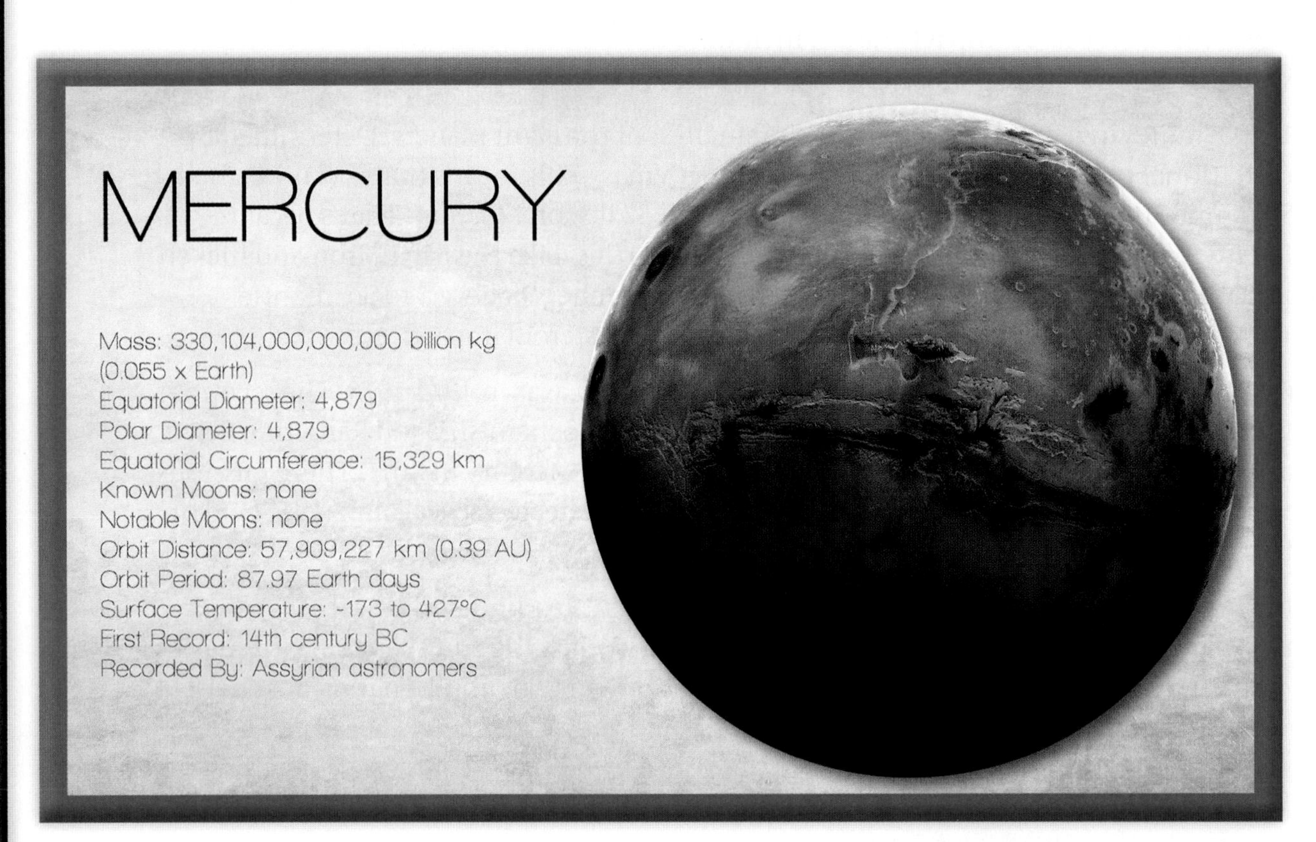

Mercury – Facts and Figures
Vadim Sadovski/Shutterstock

Did You Know?

Early researchers believed Venus would be a tropical paradise. The clouds of sulfuric acid around Venus made it impossible to view the surface, but the location, orbit, and similarities to Earth made researchers believe its environment to be favorable. In the 1960s, scientists finally observed the extreme temperatures and hostile environments of Venus. Scientists do believe that Venus once had oceans, but that they eventually evaporated as the temperature on the planet increased.

Venus

Venus was named after the Roman goddess of love and beauty. The second planet from the Sun, Venus is like Earth in size and structure. Venus is one of the easiest planets to observe from Earth because it is so bright. Except for the Sun and the Moon, Venus is the brightest object in the sky. It appears either before sunrise or after sunset and is often referred to as the Morning Star and Evening Star.

Thick clouds cover the surface of Venus and hide a harsh environment of extreme temperatures and pressures. The landscape is deserted, dry, and formed by large lava flows and tens of thousands of volcanoes. Venus is also covered in craters that are more than 0.9 to 1.2 miles across.

The diameter of Venus is 95 percent of Earth's and its mass is 82 percent of Earth's. With so much volcanic rock on its surface, scientists believe Venus must have a dense metallic core. Venus may be similar in size to Earth and has a similarly-sized core, but its magnetic field is much weaker than Earth's. So is its gravitational pull. If you weighed 150 lbs. on Earth, you would weigh 136 lbs. on Venus. Although much closer to Earth, Venus' gravitational pull is not as strong as the planet Earth.

Why is Venus's magnetic field so weak? One reason is its very slow and backward rotation. This planet is unusual because it spins the opposite direction of Earth and most other planets. It takes about 243 Earth days to spin around just once. Because it is so close to the Sun, a year goes by in just 225 days. That means, a day on Venus is just slightly longer than a year on Venus. With the day and year lengths being similar, a day on Venus is not like a day on Earth. On Earth, the Sun rises and sets once a day. On Venus, the sun rises and sets every 117 Earth days.

Made up mostly of carbon dioxide—along with nitrogen and small amounts of water—Venus's atmosphere is also very different than Earth's. The atmosphere forms large dense clouds above Venus's polar regions. These clouds contain sulfuric acid and create a strong greenhouse effect. The greenhouse effect *is the exchange of incoming and outgoing radiation that warms a planet.*

Greenhouse effect is like a car parked outside on a cold and sunny day. The Sun warms up the inside of the car. The heat is trapped inside the car when the windows are rolled up and the car doors are closed. The greenhouse effect is a similar concept. The Sun provides the heat, but the atmosphere traps the heat inside and warms up the planet. As a result of the greenhouse effect, Venus's surface temperatures are higher than 880°F.

Through a telescope, Venus shows phases that change as it orbits the Sun.
Daniel Herron/Courtesy of NASA.gov.

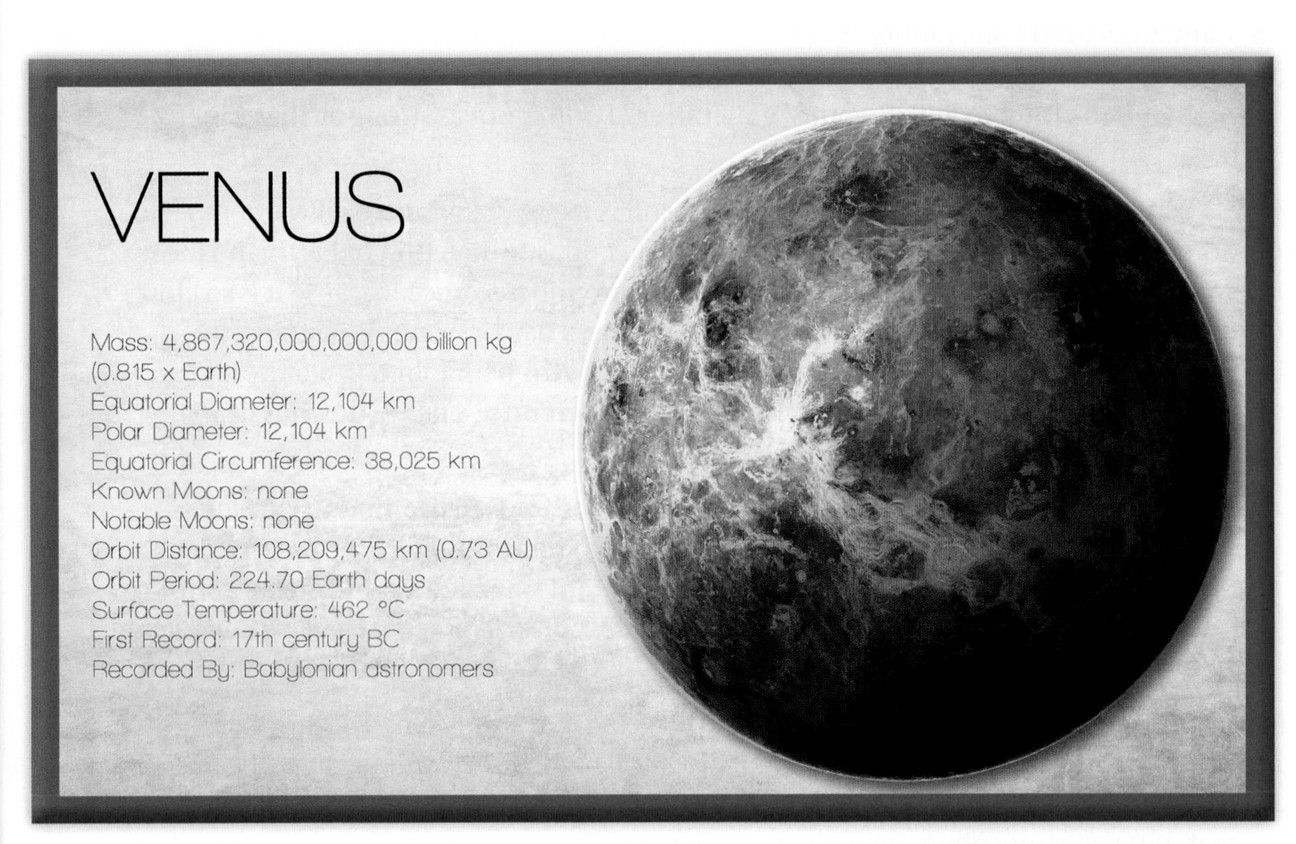

Venus – Facts and Figures
Vadim Sadovski/Shutterstock

Mars

The last of the terrestrial planets, Mars is fourth from the Sun, and the second smallest planet. Named after the Roman god of war, Ares, Mars is also called the "red planet." This is *because of the rust in the iron-filled Martian soil*. Solar winds from the Sun often cause dust storms. These make Mars's appearance change and produce large areas of sand dunes. Mars is easiest to see from Earth when it is in opposition, meaning when it is *on the opposite side of the Earth from the Sun*. Its opposition occurs about every 2.2 years.

Earth is 93 million miles from the Sun. In comparison, the average distance from Mars to the Sun is about 142 million miles. Mars also travels in an elliptical orbit. Its distance from the Sun changes from about 130 to 155 million miles. Because it is farther from the Sun than Earth, one revolution for Mars takes 687 days.

The axis of rotation for Mars is tilted very much like Earth's. Earth's tilt is about 23.5 degrees, and Mars is tilted by 25.2 degrees. Like Earth, the north pole on Mars tilts away from the Sun during winter for the northern hemisphere. Because of its irregular orbit, its northern hemisphere is closer to the Sun during the winter. This causes less difference in summer and winter temperatures. Mars rotates in the same direction as Earth, from west to east, with Mars taking only about 40 minutes longer.

The diameter of Mars is about one-half the Earth's diameter. Picture the Sun's diameter as the height of a door. Earth would appear as the size of a dime, and Mars would be about the size of an aspirin.

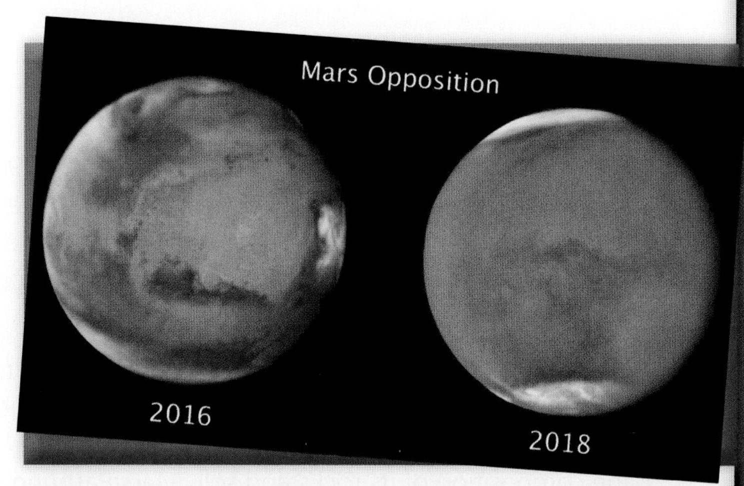

Comparison of Mars's appearance near its 2016 and 2018 oppositions. In 2016, the Martian atmosphere was clear. The 2018 image shows almost the same face of Mars. Surface features obscured by dust, the planet's cloud-hidden south pole is tilted more toward the Sun.

Courtesy of NASA, ESA, and STScI

Let's compare mass, density, and gravity on Mars and Earth. Mars's mass is only one-tenth that of Earth's. Its density is about 0.7 as great as Earth's, and its gravity is much lower, .375 of the gravity on Earth. What does this mean? A person weighing 150 pounds on Earth would only weigh about 56.5 lbs. on Mars. That person could jump higher on Mars. So, every basketball player could easily dunk the ball, even if the rim of the goal was set at 20 feet high!

Mars has some magnetic iron, as Earth does, but has no magnetic field because it does not have an iron core. Instead, its iron is evenly scattered throughout the planet. The lack of a magnetic field has caused the Sun's solar winds to blow away much of Mars's atmosphere and water vapor.

The atmosphere is about 100 times thinner than Earth's, with about 95 percent carbon dioxide and very little water vapor. Because of its cold temperatures, it does have small polar ice caps. Earth's overall average temperature is about 57^{0}F. For Mars, the average is about -81^{0}F. At its equator, the temperatures change from about 70 degrees Fahrenheit at noon to negative 225 degrees at night.

Even though Mars is smaller than Earth, it has the highest mountain of any planet in the solar system. Mt. Everest is about five miles high. Mars's volcanic mountain, Olympus Mons, is about 15 miles high with a base that is about 400 miles across. That is almost as far as the distance between Boston and Washington, DC!

The Sojourner planetary rover that was sent to Mars in 1997 onboard the Pathfinder spacecraft examined soil samples and found some iron and the same quartz material that is in our sand..

Courtesy of NASA.gov.

Did You Know?

Mars Pathfinder was launched on December 4, 1996 and landed on Mars on July 4, 1997. It was designed as a technology demonstration of a new way to deliver an instrumented lander and the first-ever robotic rover to the surface of the red planet. Pathfinder not only accomplished this goal but also returned an unprecedented amount of data and outlived its primary design life.

Both the lander and the 23-pound (10.6 kilogram) rover carried instruments for scientific observations and to provide engineering data on the new technologies being demonstrated. Included were scientific instruments to analyze the Martian atmosphere, climate, geology, and the composition of its rocks and soil. Mars Pathfinder used an innovative method of directly entering the Martian atmosphere, assisted by a parachute to slow its descent through the thin Martian atmosphere and a giant system of airbags to cushion the impact.

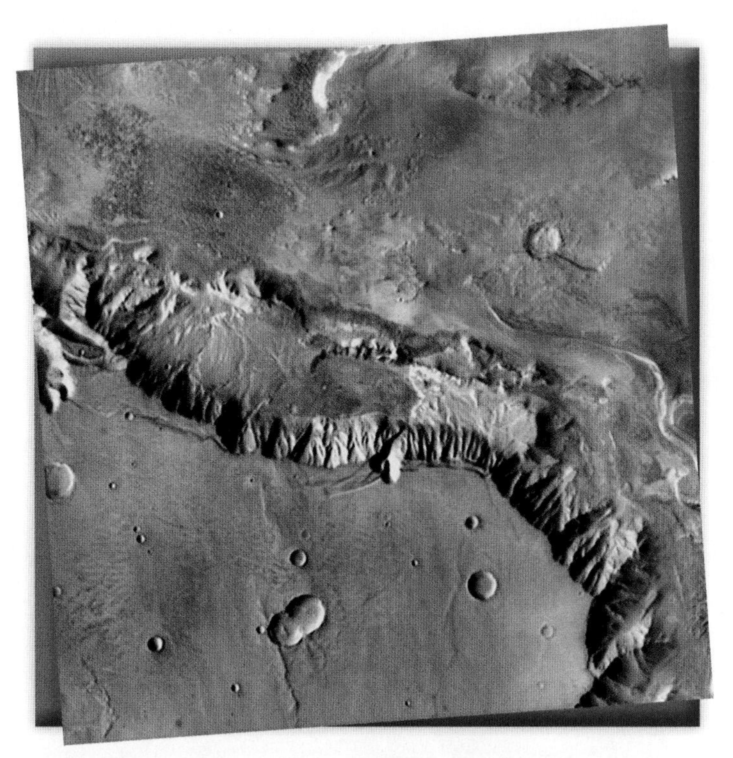

Valles Marineris is a chasm vastly larger than Earth's Grand Canyon that also has many layers of rock that serve as windows into the past.

Courtesy of NASA/JPL-Caltech/ASU

Mars also has a giant crater, Valles Marineris. This crater extends outward from the mountain for about 3,000 miles, more than 10 times longer than the Grand Canyon. Spacecraft sent to Mars have also discovered chasms or *channels that appear to have been made by water or glaciers* up to 3.5 billion years ago.

Mars has two moons, but they are very small and irregular in shape, much like an ordinary rock. The larger one is Phobos, named after the Greek word for fear, and is only about 17 miles across at its widest point. The smaller one is Deimos, named for the Greek word for terror. They both travel around Mars in the same direction that our Moon travels around Earth. We'll discuss more about the exploration of Mars in later chapters.

MARS

Mass: 641,693,000,000,000 billion kg (0.107 x Earth)
Equatorial Diameter: 6,805
Polar Diameter: 6,755
Equatorial Circumference: 21,297 km
Known Moons: 2
Notable Moons: Phobos & Deimos
Orbit Distance: 227,943,824 km (1.38 AU)
Orbit Period: 686.98 Earth days (1.88 Earth years)
Surface Temperature: -87 to -5 °C
First Record: 2nd millennium BC
Recorded By: Egyptian astronomers

Mars – Facts and Figures
Vadim Sadovski/Shutterstock

TABLE 1.1 Terrestrial Planets

	Mercury	Venus	Earth	Mars
Distance from the Sun	36 million miles	67 million miles	93 million miles	142 million miles
Orbit Length	88 Earth days	225 Earth days	365 days	687 Earth days
Length of Day	59 Earth days	243 Earth days	1 Earth day	24.6 Earth days
Diameter	3,033 miles	7,521 miles	7,917 miles	4,221 miles
Surface Temperature	-292°F to 806°F	867°F	57°F	-225°F to 70°F
Atmospheric Composition	42% Oxygen 29% Sodium 22% Hydrogen 6% Helium 1% Trace Gases	96.4% Carbon Dioxide 3.4% Nitrogen .015% Sulfur Dioxide .007% Argon .002% Water Vapor	78% Nitrogen 21% Oxygen	95.3% Carbon Dioxide 2.7% Nitrogen 1.6 % Argon

The Outer Planets and Dwarf Planets

In Chapter 2, Lesson 2, we learned that the outer planets, or Jovian planets, are large and gaseous. These include Jupiter, Saturn, Uranus, and Neptune. With lower average density, they all have ring systems and many moons. Made mostly of hydrogen and helium, we often refer to them as the "gas giants."

Jupiter

Jupiter is the largest of the gas giants. Compared to Earth, Jupiter is 11 times wider. It is about five times farther from the Sun than Earth. Jupiter takes almost 12 Earth years to cycle around the Sun, and it spins on its axis about every 10 hours.

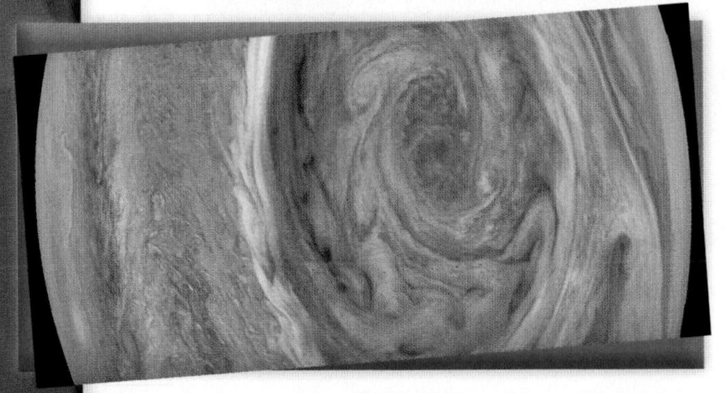

Jupiter's atmospheric bands are impacted by storms and winds of up to 388 miles per hour. The largest of the storms is the Great Red Spot. The storm itself is twice the size of Earth!

NASA/JPL-Caltech/SwRI/MSSS/Kevin Gill

Unique to Jupiter is its differential rotation. Differential rotation *is seen when a rotating object moves with different periods of rotation*. The giant planet has a system of bands that move at different speeds depending on their closeness to the equator. Jupiter is also oblate, or *somewhat flattened at its poles*. Jupiter's equator tilts three degrees from the vertical. This means it spins nearly upright and does not have seasons.

Made up mostly of hydrogen, Jupiter also contains helium and small amounts of methane, ammonia, and water vapor. Electric currents in an inner layer of metallic hydrogen create a strong magnetic field.

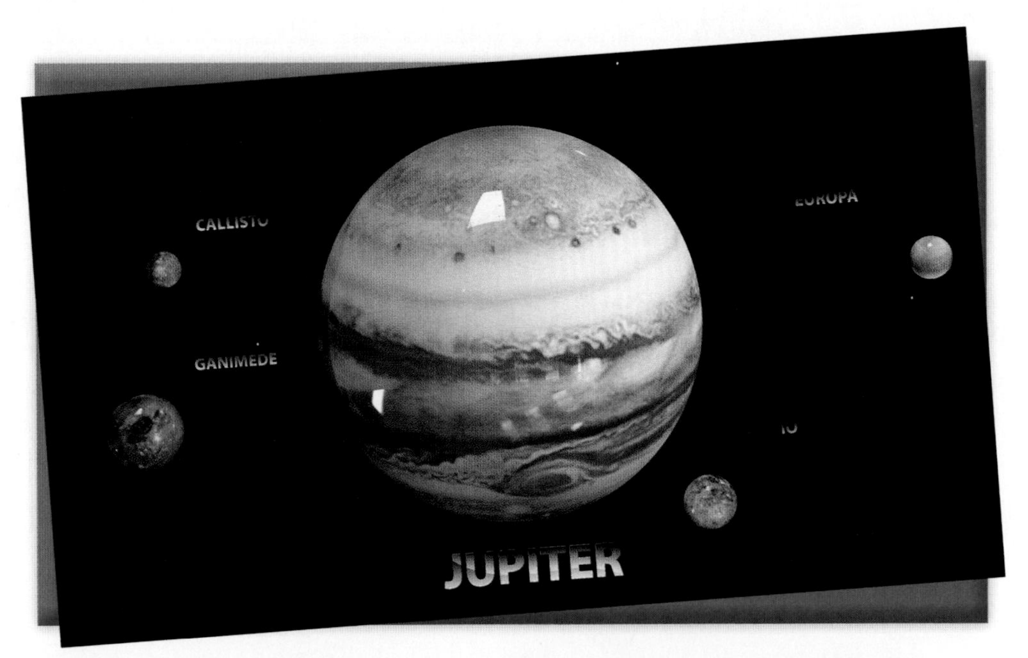

3D illustration of Jupiter's moons. Elements of this image furnished by NASA.

Victor Josan/Shutterstock

JUPITER

Mass: 1,898,130,000,000,000,000 billion kg (317.83 x Earth)
Equatorial Diameter: 142,984 km
Polar Diameter: 133,709 km
Equatorial Circumference: 439,264 km
Known Moons: 67
Notable Moons: Io, Europa, Ganymede & Callisto
Known Rings: 4
Orbit Distance: 778,340,821 km (5.20 AU)
Orbit Period: 4,332.82 Earth days (11.86 Earth years)
Surface Temperature: -108°C
First Record: 7th or 8th century BC
Recorded By: Babylonian astronomers

Jupiter – Facts and Figures
Vadim Sadovski/Shutterstock

Hydrogen is usually a gas, but with Jupiter's temperature and pressure, the hydrogen acts more like a metal and conducts electricity. At its center, Jupiter is extremely hot, up to 43,000°F. However, the temperature at the level of the planet's clouds is about -234°F.

Did You Know?

In 1979, NASA's Voyager I discovered Jupiter's faint, thin rings. Composed of small, dark particles, they are difficult to see except when backlit by the Sun. Data from the Galileo spacecraft in 1996 suggested that Jupiter's rings may have been created by dust from meteoroids crashing into the planet's inner moonlets.

Jupiter has 53 named moons, and scientists now believe Jupiter could have as many as 79 moons. The four largest moons are known as Galilean satellites, *because they were first observed by Galileo in 1610*. These moons, Io, Ganymede, Callisto, and Europa, are some of the most curious objects in our solar system. Volcanically active, Io has volcanic vents, lava flows, and lava plumes. Ganymede is the largest moon in the solar system, even bigger than the planet Mercury. Craters and ringed objects caused by asteroid impacts cover Callisto's surface. Europa just might be the most fascinating of Jupiter's moons. With a liquid water ocean beneath an icy crust, Europa may hold some of the elements needed to sustain life.

Saturn. Elements furnished by NASA
Vadim Sadovski /Shutterstock

Saturn

Saturn is the second largest and the sixth planet from the Sun. It has a small solid center or core, likely made of iron, but its core is still about 10 to 20 times larger than Earth's. Even with its dense core, we classify Saturn as a "gas planet." It is mostly made of gases, mainly hydrogen (about 96 percent) and helium (about three percent). About nine times wider than Earth, Saturn's density is so low because of these gases. If it were possible, Saturn would float in water! However, Saturn still has a strong gravitational pull, so if you weighed 150 lbs. on Earth you would weigh 137 lbs. on Saturn.

After Jupiter, Saturn is the second fastest spinning planet. It spins or rotates in about 10.5 hours. Because it spins so fast and is made of gases, Saturn also has an oblate shape like Jupiter. Although it rotates very fast, it takes about 29.5 years to revolve around the Sun because it is so far out in the solar system. Like Earth's axis of rotation tilt, Saturn's tilt is 27 degrees.

Saturn has a magnetic field about 600 times stronger than ours. That causes it to have northern and southern lights or auroras, like we have on Earth, as Saturn attracts charged particles thrown off from the Sun.

We have long been fascinated by Saturn's distinct rings. The rings were first written about by Galileo in 1610 when he observed them with a weak 20x telescope. Over the years, the rings seemed to disappear and then appear again, as Saturn's position shifted.

This image of the northern polar region of Saturn shows both the aurora and underlying atmosphere, seen at two different wavelengths of infrared light as captured by NASA's Cassini spacecraft.
Courtesy of NASA/JPL/University of Arizona

Dutch astronomer Christiaan Huygens described the rings around Saturn some 40 years later. They were thin and flat but separate from each other within a flat plane. Every 15 years or so, as the planets orbit the Sun, our Earth passes across the plane of Saturn's rings. Galileo would see them if he was looking upward or downward into that plane. When he was observing them from the edge of the plane, they were not visible.

Imagine looking at a target with a bull's-eye marked on it. From several yards away, the bull's-eye is easy to see.

However, if the target is tilted so you are looking at the edge of it, it would be hard to see it at all.

Scientists believe that Saturn's rings are made of water, ice, and rock from the size of sand grains up to about 30 feet in diameter. They are probably fragments from comets, asteroids, or broken moons. The rings are close together except for a gap. The Cassini Division is *the space between the second and third rings of Saturn.* This space may be caused by the orbits of small moons which apply a pull of gravity on the particles within the rings.

Saturn has more than 60 moons. Titan is Saturn's largest moon. It lies outside of the rings and is the second largest moon in the solar system, second only to Ganymede (Jupiter's largest moon).

NASA Cassini spacecraft captures the shadow of Saturn moon Mimas as it dips onto the planet rings and straddles the Cassini Division in this natural color image.

Courtesy of NASA/JPL/Space Science Institute

Did You Know?

In 2004, NASA's Cassini spacecraft first orbited Saturn, to reveal much about its system of rings and moons. In 2017, Cassini was purposely plunged into Saturn's atmosphere, allowing data to be collected directly from Saturn's atmosphere.

SATURN

Mass: 568,319,000,000,000,000 billion kg (95.16 x Earth)
Equatorial Diameter: 120,536 km
Polar Diameter: 108,728 km
Equatorial Circumference: 365,882 km
Known Moons: 62
Notable Moons: Titan, Enceladus, Iapetus, Mimas, Tethys, Dione & Rhea.
Known Rings: 30+ (7 Groups)
Orbit Distance: 1,426,666,422 km (9.58 AU)
Orbit Period: 10,755.70 Earth days (29.45 Earth years)
Surface Temperature: -139 °C
First Record: 8th century BC
Recorded By: Assyrians

Saturn – Facts and Figures

Vadim Sadovski/Shutterstock

Uranus

Uranus is the seventh planet from the Sun and the third largest in the solar system. It is a giant planet. Uranus has 27 known satellites. Elements of this image furnished by NASA.
NASA images/Shutterstock

Uranus is the seventh planet from the Sun, and its orbit is between those of Saturn and Neptune. It is also the third largest planet based on its radius and is four times wider than Earth. By comparison, if you imagine Earth as the size of a tennis ball, Uranus would be the size of a basketball. It has the fourth largest mass of all planets in our solar system with a mass about 63 times greater than Earth's. Of the large Jovian planets, Uranus and Neptune are the "ice giants." Large and extremely cold, temperatures on Uranus reach the coldest of any planet in the solar system.

Discovered in 1781 by William Herschel, Uranus was the first planet to be detected using a telescope. If conditions are perfect and you know exactly where to look, Uranus can just barely be seen from Earth without a telescope.

Uranus has a rocky core containing metals like iron and nickel. Like Earth, it has a second layer called the mantle, but its mantle is made of water, methane, and ammonia. Its atmosphere consists of about 83 percent hydrogen, 15 percent helium, and two percent methane. It has extremely strong winds, sometimes reaching about 500 miles per hour.

The rotation of Uranus is unusual for two reasons. Other planets have an axis of rotation that passes through their north and south poles. The axis tilts from about 0 to about 30 degrees, so the planets spin like a top. The rotation axis of Uranus is 98 degrees, so it turns more like a wheel.

The rotational axis of Uranus is tilted almost parallel to its orbital plane.

Using infrared filters, Hubble captured detailed features of three layers of Uranus atmosphere.

Courtesy of NASA/JPL/STScI

Instead of its equator always pointing toward the Sun as with other planets, the north or south poles of Uranus sometimes points toward the Sun. This affects the seasons. Its moons do revolve around its equator like those of the other planets.

The other strange thing about its rotation is its retrograde rotation. Recall that this is also true for Venus. As the other planets rotate west to east, these two spin east to west. Its rotation time is 17 hours, 14 minutes compared to our rotation of 24 hours. While Earth revolves around the Sun in one year, Uranus takes 84 Earth years for one orbit.

NASA's Voyager 2 is the only spacecraft that has traveled near Uranus. In 1986, it provided a lot of information about the outer regions of this planet. Uranus has 27 moons made of ice and rock. The largest moon is Titania, which is about one-eighth the diameter of Earth.

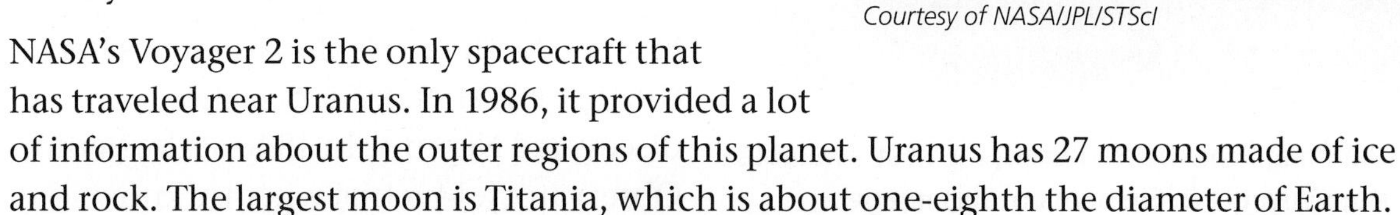

Uranus has 13 rings of unknown composition. Eleven of those are inner rings, which are dark and narrow, but the two outer rings are brightly colored. The rings have sharply outlined edges instead of being hard to see. That seems to be because they have shepherd moons, one moon on each side of the ring, which hold the rings together.

URANUS

Mass: 86,810,300,000,000,000 billion kg (14.536 x Earth)
Equatorial Diameter: 51,118 km
Polar Diameter: 49,946 km
Equatorial Circumference: 159,354 km
Known Moons: 27
Notable Moons: Oberon, Titania, Miranda, Ariel & Umbriel
Known Rings: 13
Orbit Distance: 2,870,658,186 km (19.22 AU)
Orbit Period: 30,687.15 Earth days (84.02 Earth years)
Surface Temperature: -197 °C
Discover Date: March 13th 1781
Discovered By: William Herschel

Uranus – Facts and Figures

Vadim Sadovski/Shutterstock

Neptune

At long last, we have Neptune, the eighth planet from the Sun and the most distant planet in our solar system. Like Uranus, it is also four times wider than Earth. Neptune is dark, cold, and full of wind and ice. It is mostly made up of icy materials, water, methane, and ammonia, outside its small rocky core. Of the Jovian planets, it is the densest. Hydrogen, helium, and methane form Neptune's atmosphere.

Inspired by the work of mathematician Urbain Le Verrier, Johann Galle discovered Neptune in 1846 at the Berlin Observatory. Much of what we know about Neptune today is based on NASA's Voyager 2. No other spacecraft has orbited the planet to study it up close.

Neptune spins every 16 hours and it takes about 165 years to revolve around the Sun because it is so far out in the solar system. Neptune's axis of rotation is like Earth's, with a tilt of 28 degrees. Neptune has seasons, although each season lasts over several decades. Neptune also has a strong magnetic field and its main axis is tipped over by about 47 degrees compared with its rotation axis.

Neptune has 13 moons. Triton is the only large moon in our solar system to orbit its planet in the opposite direction (clockwise) of its planet's rotation. A second major moon of Neptune is Nereid. It orbits in the same direction as Neptune's rotation, but it has the most eccentric orbit in our solar system. Neptune also has five known rings, extending from 25,500 to 40,000 miles from the planet.

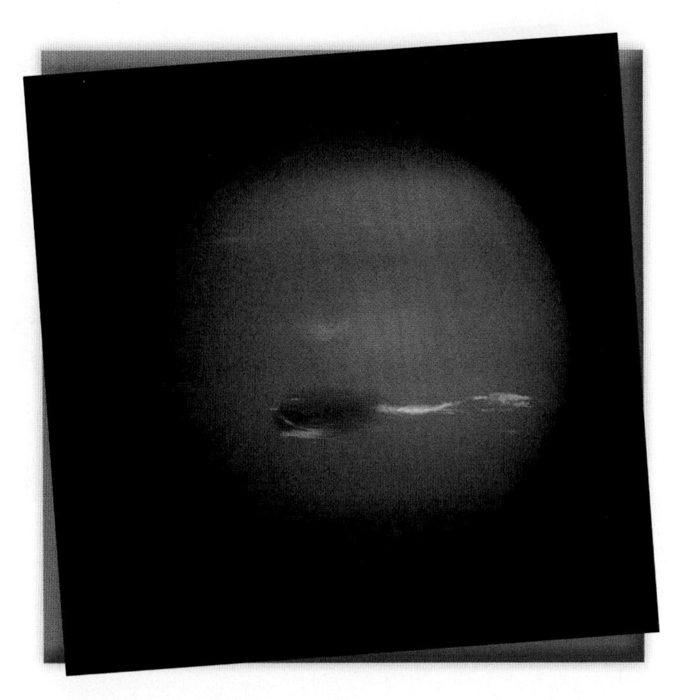

Neptune. An image constructed using the best available imagery from Voyager 2. Elements of this image furnished by NASA.

MarcelClemens/Shutterstock

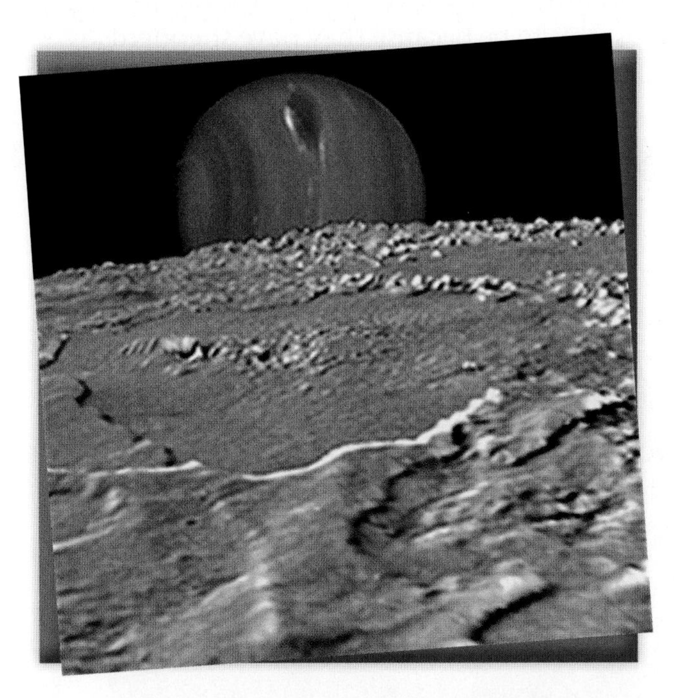

This composite illustration is of the planet Neptune, as seen from its moon Triton. Neptune's south pole is to the left; clearly visible in the planets' southern hemisphere is a Great Dark Spot, a large anti-cyclonic storm system. This three-dimensional view was created using images from the Voyager spacecraft.

Courtesy of NASA

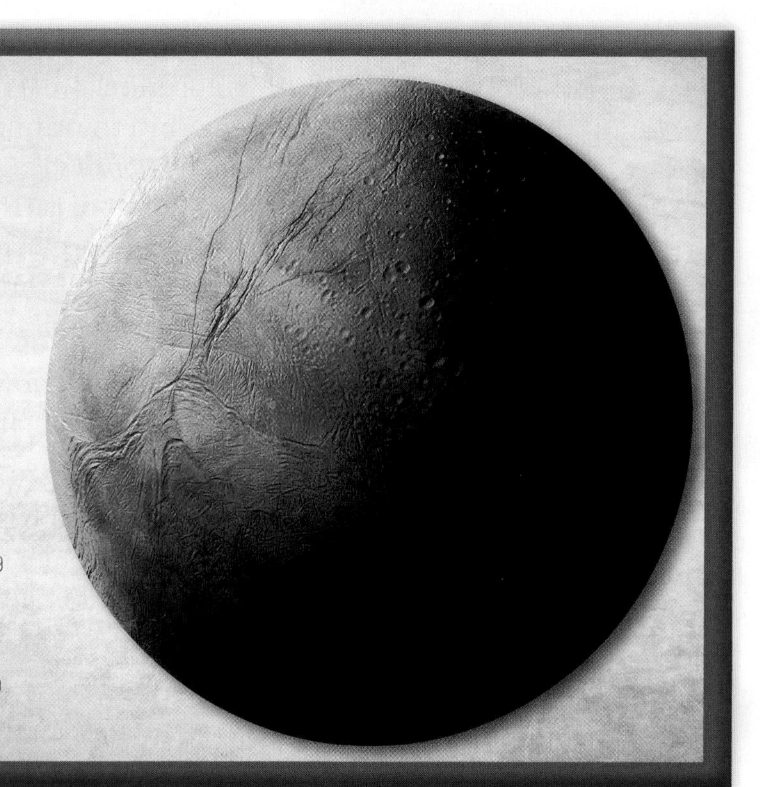

Neptune – Facts and Figures

Vadim Sadovski/Shutterstock

Pluto and Dwarf Planets

Pluto was once considered to be a planet until scientists reclassified it. Remember that a planet, by definition, meets three requirements. It orbits the Sun, it is large enough to be a spherical shape, and it is large enough to have cleared away other objects of a similar size near its orbit. Because Pluto is not large enough to have cleared away other objects, we call it a dwarf planet. Our solar system has five dwarf planets. Haumea, Makemake, and Eris are three dwarf planets that orbit around Neptune. Another dwarf planet, Ceres, orbits much closer to the Sun.

American astronomer Clyde Tombaugh discovered Pluto in 1930 at the Lowell Observatory in Arizona. Tombaugh took pictures of the night sky to see if anything moved against the background stars. Over time, he found a star that seemed to change position.

Smaller than our Moon, Pluto is covered with red snow mountains, valleys, plains craters, and its landmark heart-shaped glacier. Pluto's orbit is unique in that it is elliptical and tilted. Its average distance from the Sun is 3.67 billion miles. From 1979 to 1999, Pluto was near perihelion, bringing the planet closer to the Sun than Neptune. A day on Pluto is about 153 hours. Like Venus and Uranus, Pluto also shows a retrograde rotation.

Did You Know?

Because of Pluto's elliptical orbit, Pluto is sometimes closer to the Sun than Neptune.

NASA's New Horizons spacecraft captured this image of Pluto on July 14, 2015.

Courtesy of NASA/JHUAPL/SWRI

Did You Know?

Pluto was reclassified as a dwarf planet in 2006. Scientists are reconsidering the decision based on a lack of scientific literature to support the actual definition of a planet.

Pluto has a weak atmosphere that includes molecular nitrogen, methane, and carbon monoxide. When Pluto is at perihelion, its surface sublimates and rises briefly to form a thin atmosphere. Sublimate refers to *the direct change from a solid to a gas*. Pluto has low gravity–about six percent of Earth's. As a result, the atmosphere is much more extended in altitude than ours. Pluto becomes much colder during aphelion (when it is farthest from the Sun), causing gases to condense on the surface.

Pluto has five known moons and no rings. The largest moon is Charon. Because Charon is about half the size of Pluto the two are often called a double planet.

Remember, Pluto is one of five dwarf planets in our solar system. Although nearly the same size as Pluto, the dwarf planet of Eris is almost three times farther from the Sun. Much like Pluto, there is still debate over its status as a planet.

Next, we have the dwarf planets Haumea and Makemake. Haumea is also close in size to Pluto but is shaped like a football due to its super-fast rotation. Makemake is a little smaller than Pluto, but almost as bright.

Lastly, we have the Sun-orbiting Ceres. Ceres also happens to be the largest asteroid in the asteroid belt at 580 miles in diameter. The asteroid belt consists of the *randomly scattered asteroids between the orbits of Mars and Jupiter*.

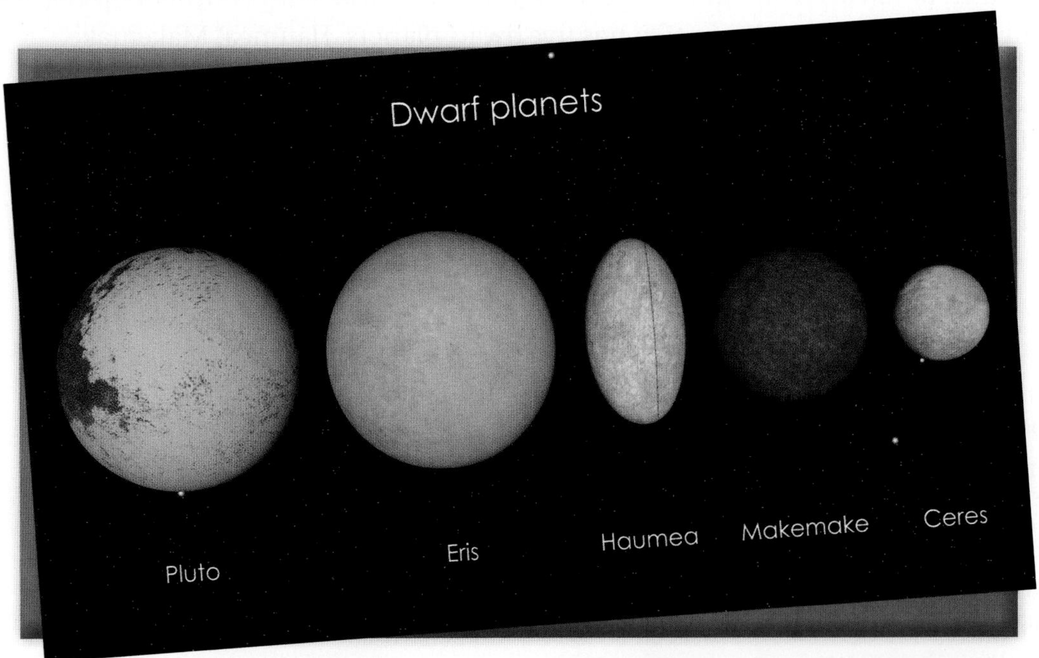

3D Illustration of Dwarf Planets.

Meletios Verras/Shutterstock

Asteroids, Comets and Meteors

Asteroids

Asteroids are large rocks ranging in size from about one mile up to a few hundred miles in diameter. There are millions of them, but only a few, like Ceres and Vesta, are large enough to have names.

It was once thought that asteroids formed when a planet exploded, but there is no theory that explains why a planet would explode, and the total mass of the asteroids would not even be as large as our Moon. Some asteroids do collide, which causes them to break into the smaller pieces that we see today.

We categorize asteroids by their position in space and by their chemical composition. Most asteroids orbit within the asteroid belt, and others, called trojan asteroids, are *asteroids that share orbits with larger planets*. There are also near-Earth asteroids that orbit near Earth. The chemical composition classifications of asteroids are C-types, S-types, and M-types. C-type asteroids consist of clay and silicate rocks. S-types include silicate materials and nickel-iron. M-types are metallic (nickel-iron).

Asteroids revolve around the Sun in elliptical orbits with unpredictable rotations. The strong gravity of Jupiter affects their orbits, creating gaps between groups of asteroids in the asteroid belt.

Second to Ceres, Vesta is the next largest asteroid at about 330 miles wide. This image is from the last sequence of images NASA's Dawn spacecraft obtained, looking down at Vesta's north pole as it was departing in 2012.

Courtesy of NASA/JPL-Caltech/UCLA/MPS/DLR/IDA

The image shows how asteroids are typically discovered by detection of their motion relative to background stars.

Courtesy of NASA/JPL-Caltech/CSS-Univ. of Arizona

Did You Know?

Double asteroids are two rocky bodies of roughly the same size that orbit each other.

Comets

Image taken in 1997 of the Comet Hale-Bopp with its two tails.
Courtesy of A. Dimai and D. Ghirardo, (Col Druscie Obs.), AAC

Comets appear in the sky at night like a ball of fire with a tail. They are chunks of ice, rock particles, and dust that orbit around the Sun but a lot farther away than the planets and asteroids.

Every comet has two visible parts, a head and a tail. The head has a small nucleus which is *the center of the comet's head made mostly of ice and frozen carbon dioxide (dry ice)*. The nucleus is surrounded by a coma, which is *a large cloud of gas and dust*. The tail is a long part of that gas and dust that is blown away from the head and always points away from the Sun. The coma and tail form when the comet comes close enough to the Sun for some of the frozen material to melt and vaporize.

The solar wind from the Sun may actually produce two visible tails. One is made of charged particles and is a light blue. The second one, made of dust, is brighter and whiter and is more visible.

Scientist believe that most comets originate either from the Kuiper Belt or from the Oort Cloud. Recall from earlier in this lesson that the Kuiper Belt is a ring of icy bodies outside of Neptune's orbit. Comets from the Kuiper Belt are called short-period comets because *they take less than 200 years to orbit the Sun.Comets that take more than 200 years to orbit the Sun* are called long-period comets. In fact, comets that move in from the Oort Cloud *a spherical shell surrounding our Sun with a distance of up to 100,000 astronomical units (AU)*, can have orbit periods of millions of years.

The Right Stuff

Halley's Comet

In 1705, the English astronomer Edmond Halley suggested that comets observed in years past were actually the same comet orbiting around the Sun every 75-76 years. Using Newton's theories of gravity and planetary motion, Halley predicted the comet would be visible again in 1758. His predictions proved correct! Sadly, he didn't get to see it because he died 16 years before that happened. Halley's Comet most recently appeared in 1986. Therefore, we should see it again in 2061.

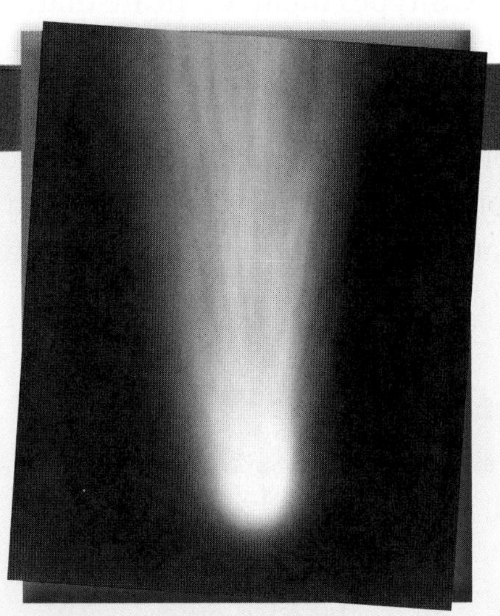

Halley's Comet in 1910.
Everett Historical/Shutterstock

Meteors

In addition to asteroids and comets, there are also meteoroids, meteors, and meteorites. Because the terms are so similar and often confused, let's look at the differences. A meteoroid is *a tiny rocky object in space usually the debris from a comet or an asteroid.* A meteor is *a very small meteoroid, usually smaller than a marble, which burns up as it enters a planet's atmosphere.* It is commonly called a shooting star or a falling star; for many, it presents a wish-making opportunity! Under perfect conditions, a person might see five or six meteors per hour. Sometimes, they occur more often in a meteor shower. A meteor becomes a meteorite if it actually *makes it through the atmosphere and strikes the surface of the planet or moon.*

The image – actually a composite of six exposures of about 30 seconds each – was taken in 2001, a year when there was a very active meteor shower.
Sean M. Sabatini/NASA

In this lesson, we have discussed the components of our solar system, including its planets, dwarf planets, the asteroid belt, icy Kuiper Belt, and the Oort Cloud. In the next lesson we will travel outside of our solar system to explore deep space and the wonders of the universe.

Lesson 3 Review

Using complete sentences, answer the following questions on a sheet of paper.

1. Describe the surface of Mercury.
2. Compare and contrast the rotation and orbit of Venus with that of Earth.
3. When is the best time to see Mars and why?
4. What is unique about Jupiter's rotation?
5. Describe the composition of Saturn's rings.
6. What is unique about Uranus's rotation?
7. What is the asteroid belt?
8. Explain the appearance of two tails on a comet.
9. What is the difference between a meteor, a meteoroid, and a meteorite?

APPLYING YOUR LEARNING

10. Given the most current research on Pluto, describe in a short paragraph whether Pluto should be a classified as a planet or remain a dwarf planet?

LESSON 4

Deep Space

Quick Write

Reflect on the story about Wylie Overstreet and Alex Gorosh's model of the solar system. What type of planning do you think was required to create an accurate model of the solar system?

Learn About

- the Milky Way galaxy
- other galaxies
- black holes, dark matter, nebulae, and pulsars
- big bang theory

The universe is so large that it can be hard to visualize. Many industries create models in order to visualize a concept. For example, architects may create models of a house or shopping mall before beginning the building process. But how do you create a model of the universe? Filmmakers Wylie Overstreet and Alex Gorosh decided it was a challenge they were up for. In their first project, To Scale: The Solar System, the duo created an accurately scaled model of the solar system. They used seven miles of dry lakebed in the Nevada desert to pull off this amazing model. In their model, the Earth was the size of a marble and the 140 million miles distance between Earth and Mars became 302 feet. The duo went a step further than just creating the model of the solar system, they used vehicles to create an accurate orbit for each planet. And when night arrived, the model came to life. Using lights attached to their vehicles, a time-lapse video was created showing an accurate model of the solar system and the orbits of each planet.

The project was so successful that they decided to move on to a larger project. Overstreet and Gorosh decided to build a scale model of time. In the project, To Scale: Time, the duo created a model to show the 13.8 billion years of the universe. Human life was a microscopic component of the model. The model itself spanned four miles, with 10 meters representing 20 million years. In the end, it took 12 minutes to film the entire 13.8 billion years of the universe's history.

The Milky Way Galaxy

A galaxy is *a group of stars, gas, and dust bound together by gravity*. The Earth and our entire solar system live in the Milky Way galaxy. And while it was originally assumed that our Sun was the center of the galaxy, our solar system is in fact closer to the edge of the galaxy. The Sun lies about 25,000 light years from the center of the galaxy. A light year is *the distance traveled by light in one year* (approximately 5,880,000,000,000 miles). American astronomer Harlow Shapley was the first to accurately determine the immense size of the galaxy and the location of our solar system. In 1917, Shapley used spatial distribution of globular clusters to determine the diameter of the galaxy. Globular clusters are *large compact spherical star clusters, typically of old stars in the outer regions of a galaxy*.

The image depicts the sun and the solar system in relation to the Milky Way galaxy. The white circle indicates the location of our solar system.

Courtesy of NASA/JPL-Caltech/T. Pyle

The Milky Way is a spiral galaxy. Think of it as a pinwheel, with a central bulge and four large spiral arms. The diameter of the galaxy is about 100,000 light years! In the spirals are glowing clouds that can be seen in the night sky. Inside the glowing clouds are gas and dust. This is where star formation occurs.

Vocabulary

- galaxy
- light year
- globular clusters
- galactic year
- Population II stars
- dwarf star
- dark matter
- interstellar medium (ISM)
- supernova
- charge exchange
- exoplanets
- elliptical galaxy
- lenticular galaxy
- irregular galaxy
- spiral galaxy
- starburst
- galaxy clusters
- superclusters
- black hole
- event horizon
- gravitational lensing
- nebulae
- emission nebulae
- reflection nebulae
- dark nebulae
- planetary nebulae
- supernova remnant
- neutron star
- pulsar
- cosmic microwave background (CMB)

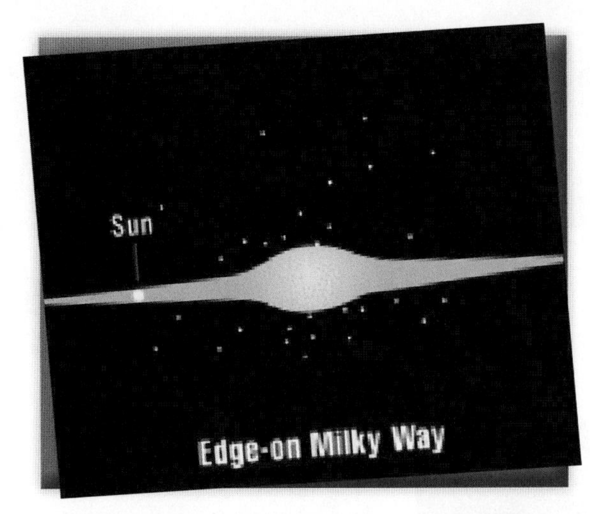

While the diameter of the galaxy is 100,000 light years, the galaxy is only about 2,000 light years thick, so it forms a thin disk.
StarDate.org

Astronomers estimate that 200-400 billion stars live in the Milky Way galaxy. On a clear night, in a rural area away from light pollution, you can see the Milky Way galaxy. You will see faint light with dark patches. The dark patches are clouds of dust and gas that are lit by the many stars in the galaxy. It looks almost like milk spilt on the night sky, hence the name the Milky Way.

While our solar system orbits the Sun, the Sun orbits the center of the Milky Way galaxy. The galaxy is so big that *an entire orbit takes 250 million years*; this is known as a galactic year.

A view of the Milky Way galaxy in the night sky.
Fabio Lamanna/Shutterstock

Did You Know?

More than half of the stars in the galaxy are over 4.5 billion years old. It is believed that galaxies go through a stellar baby boom in which massive amounts of stars are produced. It is estimated that the stellar baby boom occurred over 10 billion years ago for the Milky Way.

Structure of the Milky Way

The structure of the Milky Way galaxy is similar to other spiral galaxies. It consists of six separate parts:

- Nucleus
- Central bulge
- Disk
- Spiral arms
- Spherical component
- Halo

The nucleus is the center of the galaxy in which lies a gigantic black hole. The black hole in the center of the Milky Way is not your average black hole. It is a supermassive black hole that is 4,000,000 times the size of the Sun. This black hole may have started out small, but has grown exponentially with the ample amount of dust and gas available for it to consume. Although you cannot see a black hole, astronomers are able to locate black holes by observing the effects on nearby stars and gases. It is believed that most galaxies have a black hole in their center. We'll discuss black holes in further detail later in this lesson.

The central bulge is the area surrounding the nucleus. This area is almost perfectly spherical and contains mostly Population II stars. Population II stars are *older stars that are less luminous*. They are typically found in globular clusters.

The disk is the most difficult part of the galaxy to observe and extends out approximately 75,000 light years from the nucleus. Think of the disk as the area where the arms of the galaxy are attached. The disk contains stars and gas clouds. There are two components to the disk: the "thin disk," which contains dust, gas and the youngest stars; and the "thick disk," which contains older stars.

Omega Centauri is the largest globular cluster in the sky.

Albert Barr/Shutterstock

An illustration of the Milky Way galaxy and the spiral arms.
Courtesy of NASA/JPL-Caltech/R. Hurt (SSC/Caltech)

The spiral arms are probably the most noticeable feature of the galaxy. There are four main spiral arms: Norma and Cygnus, Sagittarius, Scutum-Crux, and Perseus. In addition to the main arms, there are several minor arms, or spurs. Our solar system is located on the Orion Spur. New stars are constantly being formed in the arms.

The galaxy is constantly moving. The material in the center of the galaxy moves faster, and, as it moves, the material in the center of the galaxy stretches out to create the spiral arms. Scientists believe the spiral arms remain as a result of density waves, or areas of greater density. The density wave model was proposed by C.C. Lin and Frank Shy in the 1960s to explain the spiral arms of the galaxy. Have you ever been in a traffic jam? Traffic slows down and you make your way past the cause of the traffic jam. But even after you have passed the cause of the traffic jam, traffic remains slower than normal. This is an example of density waves.

The spherical component of the galaxy extends above and below the disk and is really an extension of the central bulge. This region is populated by globular clusters of Population II stars and dwarf stars. A dwarf star is *a star with average or low luminosity, mass, and size.*

The halo, a spherical area that surrounds the Milky Way, is a massive component of the galaxy that may reach much farther than the 100,000 light year diameter of the galaxy. Not much is known about the halo other than its massive size. We do know that the halo only contains about two percent as many stars as found in the disk.

Dark matter is *invisible matter that can be detected through its gravitational pull* on the Milky Way's gas clouds. It is estimated that dark matter makes up 90 percent of the Milky Way. We'll discuss dark matter in further detail later in this lesson.

Did You Know?

The Milky Way galaxy gained its enormous size from devouring smaller galaxies throughout its existence.

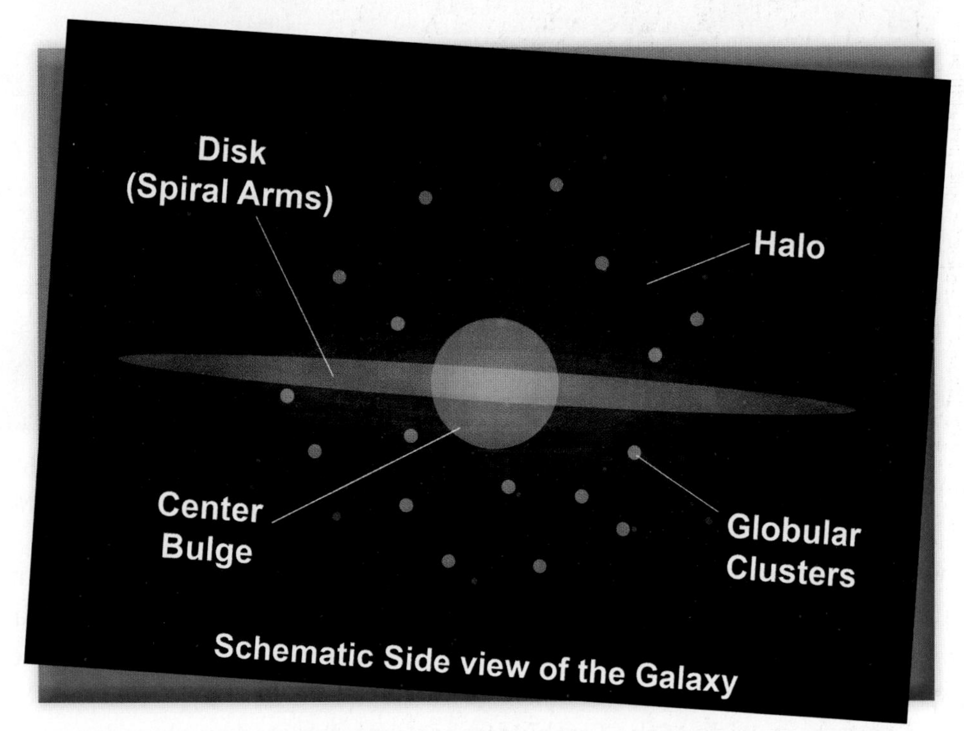

An illustration of the Milky Way galaxy and it's components.

The Local Bubble

Interstellar medium (ISM) is *the term used to describe the matter in galaxies that exists between solar systems*. This matter is made up of gases and dust. It is believed that "bubbles" exist in the ISM. Our solar system is located in one of the bubbles known as the Local Bubble. This region contains very hot gases, but is surrounded by cooler, dense air. The Local Bubble is about 300 lights years in diameter and surrounded by other bubbles.

It is believed that the Local Bubble was created by a supernova explosion over 10 million years ago. A supernova is *a massive explosion of a star*. A supernova explosion typically occurs once or twice every century in the Milky Way galaxy. However, 10 million years ago there was a plethora of supernova explosions. Think of when you make popcorn: after one kernel pops, many others start popping. This is what occurred around our solar system 10 million years ago.

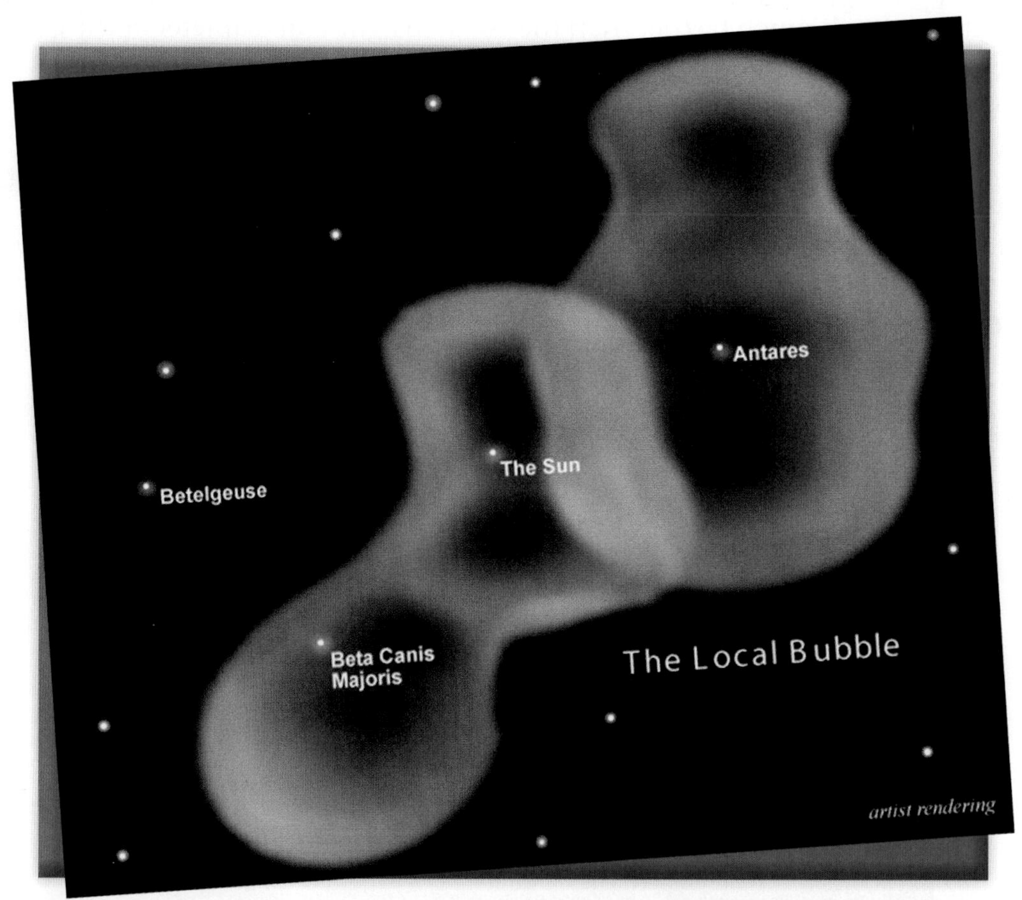

An artist's depiction of the local bubble.

Courtesy of NASA

So how did scientists discover the bubble? Scientists have always been fascinated with the interstellar medium (ISM) and continue to study the gas and dust found in deep space. However, when scientists started aiming their telescopes to study matter in our solar system they found very little of the gas, dust, and ions they found in ISM. In addition, sensors outside of the Earth's atmosphere detected an abundance of x-ray radiation. The conclusion was the formation of the bubble.

The supernova explosions were so strong they blew away the dust and gas particles in our solar system. The remnants of the supernova are the x-ray radiation in our solar system.

In recent years, the Local Bubble model has been debated. Some scientists believe that the x-ray radiation is not necessarily all coming from inside our solar system as a result of the supernova. They speculate that the radiation may come from charge exchange, which is *the passing solar wind stealing electrons and emitting radiation*. Data suggests that only 40 percent of the x-ray radiation is emitted from our solar system and that the remaining radiation is most likely coming from the hot gas walls of the bubble.

Exoplanets

Many other planets live in the Milky Way galaxy outside of our solar system, these planets are called exoplanets. Exoplanets are *planets that orbit a star outside of our solar system*. To date, over 3,700 exoplanets have been discovered and the number is constantly growing.

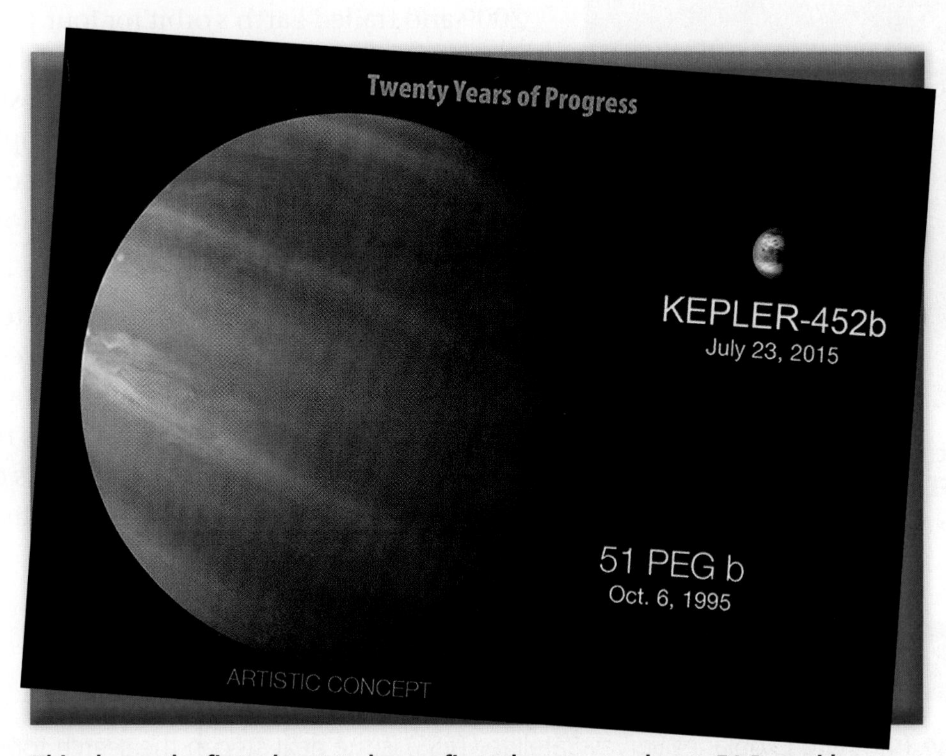

This shows the first planet to be confirmed as an exoplanet, 51 Pegasi b, and the newest Earth-sized planet confirmed.

Courtesy of NASA Ames/W. Stenzel

So how does a planet form? Scientists and philosophers have pondered this question for centuries. As scientists collect data from advanced telescopes and probes, they can speculate on the process of planet formation. But scientists have never been able to observe a planet in the process of formation. The data collected provides clues as to how planets, such as Earth, were formed. Theories suggest that planets are formed by large collisions occurring in space. The gas and dust left over from a collision are pulled together by gravity. As larger and larger pieces form together, the universe slowly creates a planet.

The Kepler spacecraft undergoing testing at Cape Canaveral, FL prior to launch in 2009.
Courtesy of NASA/Kim Shiflett

NASA has an entire division dedicated to exoplanet exploration called the Exoplanet Exploration Program (ExEP). For example, the Kepler mission is a NASA mission designed to study our region of the Milky Way for other planets that may be in the habitable zone. As a planet passes in front of a star, it is called a "transit." This event can be seen in the sky as a small black dot. The Kepler Space Telescope uses this phenomenon to detect other planets. Once a planet is detected, the orbital size and mass can be calculated to determine if the planet is in a habitable zone.

The Kepler Space Telescope was launched in 2009 and trailed Earth's orbit for four years. The data from the mission is still being analyzed and continues to reveal new planets. As a result of the four years of data, more than 2,000 exoplanets were confirmed. The Kepler Space Telescope mission ended in 2013 when two reaction wheels on the spacecraft failed.

This failure did not stop NASA's Kepler team. They had an innovative fix for the spacecraft. They used the pressure of sunlight to stabilize the axis of the telescope. The telescope was renamed the K2 and continues its path in Earth's orbit, sending data that may help scientists discover new exoplanets.

Did You Know?

The Transiting Exoplanet Survey Satellite (TESS) is the next step in the search for planets outside of our solar system, including those that could support life. The mission will find exoplanets that periodically block part of the light from their host stars, events called transits. TESS will survey 200,000 of the brightest stars near the Sun to search for transiting exoplanets. TESS was launched on April 18, 2018, aboard a SpaceX Falcon 9 rocket. The stars TESS will study are 30 to 100 times brighter than those the Kepler and K2 missions surveyed, which will enable far easier follow-up observations with both ground-based and space-based telescopes. TESS will also cover a sky area 400 times larger than that monitored by Kepler.

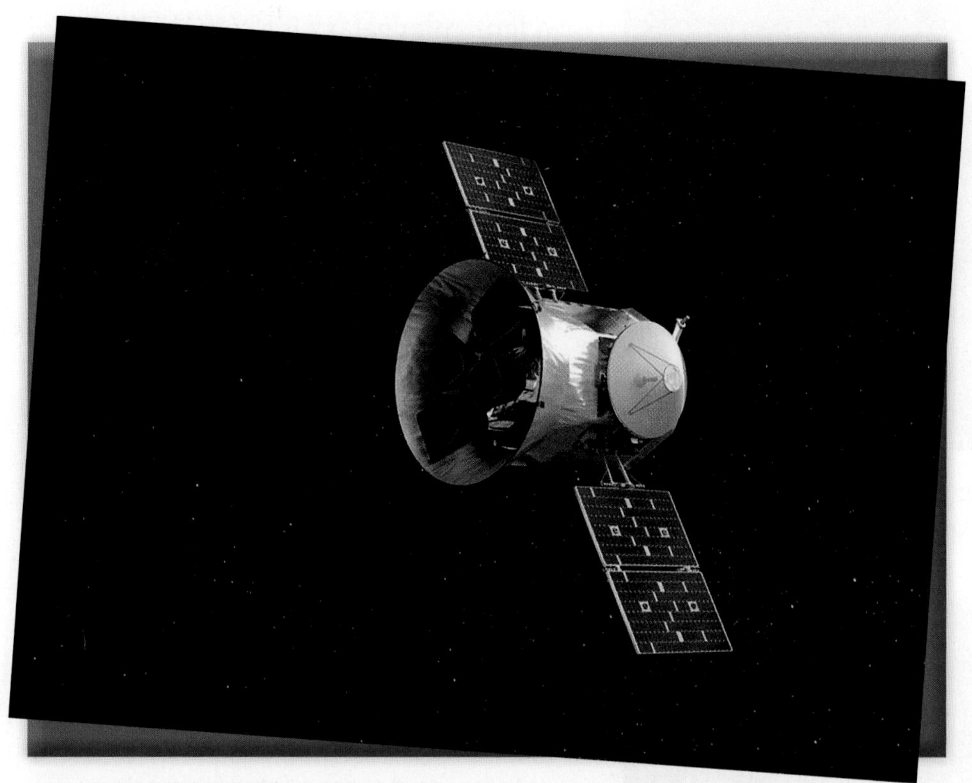

Illustration of NASA's Transiting Exoplanet Survey Satellite — TESS.
Courtesy of Goddard Space Flight Center/NASA

Other Galaxies

The universe is home to many other galaxies. Scientists theorize that galaxies are first formed with a small group of stars. The galaxy then merges with other galaxies and became a larger galaxy, like the Milky Way. Galaxies can come in different sizes and shapes. There are four known galaxy shapes: elliptical, lenticular, irregular, and spiral.

An elliptical galaxy is *elliptical in shape and does not have spiral arms or gas and dust particles as seen in the Milky Way*. Elliptical galaxies are typically home to older stars.

An elliptical galaxy appears to be mostly a big ball of light from the massive number of stars it contains.
Courtesy of Shutterstock/Diego Barucco

Figure 2.1: Lenticular galaxy NGC 4565
Courtesy of NASA/JPL-Caltech

A lenticular galaxy is a *spherically shaped galaxy with a disk of stars and gas around a nucleus*. The disk is similar to spiral galaxies, but does not have spiral arms or areas of star formation.

Figure 2.1 from NASA Galaxy Evolution Explorer shows NGC 4565, one of the nearest and brightest galaxies not included in the famous list by 18th-century comet hunter Charles Messier.

Figure 2.2: irregular galaxy IC 3583
Courtesy of NASA/JPL-Caltech

Irregular galaxies do not have a clear shape. These types of galaxies have large groupings of stars and areas of star formation.

This delicate blue group of stars in Figure 2.2 is an irregular galaxy named IC 3583 and sits some 30 million light years away in the constellation of Virgo (The Virgin).

Figure 2.3: Barred spiral galaxy M83
Courtesy of NASA/JPL-Caltech

Spiral galaxies are *spiral in shape and have long spiral arms with young stars and star formation areas*. Spiral galaxies are easily recognizable and mesmerizing to see. As discussed earlier, the Milky Way galaxy is a spiral galaxy.

Figure 2.3 shows vibrant magentas and blues in a Hubble image of the barred spiral galaxy M83. The image reveals that the galaxy is ablaze with star formation.

Observing Galaxies from Earth

You may not realize it, but you can observe many galaxies from Earth in the night sky. You just need to know where to look to locate these magnificent sites.

Andromeda Galaxy

The Andromeda galaxy is 2.5 million light years away, yet it is the closest big galaxy to the Milky Way, and you can see it without a telescope! The Andromeda galaxy is a spiral galaxy that is slightly larger than the Milky Way. It is also referred to as M31, Messier 31, or NGC 224.

When observing the galaxy in the night sky, it will appear as a long, hazy smudge of light. The Andromeda galaxy is located just north of the Andromeda constellation. It is best to observe the galaxy in a dark sky, away from urban light pollution in November in the Northern Hemisphere. The Andromeda galaxy will be about the same width as the full moon in the night sky.

In 2015, the Hubble Space Telescope took a mosaic of images of the Andromeda galaxy that resulted in an immense amount of information for scientists. Some of the key features of Andromeda are two supermassive black holes that orbit each other, and a ring of dwarf galaxies around Andromeda.

This image from NASA Galaxy Evolution Explorer is an observation of the large galaxy Andromeda, also known as Messier 31.

Courtesy of NASA/JPL/California Institute of Technology

This is a real image of the Andromeda galaxy in the night sky.

Courtesy of Shutterstock/Anze Furlan

Did You Know?

The Andromeda Galaxy is currently on a collision course with the Milky Way galaxy. In about four billion years, the two galaxies will collide, creating a new "Milkomeda" elliptical galaxy. Of course, the process of combining these two galaxies will take almost two billion years.

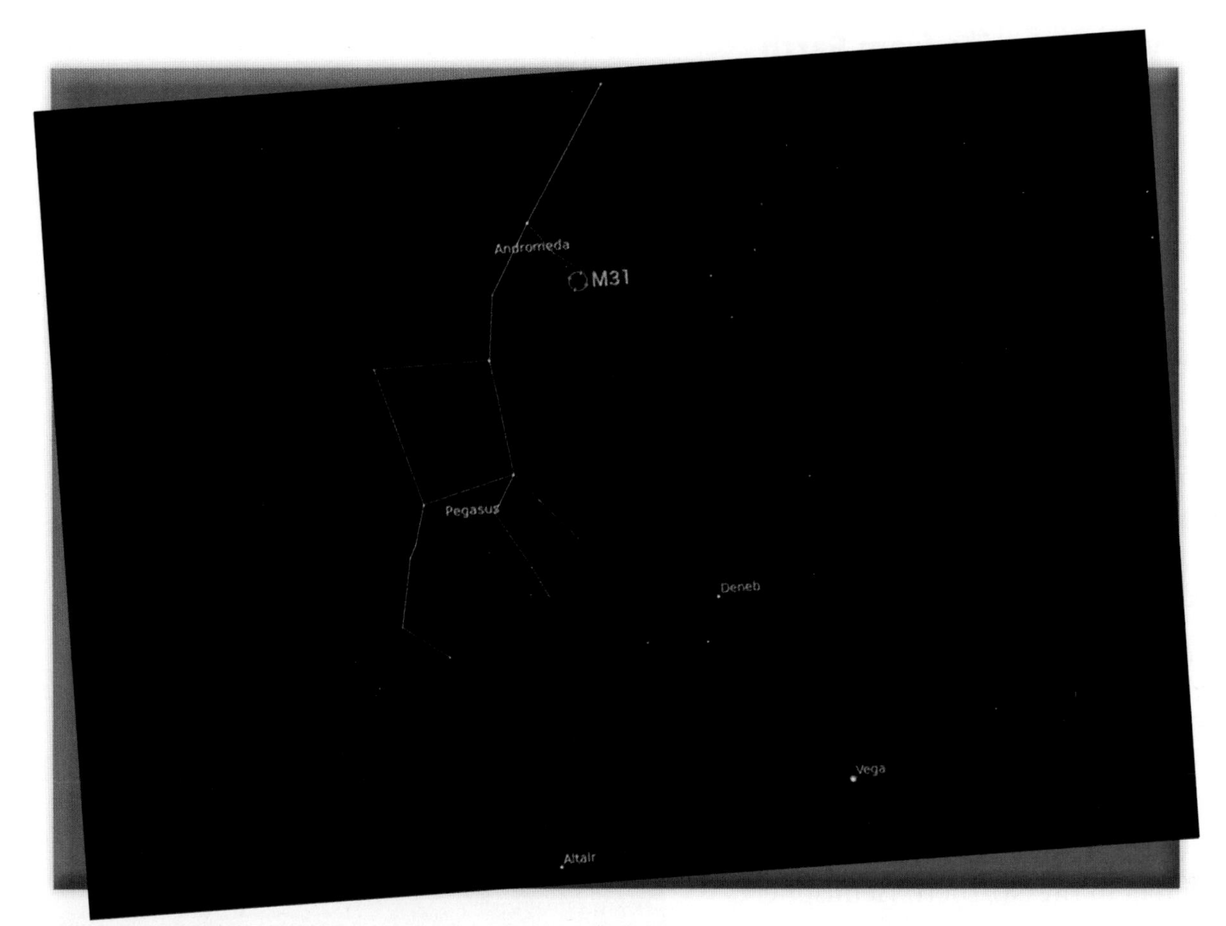

This star chart shows the location of M31 in the Northern Hemisphere in November around 10 p.m. EST.
Courtesy of Stellarium

How are Galaxies Named?

Galaxies usually have many different names. The Messier catalog of objects was maintained by Charles Messier. It was a list of 110 fuzzy objects observed in the sky. When an object is identified, it is often referred to by its Messier number.

In addition, many galaxies are given an NGC number. The NGC is the New General Catalogue that contains a list of 7,840 interesting night sky objects. As each object is identified, it may be referred to using its NGC number.

A third way of naming galaxies is based on their location in the sky in relation to constellations. For example, the Andromeda galaxy is named for the constellation in which it is located, the Andromeda constellation. But in the Messier catalog, the Andromeda galaxy is designated as M31. And in the New General Catalogue, the Andromeda galaxy is designated as NGC 224.

The International Astronomical Union maintains the official names for astronomical objects, so you cannot name a galaxy or star after yourself even if you discover it!

The Whirlpool Galaxy

The Whirlpool galaxy is a majestic spiral galaxy more than 23 million light years from the Milky Way. The Whirlpool galaxy is also known as M51 and is in the constellation Canes Venatici.

Each of the spiral arms of the Whirlpool galaxy has long lanes of stars, gas, and dust producing clusters of new stars. Some astronomers believe that the arms on the Whirlpool galaxy are more prominent because of the tidal force from its companion galaxy, NGC 5195. The Whirlpool galaxy can be spotted with a small telescope in May.

The Whirlpool galaxy is situated facing the Milky Way, which provides stunning images. This image of the Whirlpool Galaxy shows its companion galaxy, NGC 5195, to the right.

Courtesy of NASA, ESA, Beckwith (STScI) and the Hubble Heritage Team (STScI/AURA)

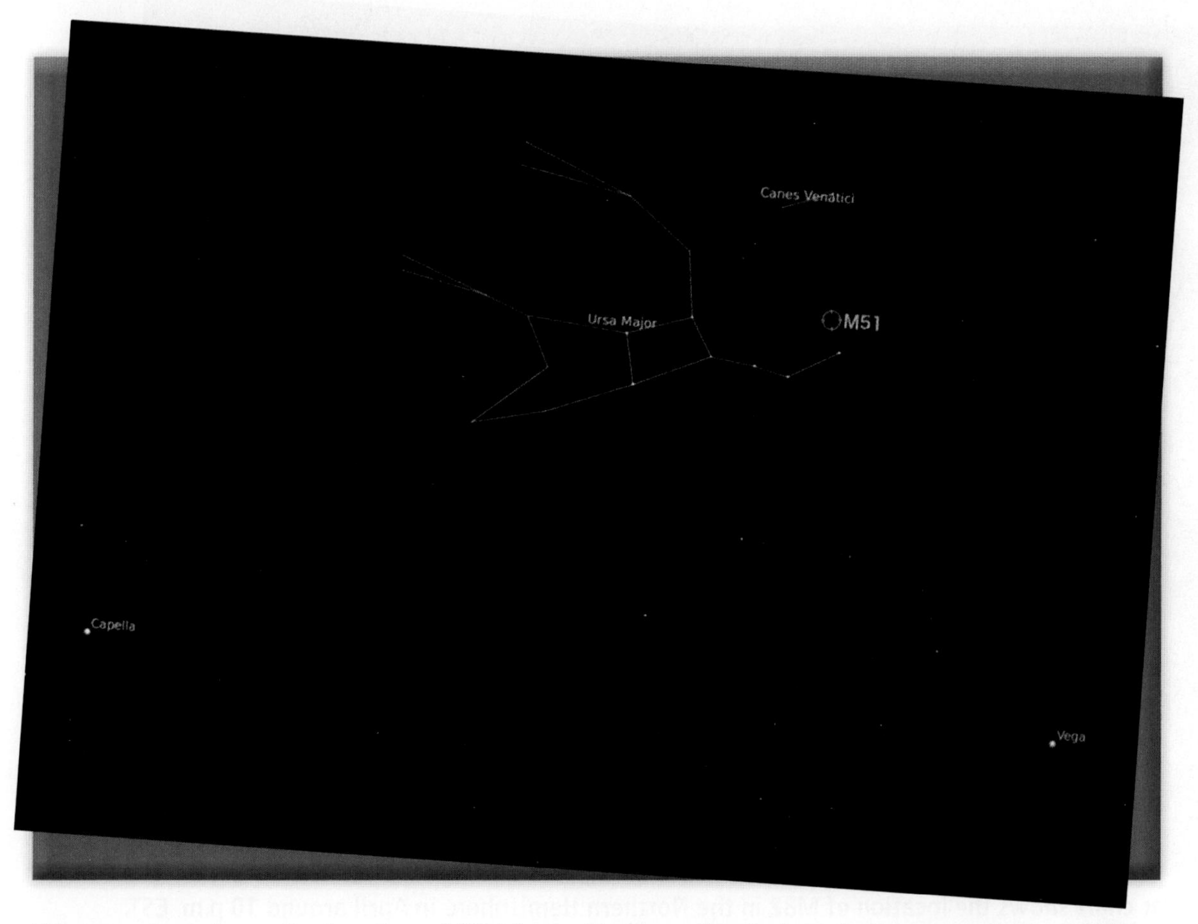

This star chart shows the location of M51 in the Northern Hemisphere in May around 10 p.m. EST.

Courtesy of Stellarium

The Cigar Galaxy

The Cigar galaxy is a spiral galaxy 12 million light years away in the constellation Ursa Major. (Ursa Major is in The Big Dipper.) It is also referred to as M82. The Cigar galaxy has *a high rate of star formation* called a starburst. This phenomenon is attributed to its galactic neighbor M81. M82 produces new stars 10 times faster than the Milky Way galaxy. M82 can be seen with a telescope or binoculars in April in the Northern Hemisphere.

In this image of the Cigar galaxy, you can see the gas and dust in pink coming from the galaxy.

Courtesy of NASA, ESA, Beckwith (STScI) and the Hubble Heritage Team (STScI/AURA)

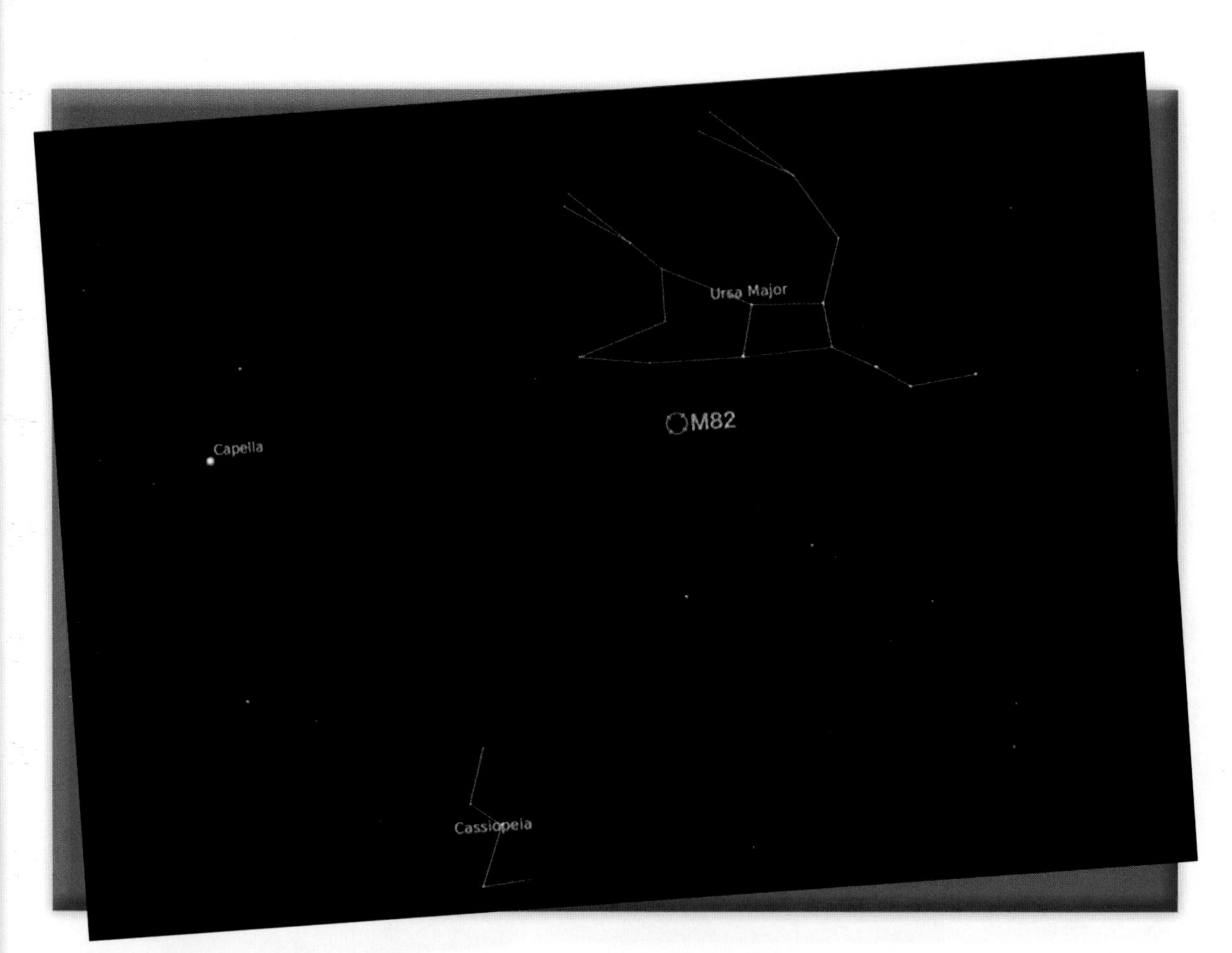

This star chart shows the location of M82 in the Northern Hemisphere in April around 10 p.m. EST.

Courtesy of Stellarium

Triangulum Galaxy

The Triangulum galaxy is a spiral galaxy about 2.8 million light years away. It is very close to the Andromeda galaxy and can also be seen unaided in the night sky away from light pollution. The Triangulum galaxy is also referred to as M33.

Triangulum galaxy, or M33.

Courtesy of Shutterstock/Antares_StarExplorer

The Magellanic Clouds

The Magellanic Clouds are a group of two irregular galaxies that live close to the Milky Way. The Large Magellanic galaxy is easy to find in the night sky and is visible to the unaided eye. It will appear as a small, faint smudge in the sky. It is often mistaken for the Milky Way. The Large Magellanic Cloud is between the Dorado and Mensa constellations. The Small Magellanic Cloud is in the Tucana constellation.

The Large Magellanic Cloud is almost 200,000 light years from Earth and contains vast clouds of gas and newly formed stars that present a magnificent display of colors in the night sky.

Central region of the Tarantula Nebula in the Large Magellanic Cloud. The young and dense star cluster R136. Retouched colored image.

Courtesy of NASA/Shutterstock

Galaxy Clusters

As you have seen from the images, many *galaxies form groups* or galaxy clusters. The Milky Way is part of a galaxy cluster called the Local Group. *Galaxy clusters can also group together* into superclusters.

The Local Group contains about 50 galaxies, including the Milky Way galaxy. The Andromeda galaxy and Magellanic Clouds are also included in the Local Group.

A computer-simulated view of a cluster of galaxies in the distant cosmos.

Courtesy of NASA, ACS Team, Rychard Bouwens (UCO/Lick Obs.)

Did You Know?

The universe is full of galaxy clusters in varying sizes. A galaxy cluster can contain a few galaxies or thousands of galaxies. They may be scattered over a large area or be packed tightly together.

Black Holes, Dark Matter, Nebulae, and Pulsars

Black Holes

Another intensely studied aspect of deep space is a black hole. A black hole is *an area of intense gravitational pull*. A black hole will suck objects in, and once inside, they cannot escape–even light can't escape a black hole.

Think of a black hole as a hungry toddler. Black holes are like toddlers in that they are a bit messy as they pull objects inside. Material being sucked into the black hole doesn't always get pulled directly in. The objects may miss the opening of the black hole, and their material is forced out at the speed of light. The black hole has powerful magnetic fields surrounding it, so when material is forced out it gets caught in the magnetic fields around the black hole. These magnetic fields act as jets to shoot the material away. In addition, they emit an enormous amount of energy, such as radio waves, visible light, and X-ray light.

There are three main types of black holes: primordial, stellar, and supermassive. Primordial black holes are extremely small black holes that have a large mass. It is thought that primordial black holes were formed early in the universe.

Stellar black holes form when a large star collapses into itself. At the final stage of a particularly massive star's life, it can detonate, or create a supernova, which is a massive explosion that scatters most of the star but leaves a cold remnant. The remnant has no energy or outward pressure to oppose the inward pull of gravity. The result is a black hole, with its immense gravitational pull.

Supermassive black holes are formed at the same time as the galaxy to which they belong. It is believed that the size of the black hole has a direct correlation to the size of the galaxy. Many galaxies have a black hole at their center, just as the Milky Way does. Black holes can grow to a massive size as they consume the matter and objects that pass by.

Did You Know?

Australian astronomers found the fastest-growing black hole known. The 20-billion-solar-mass black hole consumes the mass of the Sun every two days. If this black hole were at the center of the Milky Way, it would appear 10 times brighter than a full moon because of the massive amount of gases sucked in.

A black hole sucks in dust, gas, and even stars and planets. Objects and matter must pass fairly close to a black hole in order to be sucked into the black hole. Once *an object is pulled into a black hole to the point of no return*, the object has reached the **event horizon**. This is the point where the path into the black hole cannot be reversed.

While black holes are very powerful, they are also very small. If you had a black hole the size of our Sun, the black hole would be approximately 2 miles in diameter. Because of their size and distance from Earth, it is impossible to directly observe a black hole. However, scientists are able to measure the mass in a specific area of the sky. When they locate dark mass, this equates to a black hole.

An artist's rendition of a black hole.
Jurik Peter/Shutterstock

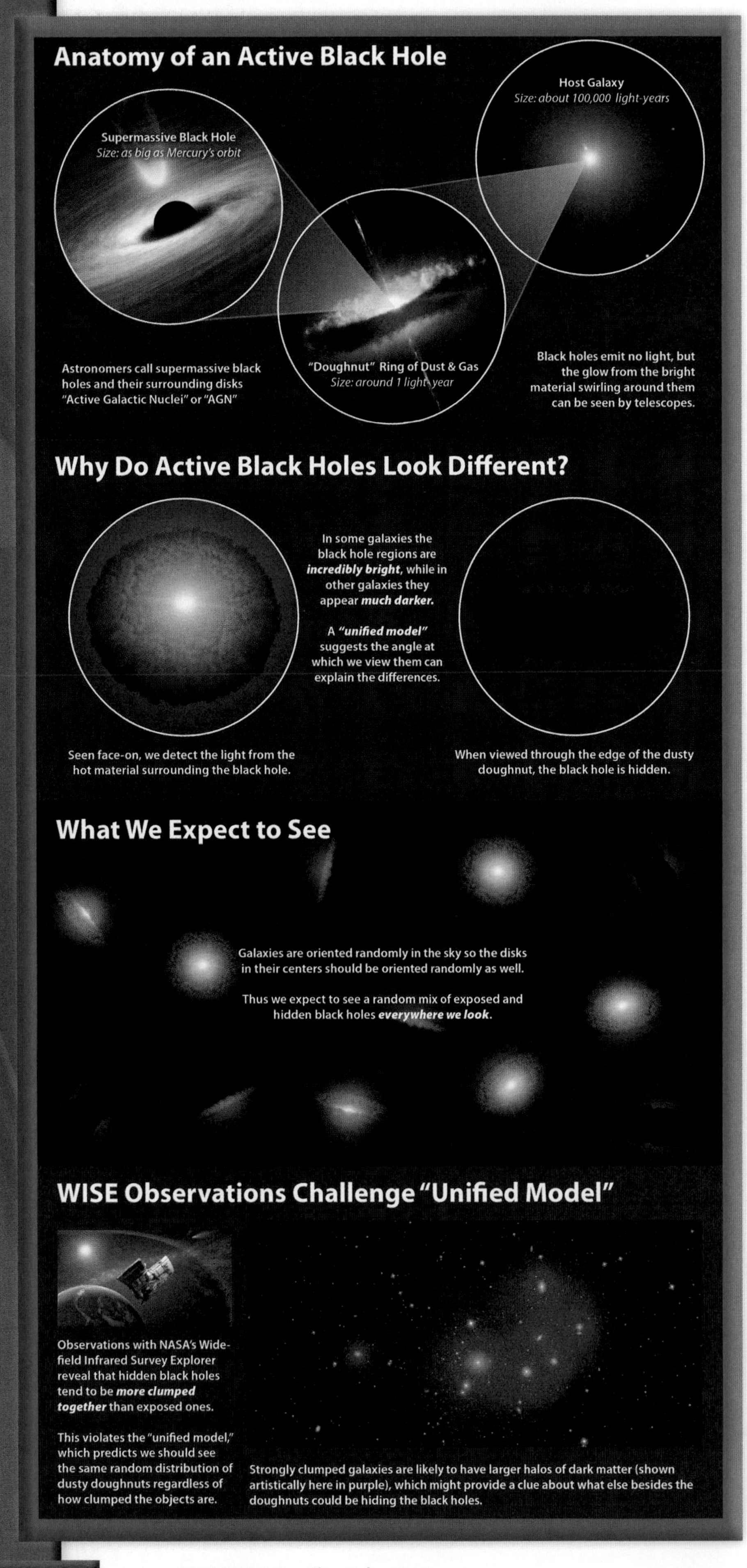

There are still a lot of unknowns about black holes. However, much was learned by theoretical physicist, cosmologist, and author Stephen Hawking, who dedicated his time to researching them. Hawking developed a solid mathematical backing for Albert Einstein's Theory of Relativity and the concept of black holes proposed in Einstein's theory. Hawking applied the area of quantum mechanics to black holes and discovered that black holes emit radiation. The source of the radiation is virtual particles that jump in and out of the black hole. If a black hole consumes a negative particle, the black hole will shrink by a miniscule amount and radiation will be emitted. This is known as Hawking radiation. With the application of this theory, all black holes will eventually shrink away when there is no matter left for them to consume. This process would, of course, take an immensely long time to occur. Although no emissions have been spotted, most physicists do believe they exist.

The anatomy of a black hole.

Courtesy of NASA/JPL-Caltech

Dark Matter

What is dark matter? Dark matter may be the biggest mystery of space. We have talked about galaxies containing dark matter, but what exactly is it? Well, astrophysicists don't really know. We know that it does not emit radiation like other matter. And we also know that it cannot be seen.

This illustration shows Jupiter surrounded by filaments of dark matter called "hairs," which are proposed in a study in the Astrophysical Journal by Gary Prézeau of NASA's Jet Propulsion Laboratory, Pasadena, California. A hair is created when a stream of dark matter particles goes through the planet.
Courtesy of NASA/JPL-Caltech

So, if you can't see dark matter, how do we know it exists? Let's first review how galaxies form. Over time, stars form and collect into galaxies. The galaxies then cluster together. Between the stars, galaxies, and solar systems, there is matter smashed together. Gravity is the glue that holds everything together. The matter between everything is filled with gas, dust and "stuff" that we just can't detect. This "stuff" is dark matter. It is the invisible matter between planets, solar systems, and galaxies.

Dark matter was discovered when scientists were studying the mass of objects in space. Using the observable light of a galaxy, they could estimate the mass of the galaxy. However, when looking at the weight, position, and speed of rotation, scientists learned that the mass calculation was far greater than the observable light would indicate. This finding established the existence of dark matter. Dark matter has a mass that contributes to the overall mass of the object, even though we cannot observe dark matter.

Because scientists cannot see dark matter, how do they go about studying it? They use gravitational lensing, which is *a technique used to study dark matter*. When light is passed through a gravitational lens, it will bend. The amount of bending can help scientists study dark matter.

Another way that scientists study dark matter is by using the Fermi Gamma-Ray Space Telescope. The telescope detects gamma rays, the highest energy form of light. It is believed that when dark matter particles collide they release gamma rays. The burst of gamma rays would be detected by the Fermi telescope. The Fermi Gamma-Ray Space Telescope was launched in June 2008 and maps the entire sky every three hours. It is still considered a new technology, and data from the telescope is continually being analyzed.

For now, dark matter continues to be a great mystery. Many theories surround dark matter, and little is known about it. But, through studies, scientists have been able to determine what dark matter is "not," thus getting them closer to discovering the mystery of dark matter.

Nebulae

A beautiful nebula and galaxy.

Courtesy of NASA/Shutterstock

We have discussed how dust and gas are present in the space between stars and planets. In some areas, the *dust and gas are very dense and create clouds* known as nebulae (nebula for singular). Within the nebulae clouds, star formation occurs. The combination of the dust, gas, and star formation produces magnificent displays of light from the nebulae clouds. There are many types of nebulae that are found in the universe, including emission nebulae, reflection nebulae, dark nebulae, planetary nebulae, and supernova remnant.

As young stars are born, they create strong stellar winds and interstellar gas. The solar winds create bubbles and canyons in the nebula. The interstellar gas creates light to form emission nebulae, which are *nebulae clouds lit up by interstellar gas*. With binoculars or a small telescope, you can witness emission nebulae in the night sky.

A reflection nebula in the star cluster Pleiades.

Courtesy of Mironov/Shutterstock

Some nebulae do not emit light but do reflect it. They are called reflection nebula. Reflection nebulae are *scattered nebula clouds that reflect light from the nearby stars*. The reflected light typically appears as blue starlight.

Then we have *nebulae that do not emit or reflect light*; these are dark nebulae. Dark nebulae can be seen, as they are silhouetted against brighter backdrops.

Another type of nebulae is a planetary nebula. A planetary nebula *occurs when a star reaches the end of its life*. The star will push its atmosphere into space and create a magnificent planetary nebula. The core of the star remains, lighting up the surrounding material.

When a massive star dies in a supernova explosion, remnants of the star remain. The explosion sends glowing material into space. As the material crashes into the surrounding gas, it expands over time. This type of nebula is called a supernova remnant. A supernova remnant is *the remains of a star that has expanded to create a nebula*.

The Right Stuff

Nebulae

Nebulae are fascinating features of the night sky, Using binoculars or a small telescope, you can easily observe some common nebulae. The Orion nebula is one of the easiest nebulae to observe and is located within the Orion constellation. It is an emission nebula that is 1,350 light years away.

Orion Nebula

Courtesy of NASA/ JPL-Caltech/Univ. of Toledo

Carina Nebula

Courtesy of NASA Goddard

The Carina Nebula is 7,500 light years away in the Carina constellation. The Carina Nebular is a spectacular sight and produces many stars.

Crab Nebula is a supernova remnant. It is located in the Taurus (the Bull) constellation and is thought to have occurred in 1054. To locate this nebula, locate the Hyades star cluster at the head of Taurus. Follow the bull's horn to the Zeta star. The Crab Nebula is just a degree away from Zeta.

Crab Nebula

Courtesy of NASA/JPL

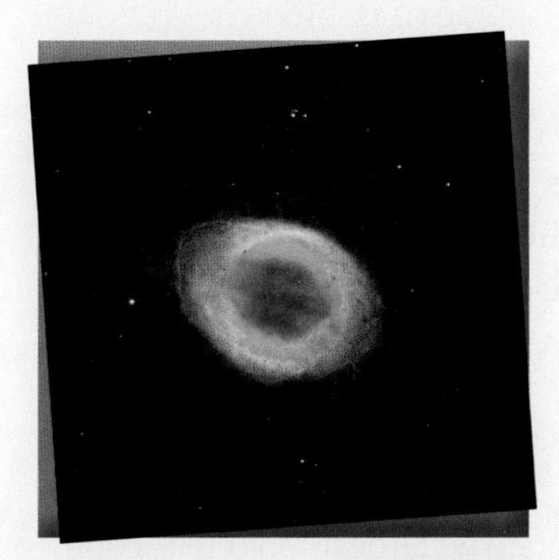

Ring Nebula

Courtesy of NASA Goddard

The Ring Nebula is a planetary nebula that is a must-see. It This small, glowing ring of light is the remnants of a star similar to the Sun. It is located in the constellation Lyra between the stars Gamma and Beta. Using a small telescope, you can easily view this wonder.

Pulsars and Neutron Stars

During a supernova explosion, the protons and electrons of a star are crushed together creating neutrons. This process creates *a ball of neutrons* called a neutron star. Think of an ice skater spinning on the ice. As they move faster, their arms and legs are tightly tucked against their body. A neutron star is similar to this concept; as the neutrons spin faster, they are pulled closer into a neutron star. A neutron star is about 20 miles in diameter, yet has approximately the same mass as the original star. If a neutron star the size of a sugar cube were on Earth, it would weigh over 100 million tons!

Neutron stars are very small, strange objects that have an extremely strong magnetic field. Neutron starts can become pulsars over time. A pulsar is *a rapidly spinning neutron star that emits radio waves in pulses*. The pulsar will emit a regular pulse at specific intervals. While some pulsars are extremely fast with a pulse period of 1.56 milliseconds, others are slow pulses at .715 seconds. As the pulsar emits a pulse, beams of radiation and light are pushed out.

Illustration of a pulsar.
Courtesy of Jurik Peter/Shutterstock

The Crab Nebula has a pulsar inside it. The pulsar pulses 33 times per second, lighting up the nebula as it pulses. The pulse of a pulsar is incredibly consistent. In fact, the timekeeping of a pulsar is more accurate than an atomic clock.

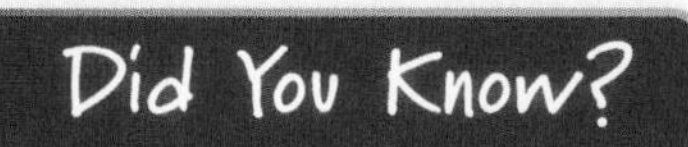

Scientists use large radio telescopes to locate pulsars. The largest radio telescope in the world is located in Arecibo, Puerto Rico. It is used to search for pulsars.

Big Bang Theory

By this point, you have discovered that there is a lot of action happening in space! But how was the universe itself created? In 1927, astronomer Georges Lemaitre had an idea. He speculated that the universe began as a single point, and over time it expanded. His theory was that the universe would continue to stretch and expand over time. In 1929, the astronomer Edwin Hubble observed that other galaxies were moving farther away from Earth. And the galaxies that were the farthest away moved faster than those that were closer to us. Just as Lemaitre speculated, the universe is still expanding, and a long time ago everything was much closer together.

In the beginning, the universe was made of hot, tiny particles, light, and energy. The particles began to group together to form atoms. Then atoms began to group together to form stars and galaxies. Stars, galaxies, and molecules continued to group together. Asteroids, comets, black holes, and planets were forming. This entire process took 13.8 billion years. The theory of how the universe was created is called the Big Bang theory.

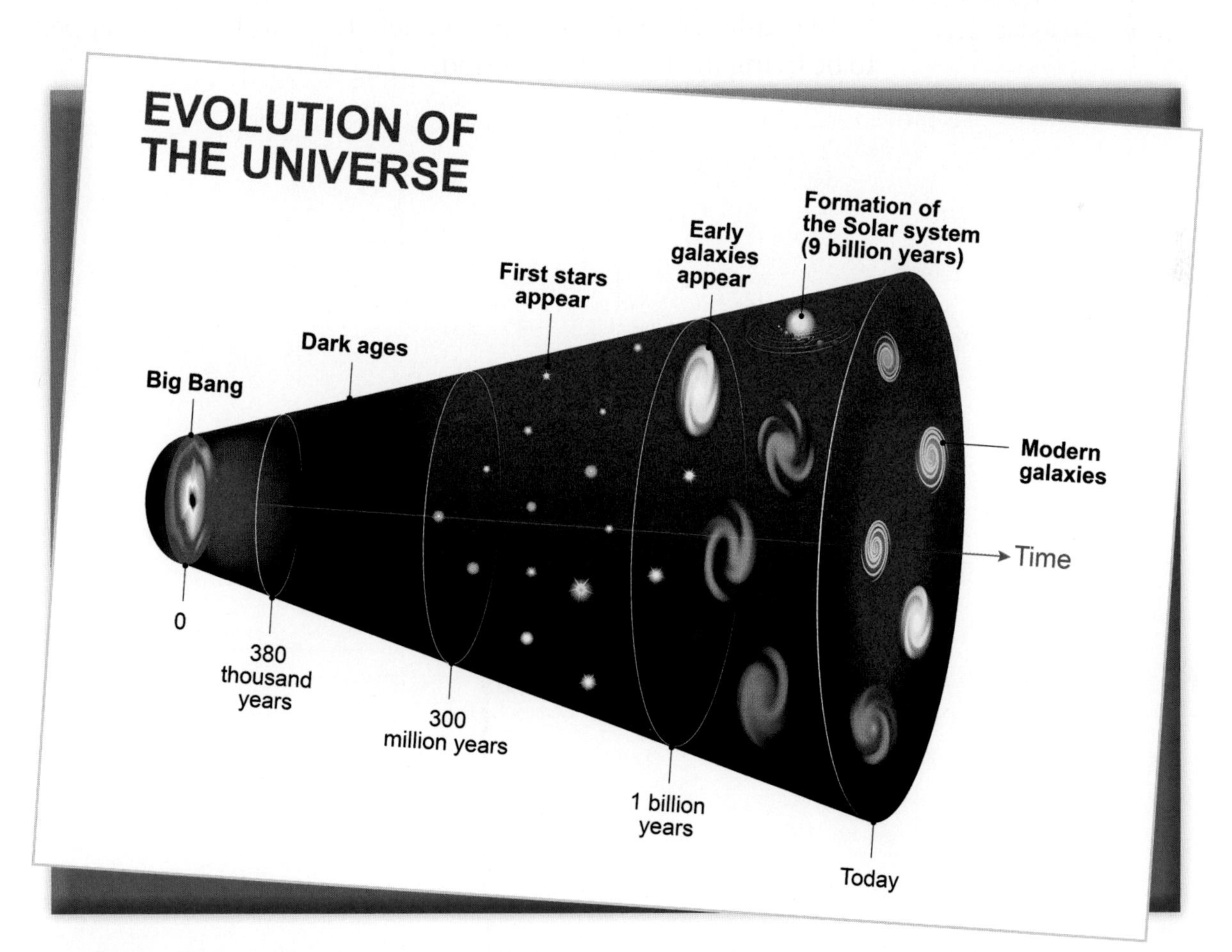

Evolution of the Universe. Cosmic Timeline and evolution of stars, galaxy and Universe after Big Bang.

Courtesy of Designua/Shutterstock

The Big Bang theory is still a controversial theory, as it is difficult to prove. Scientists rely on mathematical formulas and models to try to prove the theory. But astronomers are able to witness the universe expanding through "cosmic microwave background." The cosmic microwave background (CMB) is believed to be *the radiation leftover from the Big Bang*. This radiation is everywhere, but it is so cold (only 2.725^0 F above absolute zero) it is impossible to see with the human eye. CMB is visible in the microwave part of the electromagnetic spectrum.

CMB was first discovered by accident in 1965 by two researchers with Bell Telephone Laboratories. Arno Penzias and Robert Wilson were creating a radio receiver and confused by a signal they picked up coming from all over the sky. They had accidently found CMB.

Did You Know?

The term "Big Bang theory" has always been well known by astrophysicists, but it wasn't until the launch of the popular sitcom series in 2007 that the term became more mainstream.

Although the Big Bang theory is regarded as fact among the majority of scientists, some skeptics have alternative theories as to the creation of the universe. In addition, some scientists believe that while the Big Bang explains our universe, it is not the first time the universe expanded. Some believe that the universe has a regular cycle of inflation and deflation, and we just happen to be living in an inflation period.

✔ CHECKPOINTS

Lesson 4 Review

Using complete sentences, answer the following questions on a sheet of paper.

1. Who was the first to accurately determine the immense size of the galaxy and the location of our solar system?
2. How long is a galactic year?
3. What are the names of the four main spiral arms in the Milky Way galaxy?
4. What percentage of the Milky Way is composed of dark matter?
5. How was the Local Bubble created?
6. What are the four types of galaxies?
7. Which galaxy can be seen without a telescope?
8. What are the three ways galaxies are named?
9. How do scientists locate black holes?
10. How do nebulae form?
11. Which nebula is the easiest to observe?
12. How does a neutron star form?
13. What astronomer first suggested that the universe started with one single point?
14. How was cosmic microwave background (CMB) discovered?

APPLYING YOUR LEARNING

15. The Big Bang theory is still a controversial theory for the beginning of the universe in the scientific community. Based on what you have read and know, describe why you agree or disagree with this theory.

CHAPTER 3

This artist rendering shows an aerial view of the liftoff of the 70-metric-ton (77-ton) lift capacity configuration Space Launch System (SLS) from the launch pad.

Image courtesy of NASA.gov.

Space Exploration

Chapter Outline

"We are at the dawn of a golden age of space exploration which will transform our relationship with the Earth and with the cosmos."

Sir Richard Branson

LESSON 1

Why Explore Space?

Quick Write

The Humanity Star was sent to space to help humans look to the stars and reflect on our place in the universe. If you had the opportunity, what would you send into space that would generate interest in space exploration?

Learn About

- the benefits of space exploration
- NASA's plan to explore space
- private industry in space

On January 21st, 2018, Rocket Lab, a commercial space company, launched its Electron rocket into orbit. Unbeknownst to the world, Electron carried a surprise for humankind: the Humanity Star The Humanity Star was a highly reflective satellite that blinked across the night sky and could be seen with the naked eye! Think of a disco ball in space! Every 90 minutes the Humanity Star orbited Earth and could be seen anywhere in the world. Rocket Lab founder and CEO, Peter Beck created the Humanity Star as a bright symbol to remind us that we have a fragile place in this universe The Star also encouraged us to look up into the stars to be inspired to explore the universe.

The Humanity Star was designed to orbit Earth for a few short months; however, it only lasted a few weeks. Due to the small mass of the satellite, it deorbited faster than expected. In March 2018, it descended back to Earth and burned up during re-entry. No evidence of the Humanity Star survived the descent.

The company has no plans to send another Humanity Star to space. This was a one-time experience.

The Benefits of Space Exploration

Vocabulary

- low earth orbit (LEO)

Mankind has an immense amount of curiosity and a desire to explore the unknown. Space is no different. Mankind is compelled to explore unknown worlds and expand their scientific and technical knowledge. By exploring space, humans can gain further understanding about our solar system and universe. Through this exploration, we can discover our place in the universe.

For more than 60 years, NASA (National Aeronautics and Space Administration) has been the premiere organization for space travel in the United States. The knowledge gained from the research, space missions, and experiments conducted at NASA have provided many benefits to the human race.

The NASA logo in front of the Kennedy Space Center in Cape Canaveral, Florida.

Alexanderphoto7/Shutterstock

Knowledge

The greatest benefit that can be achieved by space travel is knowledge. The technical knowledge gained by space travel has provided many benefits to the public. For example, communications satellites, which are possible because of space travel, provide the means to make cell phones possible. As new products and services are developed, the economy is positively impacted for those countries investing in space travel.

Space travel provides direct and indirect benefits for society. Direct benefits are benefits that you can see immediately, like scientific knowledge or creation of new products. Indirect benefits are benefits that take time to develop and include items such as economic prosperity and health benefits. See Figure 3.1 for a list of direct and indirect benefits of space travel, as outlined by NASA.

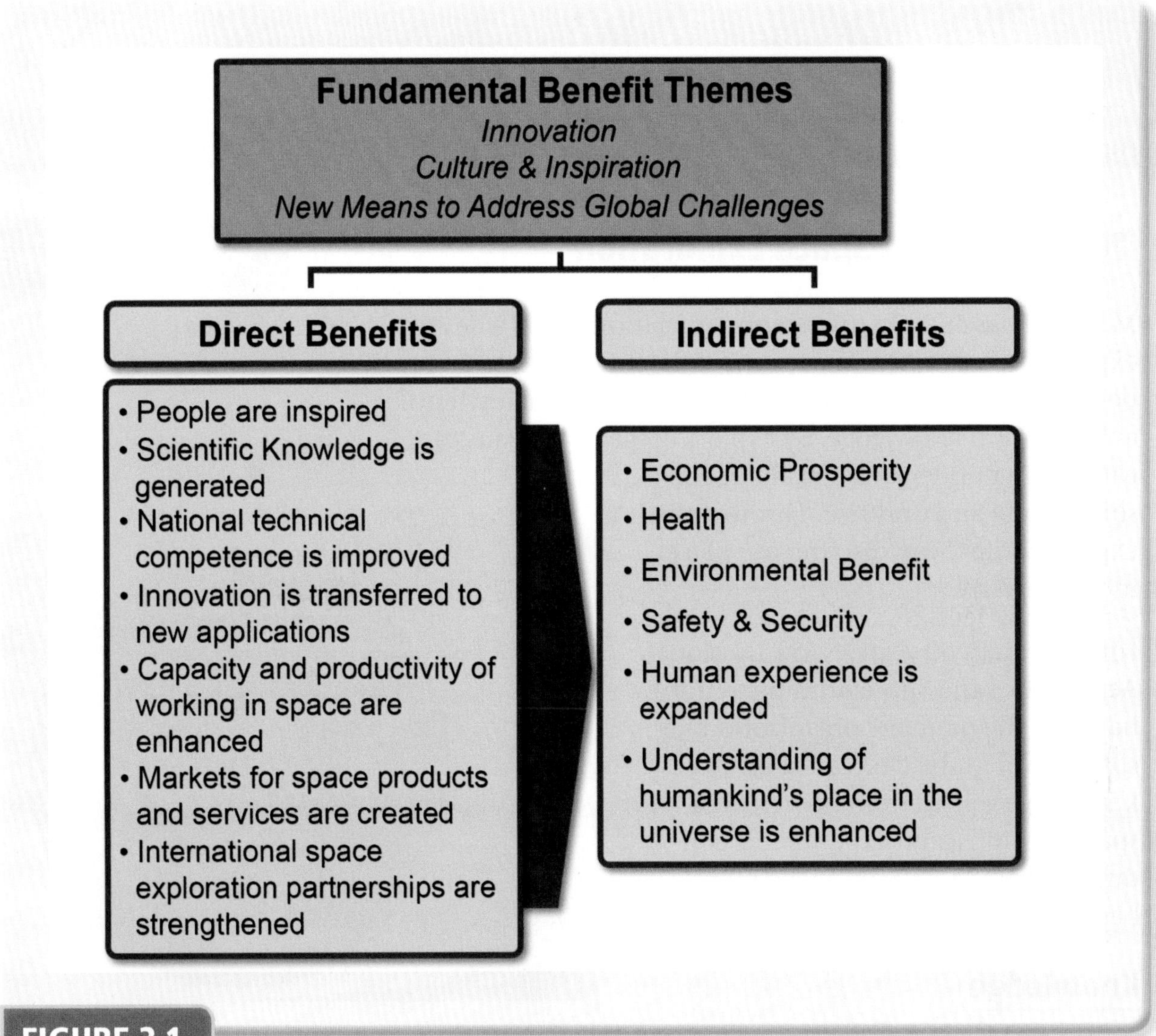

Fundamental Benefit Themes
Innovation
Culture & Inspiration
New Means to Address Global Challenges

Direct Benefits

- People are inspired
- Scientific Knowledge is generated
- National technical competence is improved
- Innovation is transferred to new applications
- Capacity and productivity of working in space are enhanced
- Markets for space products and services are created
- International space exploration partnerships are strengthened

Indirect Benefits

- Economic Prosperity
- Health
- Environmental Benefit
- Safety & Security
- Human experience is expanded
- Understanding of humankind's place in the universe is enhanced

FIGURE 3.1

The direct and indirect benefits of space travel.
Courtesy of NASA.

Space exploration is a challenging feat that requires humans to develop innovative solutions to meet lofty space exploration goals. In addition, space travel stimulates the advancement of technology in areas outside of space exploration and throughout the scientific and technical community. For instance, innovations developed for space travel initiate progress in health care, robotics, automation, and many other industries.

Some specific examples of technological benefits of space travel include cordless tools, cameras in cell phones, biomedical technologies, implantable heart monitors, and solar panels. Scientific discoveries are ongoing with the use of the International Space Station (ISS). The ISS conducts experiments in space on human physiology, plant biology, materials science, and physics.

Education

Space travel and research also have a direct benefit on the education of youth in science and technology. As space travel continues to expand, a new generation is inspired to dream of ways to go farther into space. In the 1960s, the Apollo space mission to the Moon provided a large increase in the technical knowledge of US students at the time. The recent developments in commercial space travel are expected to provide the same benefits.

Mathematics and science scores in the US have been falling behind other nations. STEM (science, technology, engineering, and mathematics) is an initiative to provide science, technology, engineering, and mathematics knowledge to students throughout their learning. Toys for preschoolers are even being developed to help increase the STEM skills in youngsters. STEM education not only increases science knowledge, it also develops critical thinking skills and promotes innovation.

Teacher with students building robotic vehicle.
Monkey Business Images/Shutterstock

STEM education is growing in classrooms and will continue its upward trend. According to the U.S. Department of Commerce, STEM occupations are growing at a 17 percent rate annually. The average growth rate for other highly sought-after occupations is only 9.8 percent. In addition to the growth rate of STEM positions, students who hold STEM degrees achieve a higher pay rate. As you can see, STEM is a major factor in the growth of the US economy and also helps the US secure its future in STEM development.

Social Development

Space travel has a profound impact on a nation's cultural and social development. Space travel can help answer questions such as "Are we alone?" or "What lies beyond our solar system?" Space exploration has helped humans develop a new perspective of the Earth and space. Space travel and images from space allow us to understand the fragility of the Earth and the need to protect the Earth and its environment.

In addition, space travel has inspired movies, songs, books, photographs, and paintings. These cultural aspects further the social development of the human race.

International Partnerships

Space travel can also serve as a catalyst for international partnerships. Together, nations can advance their goals for space exploration, share knowledge gained during space travel, and conduct joint missions in space. The ISS is a partnership between 15 nations. Over the years, 68 nations have participated in ISS activities. This partnership allows us to invest in life in space, which would not be economically feasible as a single nation.

An illustration of the International Space Station (ISS) passing over Florida.

Courtesy of NASA.

NASA's Plan to Explore Space

NASA 2018 Strategic Plan Framework		
Theme	**Strategic Goal**	**Strategic Objective**
DISCOVER	EXPAND HUMAN KNOWLEDGE THROUGH NEW SCIENTIFIC DISCOVERIES.	1.1: Understand the Sun, Earth, Solar System, and Universe. 1.2: Understand Responses of Physical and Biological Systems to Spaceflight.
EXPLORE	EXTEND HUMAN PRESENCE DEEPER INTO SPACE AND TO THE MOON FOR SUSTAINABLE LONG-TERM EXPLORATION AND UTILIZATION.	2.1: Lay the Foundation for America to Maintain a Constant Human Presence in Low Earth Orbit Enabled by a Commercial Market. 2.2: Conduct Exploration in Deep Space, Including to the Surface of the Moon.
DEVELOP	ADDRESS NATIONAL CHALLENGES AND CATALYZE ECONOMIC GROWTH.	3.1: Develop and Transfer Revolutionary Technologies to Enable Exploration Capabilities for NASA and the Nation. 3.2: Transform Aviation Through Revolutionary Technology Research, Development, and Transfer. 3.3: Inspire and Engage the Public in Aeronautics, Space, and Science.
ENABLE	OPTIMIZE CAPABILITIES AND OPERATIONS.	4.1: Engage in Partnership Strategies. 4.2: Enable Space Access and Services. 4.3: Assure Safety and Mission Success. 4.4: Manage Human Capital. 4.5: Ensure Enterprise Protection. 4.6: Sustain Infrastructure Capabilities and Operations.

NASA 2018 Strategic Plan

Courtesy of NASA.

Every four years, NASA produces a strategic plan for space exploration. The most current plan was published in 2018. The 2018 Strategic Plan identifies four long-term goals and how they plan to achieve them:

- DISCOVER
- EXPLORE
- DEVELOP
- ENABLE

Strategic Goal #1: Discover

The "Discover" element of NASA's strategic goal is meant to expand human knowledge through new scientific discoveries. The specific aims of this goal are to better understand the Sun, Earth, solar system, and universe and learn more about the responses of physical and biological systems to spaceflight.

NASA's vision is to discover the secrets of the universe. They plan to accomplish this goal by studying the Sun, the Earth, other planets and solar systems, interplanetary environment, and the space between stars. One tool to help NASA enhance its knowledge is the James Webb Space Telescope.

The James Webb Space Telescope (Webb) will be launched in 2021. The Webb will be a premiere observatory in space available to astronomers across the world. The telescope is designed to study the history of the universe, which includes the first light after the Big Bang, formation of solar systems that can support life, and the evolution of our solar system. Webb is an international collaboration between NASA, ESA (European Space Agency), and the CSA (Canadian Space Agency).

The James Webb Space Telescope during the construction process.
Courtesy of NASA/Chris Gunn.

In addition, as part of its "Discover" goal, NASA would like to explore the fundamental question, "Are we alone?" Such exploration would involve searching for habitable zones both in our solar system (Mars, moons, etc.) and outside of it.

Did You Know?

How will we know if we find life? Can you believe a rainbow may be the answer? Sir Isaac Newton discovered that white light shot through a prism (or through curtains of mist seen with the sun at your back) is exposed for what it really is: a band of color, spanning from violet to red, and characterized by "wavelength." Chemicals and gases in the atmospheres of planets can absorb certain slices of this band, called a spectrum, and leave behind a narrow black gap. As light from a star is analyzed it looks almost like a bar code. The pattern of black gaps on the bar code can identify the gases present in the atmosphere, like oxygen or methane. The bar code we see may help identify life on other planets.

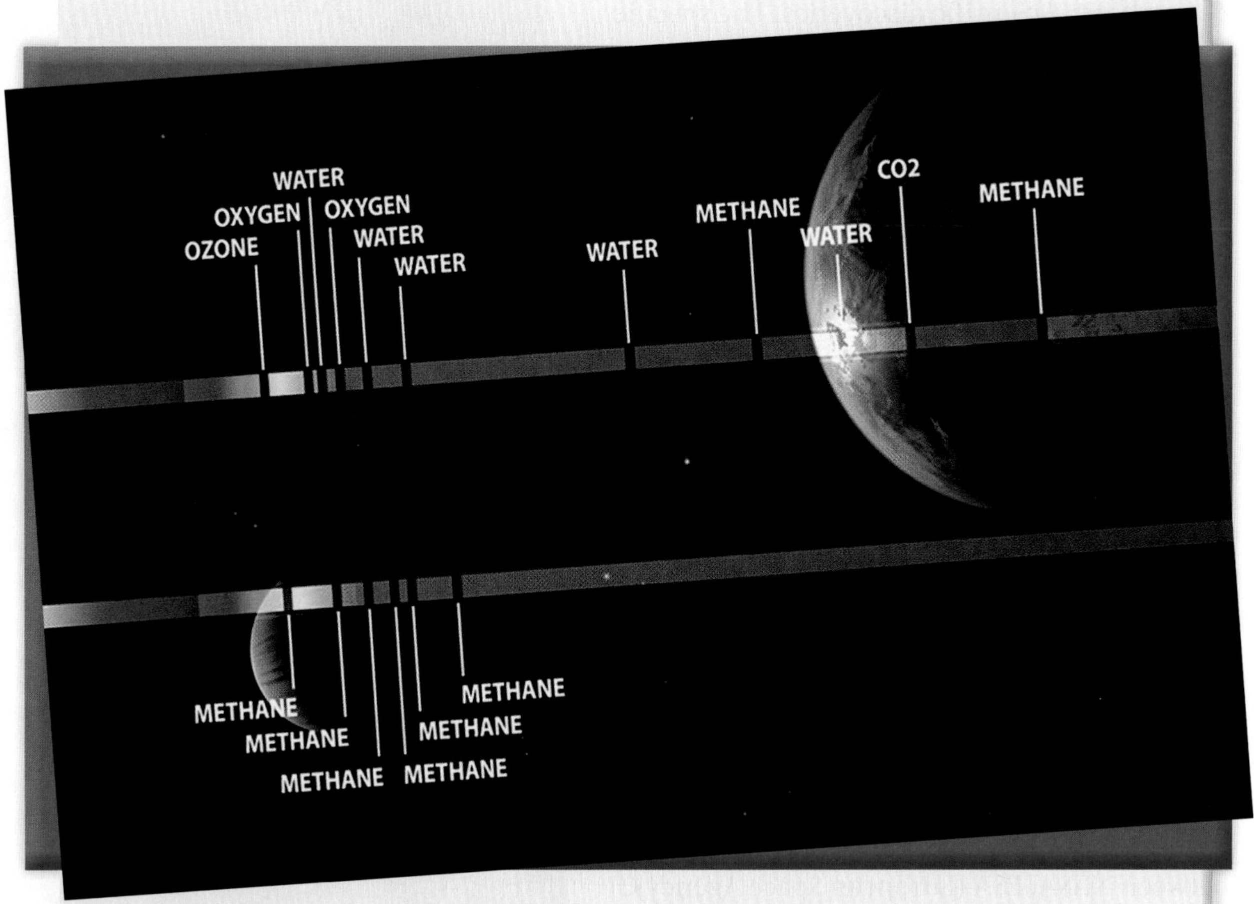

Different colors correspond to different wavelengths of light. Missing colors show up as black lines, indicating specific gases are present, because each gas absorbs light in a specific wavelength (or color).

Courtesy of NASA.gov

Finally, NASA aims to safeguard and improve life on Earth by investigating the hazards to life from the solar system, Sun, and Earth itself. NASA works to predict the paths of asteroids and comets that could harm life on Earth. They study severe weather events to help create more predictability, so that humans can prepare when a weather event is going to hit. To safeguard life on Earth, NASA works with NOAA (National Oceanic and Atmospheric Administration), the US Geological Survey, and the Environmental Protection Agency.

Astronaut Sunita L. Williams, Expedition 15 flight engineer, exercises on the Cycle Ergometer with Vibration Isolation and Stabilization (CEVIS) system (CEVIS) during a Periodic Fitness Evaluation with Oxygen Uptake Measurement (PFE-OUM) on the International Space Station.

Image courtesy of NASA.gov/ Rodney Grubbs.

NASA's exploration does not end with Earth itself. One of the objectives of the "Discover" goal is to research the physical and biological responses to spaceflight. The human body, and other organisms, react differently when living in space. NASA will continue its experiments and research in the ISS with long-duration missions to study the effects of space life.

Strategic Goal #2: Explore

When you think of space travel, you most likely think of the exploration aspect. NASA's second strategic goal focuses on exploration and, specifically, human presence in space and on the Moon for long-term exploration. So, what does this mean?

Currently, NASA is laying the foundation for the US to sustain a commercial, human presence in space. Plans are in the works to return humans to the Moon. Space travel to the Moon is seen as a stepping stone to reaching Mars. By creating a lunar space community that can be used for training, and developing the space program on the Moon, humans are setting the stage for deep space travel. It may sound like a scene from a movie, but the reality is that future generations could choose to live on the Moon, if NASA accomplishes its goals.

NASA does not plan to accomplish this mission alone. The US will seek international partnerships and rely on the US private sector for assistance. This will be discussed further in a later lesson.

Astronauts in space suits viewing the Milky Way galaxy.
Gorodenkoff/Shutterstock

Strategic Goal #3: Develop

NASA's third goal strives to develop aeronautical advancements that secure the US's status as a forerunner in space travel. The planned demonstrations and research are designed to advance flight well into the 21st century. It could mean the return of supersonic flight, new airliners that use less fuel, or "self-driving" airplanes. To achieve this goal, NASA is dedicated to recruiting students in STEM fields, with hopes of inspiring future generations to pursue STEM degrees, as well. Public interest in NASA is growing, and missions are attracting not only potential scientific minds but interested observers. In 2015, NASA's mission to Pluto prompted 10 million views of the NASA page.

The window in the sidewall of the 8-by-6 foot supersonic wind tunnel at NASA's Glenn Research Center shows a 1.79 percent scale model of a future concept supersonic aircraft built by The Boeing Company
NASA/Quentin Schwinn

In addition to expanding the potential science community, NASA seeks to develop revolutionary technologies that ensure the possibility of space exploration. Investments have been made in large-scale industrialization of space and projects to enable humans to live and work in space. These technologies will also focus on commercial aviation, including supersonic aircraft, efficient aircraft, real-time safety, and autonomy in aviation.

Strategic Goal #4: Enable

Finally, we have NASA's goal to focus on enablement. This goal means NASA will optimize its programming and resources (including people) to concentrate on activities and projects that push toward achievement of its long-term goals. To reach those goals, NASA must partner with private agencies and academic organizations to become the leader in space travel. NASA will continue using international and private partnerships to enhance the priorities of the nation. The agency also will use private and government resources to deliver people and goods to space. Later in this lesson, you will read about the boom of private organizations in space travel.

Cost of Space Travel

As you just read about NASA's strategic goals, you probably wondered, "How do we pay for this? Space exploration is expensive!" NASA does have lofty goals for sending humans back to space and ensuring a future space exploration program. But truth be told, our current space program fell behind for some time due to the exorbitant cost of space travel.

In 1972, when NASA's space shuttle program was announced, it was estimated to cost $20 million per flight. This estimate was massively under budget. In 2010, a single space mission cost $1.6 billion. In all, NASA spent $209 billion dollars on the shuttle program. The program came to an end on August 31, 2011.

Space Launch System (SLS) solid rocket booster prime contractor Grumman Innovation Systems (formerly known as Orbital ATK) recently completed work at its Utah facilities on the booster nozzles for Exploration Mission-1 (EM-1), the first flight of SLS and the Orion spacecraft.
NASA/Orbital ATK

NASA was not designed to make money, so cannot fund its budget on its own; instead, it is funded by the government. In recent years, NASA funding flatlined, and budget cuts led to delays in programs and new technology. NASA's 2019 budget provides $19.9 billion to lead an innovative space campaign and return humans to the Moon for long-term exploration.

Private Industry in Space

For decades, NASA has relied on private companies to supply components and equipment for space travel. In 1962, the Telstar 1, a communications satellite, became the first object built entirely by a private company.

In 1984, Congress passed a law that allowed private companies to launch rockets into space. Then, in 1990, Congress passed another law that required NASA to use private companies to launch supplies into space. The private space industry was born.

In 2004, Congress passed the Commercial Space Launch Amendments Act to make private space exploration legal. Until then, private organizations could launch satellites and supplies into space, but they could not actively explore space or send humans into space. With the shutdown of the NASA shuttle program in 2011, private companies have now entered the space exploration industry. NASA plans to use private space companies to launch humans and supplies into space.

Some of the businesses currently working on plans for space exploration are SpaceX, Rocket Lab, Blue Origin, Sierra Nevada Corporation, and Virgin Galactic.

SpaceX

SpaceX currently has the most advanced space travel plans. SpaceX was founded in 2002 by Elon Musk with the goal of revolutionizing space technology. Elon Musk is also the co-founder and CEO of Tesla. In addition, he co-founded PayPal. SpaceX designs, manufactures, and launches rockets and spacecraft. The ultimate goal of SpaceX is to provide the technology for humans to live on other planets.

SpaceX's Dragon spacecraft was the first commercial spacecraft to deliver supplies to the ISS in 2012. The Dragon spacecraft was designed to transport humans to space, and SpaceX is well on its way to making that a reality.

Kennedy Space Center, Florida — February 19, 2017: A SpaceX rocket launches on the CRS-10 resupply mission to the International Space Station.

John Huntington /Shutterstock

SpaceX is also responsible for designing and manufacturing the world's most powerful rocket, Falcon Heavy. The Falcon Heavy rocket can lift twice the payload of its competitors and will be used to launch satellites and spacecraft.

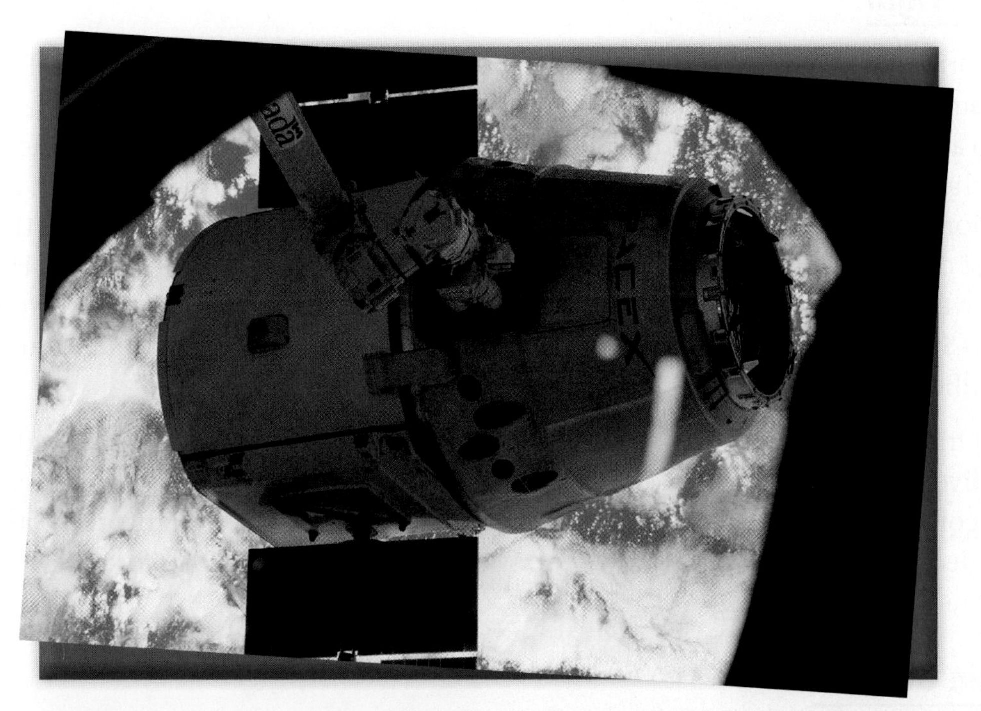

The SpaceX Dragon is pictured from inside the seven-windowed cupola moments before ground controllers remotely commanded the Canadarm2 to release the resupply ship from its grips.
Mark Garcia/NASA

Did You Know?

Japanese billionaire Yusaku Maezawa, will be the first paying customer to travel around the moon on SpaceX's yet-to-be-developed Big Falcon Rocket (BFR). In a presentation at SpaceX headquarters in Hawthorne, CA, Maezawa — said that his flight will occur in 2023 and that he wants to invite six to eight artists to accompany him. He called his trip around the moon a "lifelong dream."

Rocket Lab

In January 2018, Rocket Lab became the second commercial company to achieve orbit. Its Electron rocket is one of the cheapest rockets that can be used to launch into space, due to its lightweight design; the rocket is made entirely of carbon fiber. Rocket Lab is now using the Electron to deliver small satellites into low Earth orbit (LEO), which means that the *objects orbit at between 99 to 1,200 miles above the Earth's surface.*

Blue Origin

Blue Origin was founded by Jeff Bezos. In case you have not heard of him, Jeff Bezos is also the founder and CEO of Amazon. Blue Origin was the first commercial organization to equip a space rocket with landing gear. The first rocket by Blue Origin, New Shepard, was launched into space, completed its orbit, and then landed back on Earth in an upright position, ready to take off again. And two months later it did! Bezos estimates that the largest cost of space travel is the unwanted waste or debris created by a rocket. He aims to reduce the cost of space travel with reusability. He estimates that a launch currently costs $60 to $100 million, but if reusable rockets were deployed, the cost would be just under $1 million per launch.

Blue Origin is currently working on a new reusable rocket, New Glenn, that is designed to deliver humans and payloads to space on a regular basis. Blue Origin's plan is to offer commercial trips to space for those willing to pay. Their test flights thus far have included human dummies. However, they estimate it will be 2019 before they can offer commercial service.

Virgin Galactic

Virgin Galactic was founded by Virgin Atlantic founder Sir Richard Branson. The company's reusable SpaceShipTwo is the first commercial spacecraft built to carry passengers. Yes, that's right! Virgin Galactic has a goal to offer space travel to all humans, not just astronauts. One of the first passengers into space will be Sir Richard Branson himself.

The futuristic Virgin Galactic reusable, suborbital spacecraft on static display at the Farnborough International Airshow, UK, on July 15, 2012.

Steve Mann /Shutterstock

Benefits of Private Space Travel

Private space travel provides an enormous benefit when it comes to cost. By using commercial launches, NASA can drop the per-launch cost to send satellites into space from $4 billion to less than $50 million. In addition to the cost benefits, the introduction of private space exploration brings innovation. When you bring multiple companies and organizations into the mix, they begin to challenge each other to develop new technologies.

Of course, with benefits, also comes the consequences of space travel. As with any private organization, profit is the bottom line. Space travel technology will only advance if space travel is profitable for these organizations.

Cost of Private Space Travel

Are you ready to go into space? Virgin Galactic and Blue Origin are both primed to offer commercial space travel, but you'd better start saving. A trip to space on a commercial company's spacecraft will cost approximately $250,000 per seat. That gets you a round-trip ride into space, during which you will experience a few minutes of weightlessness before the return trip to Earth. If you want to orbit the Earth during your time in space, it will cost you about $60 to $70 million. Believe it or not, there may come a day when your annual vacation could be a trip into space!

Lesson 1 Review

Using complete sentences, answer the following questions on a sheet of paper.

1. What is the greatest benefit that can be achieved by space travel?
2. What is STEM education?
3. What are the four 2018 strategic goals for NASA?
4. What are the goals of the James Webb Space Telescope?
5. What did a single space mission cost in 2010?
6. What was the first space object built entirely by a private company?
7. What does Blue Origin see as the key to commercial space travel?
8. What are the benefits of private space travel?

APPLYING YOUR LEARNING

9. Provide a brief description of why the benefits of commercial involvement in space exploration would outweigh any negative consequences.

LESSON 2

Assembling a Space Mission

Quick Write

Read the vignette about astronaut Leland Melvin. Now, consider the importance of being prepared for an alternative career path. What career paths and options have you identified for your future?

Learn About

- the essential components of a space mission
- the selection and training of astronauts

Leland Melvin, NASA Associate Administrator for Education (2007)
Mark Sowa/JSC NASA

Imagine, you spend years of blood, sweat, and tears training for the big leagues. Then, while playing for the National Football League (NFL), you suffer a hamstring injury that ends your football career forever. For many, it ends any career plans altogether. For others, it presents new professional opportunities in other areas. For Leland Melvin, it meant a bigger, better, and literally out-of-this-world endeavor. He joined the National Aeronautics and Space Administration (NASA)!

After his football injury in the NFL, Melvin went back to school to build on his undergraduate education in chemistry. In graduate school, he received a Master of Science degree in materials science engineering. In 1998, thanks to a persistent recruiter and NASA's efforts to diversify, Melvin began training with the NASA Astronaut Corps.

Melvin suffered hearing loss during underwater training that grounded him from going to space. Eventually, his hearing returned, and in 2008 and 2009, Melvin flew two missions on the space shuttle Atlantis. Both missions involved construction on the International Space Station. The International Space Station (ISS) *is a large spacecraft in orbit around Earth*. It houses astronauts and cosmonauts from around the world and acts as a science laboratory. Cosmonauts refer to *individuals trained and certified by the Russian Space Agency*.

After Melvin left the astronaut corps, he became NASA's associate administrator for the Office of Education until his retirement in 2014. His experience as an astronaut, athlete, and scholar has inspired NASA's education programs. It has also raised public awareness about NASA strategic goals.

The Essential Components of a Space Mission

NASA's Mission

Since 1958, NASA has been leading the United States in space exploration and has experienced amazing results. As you read in the previous lesson, NASA has been joined by commercial space exploration organizations that will help the US lead the way in exploring the vast space beyond Earth. NASA's vision is to discover and expand knowledge for the benefit of humanity.

NASA strives to discover new information about scientific principles in our solar system, and beyond—and to apply those facts to make space exploration safer and more convenient. We use much of our knowledge about space to create new products here on Earth, especially in the transportation industry. Automobiles and passenger planes are more comfortable, safer, and more fuel efficient than ever before.

To continue forward progress, we need deeper exploration of the regions of space that have already been studied, such as the Moon and Mars. We also need to gain more information about other places that might benefit us on Earth. We have put humans on the Moon and brought them back home safely. New explorations may enable us to build permanent laboratories, or even normal living spaces, there and beyond. Those explorations would begin with long-term studies of the Moon and Mars to gather more detailed information, instead of just short-term visits.

Astronaut on lunar (moon) landing mission. Elements of this image furnished by NASA.

Courtesy of Castleski/Shutterstock

Vocabulary

- International Space Station (ISS)
- cosmonauts
- mission directorates
- heliophysics
- astrophysics
- aeronautics
- launch vehicle
- Tracking and Data Relay Satellite System (TDRSS)
- microgravity
- underwater training
- flight simulators
- g-force

NASA's Langley Research Center has been providing NASA with aeronautical research for over 100 years.
Courtesy of NASA/Sandie Gibbs

We use the data gathered from our studies in space to develop new technology for space travel. We also use it to create new and better products in our everyday lives. Consider the fields of transportation, communications, construction, and food processing. Many of the new products we enjoy today resulted from NASA research. For example, we have memory foam, freeze-dried food, and parachute systems on aircraft. We benefit from this technology across the globe. It helps to bring us all together as one large, human society.

NASA aims to improve and continue beneficial space operations. For example, we rely on the ISS for ongoing studies on human safety in space as we advance our explorations. Remember the early pioneers in our own country? These settlers gradually moved forward from the eastern settlements to the far western regions. Our space travelers of today are truly pioneers in their own right! As we continue to explore space, we enable greater advances in engineering, design, and construction of products at home.

NASA's Mission Directorates

Several organizations work together to achieve the mission and goals that NASA has set forth. NASA's organizational structure is built on mission directorates. *These are groups that study space from Earth, and in space.* Each group serves a specific purpose in contributing to the knowledge base of understanding air and space. The directorates include the following:

- Science Mission Directorate (SMD)
- Aeronautics Research Mission Directorate (ARMD)
- Space Technology Mission Directorate (STMD)
- Human Exploration and Operations Mission Directorate (HEOMD)
- The Mission Support Directorate (MSD)

The Science Mission Directorate (SMD) develops the knowledge base of Earth science, heliophysics (*the study of the Sun and its effects on space*), planetary science, and astrophysics. Astrophysics is *science that applies the laws of physics and chemistry to explain the birth, life, and death of stars, planets, galaxies, and nebulae.* The SMD seeks to answer some of the most important questions about our changing Earth. It uses advanced scientific methods and the latest Earth-based technologies. It also uses space technologies, such as satellites and probes. The SMD determines the needs of the space mission, including research needs and technology.

Did you know that every U.S. commercial aircraft and U.S. air traffic control tower has NASA-developed technology to help improve efficiency and safety? This is due to the efforts of the next directorate of NASA, the Aeronautics Research Mission Directorate (ARMD). The ARMD's primary goal is to use advanced aviation tools and technologies to improve flight operations and aviation safety, and to contribute to aeronautical research. We define aeronautics as *the study of the science of flight*. The ARMD also aims to reduce the overall environmental impact of flight on Earth.

The ARMD's Ikhana aircraft is the agency's first large-scale, remotely-piloted aircraft flight in the national airspace without a safety chase aircraft.

Courtesy of Ken Ulbrich/ NASA

The next directorate is the Space Technology Mission Directorate (STMD). The STMD exists to maintain important space-based technological gains. It also serves to counteract technologies that could endanger future missions or US national security. This mission directorate also provides technical and flight support for the rest of NASA.

The Human Exploration and Operations Mission Directorate (HEOMD) is most vital, as it provides NASA with leadership and oversight for operations related to human exploration in and beyond low-Earth orbit. The ISS is an example of space operations in and beyond low-Earth orbit. HEOMD manages operations related to launch services, space transportation, space communications, and human research programs.

The Human Research Program (HRP) was developed as a result of NASA refocusing the space program on exploration in early 2004. The program uses research findings to develop procedures to lessen the effects of a space environment on the health and performance of humans working in that setting. With the goal of traveling to Mars and beyond, the program is using ground research facilities and the International Space Station to develop these procedures and to further research areas that are unique to Mars.

The Mission Support Directorate (MSD) manages institutional services and capabilities. It oversees mission processes and operations to ensure efficiency, consistency, and standardization.

Components of a Space Mission

For NASA, a space mission looks a lot like a research project. Starting with a research goal, the MSD forms a scientific plan to accomplish that goal. Then, the appropriate research-gathering components are identified and assembled. These components include spacecraft, astronauts, advanced technology, ground support, and communications.

Artist concept of NASA's first Space Launch System (SLS), Block 1, on the launchpad. SLS is an advanced launch vehicle that provides the foundation for human exploration beyond Earth's orbit. It is the only rocket system that can send the Orion spacecraft, astronauts, and large cargo to the Moon on a single mission.

Courtesy of NASA/MSFC

The next important step involves building a spacecraft designed specifically to accomplish the goals of the mission. For example, manned spacecraft that go into deep space require specific features to ensure the safety of the spacecraft and the astronauts inside. The distance and duration of the mission also determine how the spacecraft is designed and built.

Along with the spacecraft, a launch vehicle is built. A launch vehicle is *a rocket that is used to propel the spacecraft into orbit*. The launch vehicle design and construction are also based on the goals of the mission. The weight of the spacecraft and the payload are careful considerations of the plan too. Before the launch vehicle propels the spacecraft into orbit, the spacecraft is tested and retested on the ground.

With a spacecraft and launch vehicle built and tested, it is time for lift off! Carefully planned communication systems, such as the Tracking and Data Relay Satellite System (TDRSS), and ground support operations, are essential components before, during, and after lift-off and landing. The Tracking and Data Relay Satellite System *provides near-constant communication links between the ground and the orbiting satellites*. On-ground support, often referred to as Mission Control, oversees the most important aspects of manned space missions. Its experts gather and analyze data from the spacecraft and launch sites to direct the mission and make time-sensitive decisions when needed.

Tracking and Data Relay Satellite-7 (TDRS-7) is placed in orbit by the Space Transportation System 70 (STS-70) Space Shuttle mission.

Courtesy of Thuy Mai/ NASA

Did You Know?

"There are thousands of people all around the country supporting the astronauts. Mission Control is the visible part, but it's less than one percent of the support team..." – William Foster (NASA Mission Control Center Ground Controller)

An illustration of the third generation Tracking and Data Relay Satellite in orbit.
Courtesy of NASA

Funding of Space Missions

As one of many government agencies, NASA submits a yearly budget request to the federal government for operating funds needed in the following year. They send their request to the US President's office, where it is included with those received from other government agencies. The President, along with staff and advisors, put that information together and decide on an itemized list for the total federal budget. They submit that document to the US Congress for adjustments and approval. Congress has the power to increase or decrease any of those requests. In some years, Congress has cut the amount requested by NASA; on other occasions, it has added more to NASA's requests.

The 2019 budget for NASA is expected to be approximately $19.9 billion. Although that seems like a very high figure, it usually represents less than one percent of the total federal budget.

NASA needs these funds for the construction of launching sites and rockets for space exploration and manned missions. In addition, NASA uses the funds for research, technology, offices, employees, and for joint projects with other public and private agencies.

NASA passes on much of the information gained by research and explorations to benefit our military, weather research, and communications companies and their satellites. In turn, large companies assist smaller companies that have contracts with them.

Directly or indirectly, NASA's projects affect each of the 50 states in a positive way. For this reason, members of Congress from all states usually support funding to NASA.

Did You Know?

Before their first mission, astronauts typically train for a combined total of 300 hours in flight simulators.

The Selection and Training of Astronauts

When asked, "What do you want to be when you grow up?" most young people will respond with, "a firefighter," "a nurse," or "an astronaut." Space travel makes the bucket list for many adults too. How many other jobs allow you to float around the office all day with a view of the entire Earth? The path to becoming an astronaut, however, is challenging. The selection process itself is strict, with a very tiny group of astronaut candidates chosen for nearly two years of extensive training. Training to be an astronaut is also hard work. Not only is there intensive classroom training, but also microgravity and survival training (in water and on land). Microgravity *is the condition in which people or objects appear to be weightless.*

Qualifications

NASA's astronaut program began in 1959 with seven astronauts selected from the military. At the time, required qualifications included flight experience in jet aircraft, a background in engineering, and knowledge and understanding of science. Because of the small space inside the first vehicles, astronauts were limited to a height of 5'11" or less. The height requirement has been changed to 5'2" up to 6'3," and applicants must complete advanced studies in science.

The education requirements today include having a bachelor's degree in engineering, biological science, physical science, chemistry, computer science, math, or medicine. A doctorate in one of those fields is considered valuable.

Applicants must have three or more years of professional experience beyond their bachelor's degree, or 1,000 hours or more of pilot-in-command experience in jet aircraft. They must also pass a long-duration astronaut physical and have near and distant vision correctable to 20/20 in each eye and a blood pressure of 140/90 in a sitting position. They must be US citizens or have dual citizenships and must complete a thorough background check and military water survival training. The first seven astronauts were all men. Today, NASA does not select astronauts based on gender or race.

Most importantly, NASA expects all applicants to have strong skills in leadership, teamwork, and communication.

Selection

When you apply for most any job, you are likely part of a pool of applicants with a variety of skills and backgrounds. The pool may be relatively small, consisting of only a few candidates, or the pool may be the size of a hundred or more. Similarly, NASA's astronaut selection process starts with applicants, but the pool of want-to-be astronauts is typically in the tens of thousands. Astronaut applicants represent a variety of backgrounds—ranging from teachers to scientists, and, yes, even former NFL players!

From the thousands of astronaut applicants, human resources personnel screen out those without the basic qualifications discussed earlier. After that, applicants are grouped by field of study and interviewed by review boards. From there, the pool shrinks to a few hundred individuals.

After first and second interviews, medical testing, and meetings with current astronauts, a very small group of astronaut candidates are selected for NASA's two-year spaceflight training.

Did You Know?

NASA selects fewer than 0.08 percent of applicants to become astronauts. In 2017, NASA selected 12 new astronauts to begin training out of 18,300 applicants.

The new astronauts will train for deep space missions on NASA's new Orion spacecraft and SLS rocket.
Courtesy of NASA/Rob Markowitz

Training to Become an Astronaut

Ask most any astronaut what it takes to fly in space and they will tell you: years of training—training before being selected for training; astronaut candidate training; and advanced on-going training. The first phase of NASA's two-year spaceflight training involves general instruction on science, space technology, foreign languages, and public speaking. Astronaut candidates also go through challenging survival training (in water and on land), along with microgravity and medical training.

Underwater Training

Candidates train underwater in a deep tank called the Neutral Buoyancy Laboratory (NBL) at NASA's Johnson Space Center in Houston, TX to prepare for space walks. Underwater training *simulates conditions similar to the weightlessness in space.* One of the first training tests for astronaut candidates is a swimming test where they swim 75 meters and tread water nonstop for 10 minutes in a flight suit and tennis shoes.

NASA astronaut Peggy Whitson trains underwater for a spacewalk at the Neutral Buoyancy Laboratory (NBL) at Johnson Space Center in Houston. Whitson launched to the International Space Station in late 2016 as part of Expedition 50/51.
Courtesy of NASA

Did You Know?

Established in 1995, the Neutral Buoyancy Laboratory (NBL) holds 6.2 million gallons of water and contains a model of the ISS underwater.

Underwater training also occurs at the Aquarius Reef Base that is located six miles off the coast of Florida. The Aquarius Reef Base is 62 feet underwater and is operated by Florida International University. Astronauts spend 10 to 14 days at the facility with an average of nine hours a day diving outside the facility to simulate the experience of living in space.

Flight Simulation

Besides underwater training, astronaut candidates experience the microgravity of space by using flight simulators. Flight simulators are *safe and cost-effective alternatives to actual flights to gather data, and provide facilities for practice and training*. Pilot astronauts maintain flying proficiency by flying 15 hours per month in NASA's fleet of T-38 training aircraft. Mission specialist astronauts fly a minimum of four hours per month.

Centrifuge Training

Astronauts also travel to NASA's Ames Research Center in Silicon Valley, CA to experience the 20-G Centrifuge. Most astronauts are not a fan of this portion of the training. Centrifuge training helps astronauts prepare for the g-forces they will experience during launch and re-entry. G-force is *the gravitational force that is put on a body during acceleration*. For example, 1 g is the force of gravity at Earth's surface. This is what we all experience every day. During launch, astronauts will experience 3 g, which is about three times the normal acceleration. Think of your body weighing three times more and the pressure of that weight being pushed into your seat. During re-entry, the g-forces are much more extreme. Astronauts experience 8 g for 30 to 60 seconds during re-entry.

The 20-G Centrifuge helps astronauts prepare for the g-forces they will experience during space travel. The centrifuge is a large rotating arm that rotates up to 50 times per minute. It creates g-forces up to 12.5 g for astronauts.

The 20-G Centrifuge at NASA's Ames Research Center.

Courtesy of NASA

Vomit Comet

The nickname of this training makes it sound like a fun experience, right? The "Vomit Comet," as astronauts fondly call it, is a hollowed-out Boeing 727 that flies in parabolic arcs over Las Vegas, NV to simulate weightlessness. NASA began using the Boeing 727 owned by Zero Gravity Corporation after canceling its Reduced Gravity Research Program in July 2014. Parabolic arcs refer to the flight pattern of an aircraft that makes a steep incline, briefly levels out, and then makes a steep decline to create a brief period of weightlessness.

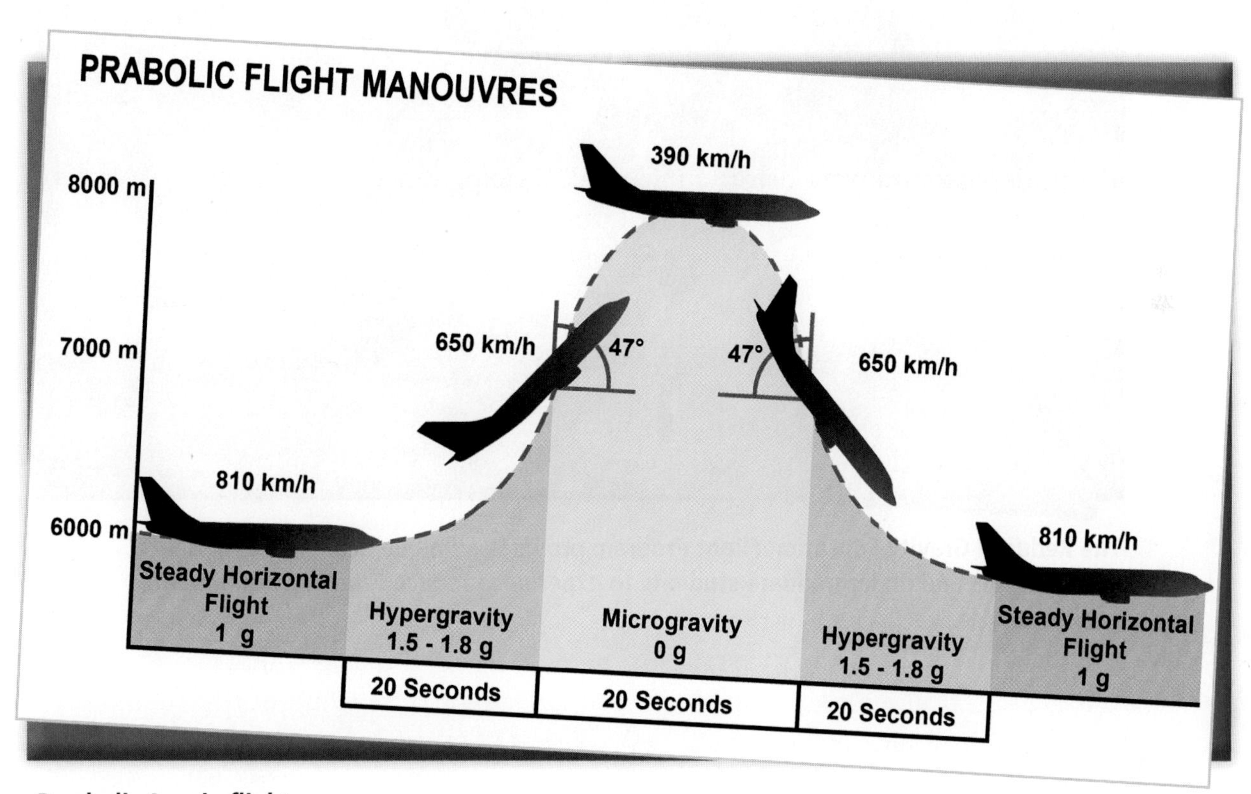

Parabolic Arcs in flight.

Astronauts take off and wait about 30 minutes to reach an altitude of 24,000 feet. The pilot then climbs to 32,000 feet at a sharp 47-degree angle. This allows astronauts to reach 1.8 g. The plane then levels out and astronauts experience 20-30 seconds of weightlessness before they dive back down to 24,000 feet. The flight repeats this pattern at least 15 times to give astronauts enough practice at 0 g. (0 g is the value for weightlessness.)

After the initial phase of spaceflight training, astronaut candidates go on to learn more about the finer details of the ISS, the experiments, special mission-specific activities, the transport vehicles, and the role of ground control. Part of training at this stage involves supporting current astronauts in space and astronauts from other training facilities.

> **Did You Know?**
>
> For $5,000, Zero G Corp. will provide you the same experience. You can take a ride on the Vomit Comet and experience weightlessness.

The Reduced Gravity Education Flight Program provides a unique academic experience for educators and undergraduate students to experience reduced-gravity experiments of their choice.
Courtesy of NASA

Duties

Astronauts hold several different titles and responsibilities on each space mission. As the person in charge, the mission commander handles everything from the vehicle itself to the crew and their safety. Successful completion of the mission is paramount. Commanding a space flight may also involve spacecraft deployment and retrieval of satellites, using payload equipment. "Payload" refers to cargo such as instruments and space equipment that is carried into space to accomplish a mission. Payload cargo even includes the astronauts themselves!

Just as a commercial flight captain has a first officer or co-pilot, every space mission commander has a supporting pilot, or next-in-command. Spacecraft support pilots assist the commander and have primary responsibility for operating the spacecraft. They perform many of the same duties as a commercial pilot and use a mechanical arm or other payload equipment to release and retrieve experiments conducted outside the spacecraft. Pilots may also be involved in scheduled spacewalks.

Working closely with commanders and pilots, mission specialists direct the daily activities. These activities include distribution of food and water to other crew members, monitoring of fuel, and carrying out experiments and payload activities. Mission specialists are experts who oversee the operation of all systems aboard the craft, including experiments, spacewalks, and payload-handling duties performed during a mission.

Dr. Ellen Ochoa became an astronaut in July 1991 and was the first Hispanic woman to go into space. She is a veteran of four space flights and has logged over 978 hours in space. She currently serves as director of the Lyndon B. Johnson Space Center in Houston, Texas.

Courtesy of NASA

One of the more recent astronaut roles is that of the payload specialist. Payload specialists are scientists or engineers selected for their expertise. They perform special functions during a mission, beyond the normal routine for standard flights. These functions may include operating unique spacecraft equipment or carrying out experiments based on their expertise. Unlike pilots and mission specialists, payload specialists do not appear on most missions.

With more missions by NASA planned for the future, the demand for the number of astronauts needed may increase. Missions include more studies in the International Space Station and a return to the Moon for additional research. Manned missions to Mars for short-term or even long-term projects may be possible. Depending on those results, we hope to see astronauts venture even farther into space than the Moon and Mars!

Do you have dreams of becoming an astronaut? Has completing this lesson sparked an interest in possibly becoming an astronaut? Although the path to becoming an astronaut is very competitive and challenging, the opportunities in fields of study related to astronomy, engineering, and medicine are endless. Find an interest in something that ignites your passion and plan a career path that will take you there. Plan for detours along the way, and the sky just may not be your limit!

The Right Stuff

Day in the Life of an Astronaut on the ISS

It's 6:00 in the morning, and it's time for breakfast and another day at the office--unless you are on a mission in the ISS. Astronauts in the ISS start their day a little differently. Getting showered, dressed, and ready in microgravity is a bit more challenging. For example, the shower is a soothing wake-up call for most of us. Showering aboard the ISS is more like a sponge bath with very little water and with rinseless shampoo. After toweling dry, an airflow system evaporates the rest of the water.

After a morning "shower," astronauts have the first of three daily meals. Space meals are much like MREs (Meals Ready to Eat) for soldiers. Without refrigeration on the ISS, space food must be spoil-proof. It must also be easy to prepare and store. You also need to consider microgravity. For example, salt and pepper are not something you find on the ISS. Imagine trying to sprinkle salt on your food, just to have it float away!

How astronaut Scott Kelly enjoys a snack aboard the ISS.

Courtesy of NASA/JSC

When astronauts are not working, they spend free time exercising, relaxing, or just having fun. Exercise is actually a requirement in space. Daily exercise, up to two and a half hours, helps to prevent bone and muscle loss. A major difference in working out in space, of course, is the amount of weight you are able to lift just to feel some resistance.

It's now 9:30 p.m. and time to strap into a sleeping bag for a good night's sleep without floating around and bumping into things. When the sun rises, it will not be time to wake up for another day, though. Keep in mind, the sun rises 16 times in a day aboard the ISS, so astronauts take special measures to avoid the light while sleeping.

Lesson 2 Review

Using complete sentences, answer the following questions on a sheet of paper.

1. What is NASA's vision for exploring space?
2. What are NASA's mission directorates?
3. What are the components of a space mission?
4. How is NASA funded?
5. What are the basic qualifications of astronauts?
6. How do astronaut candidates train for microgravity?
7. What is involved in centrifuge training?
8. What is a payload specialist responsible for?

APPLYING YOUR LEARNING

9. NASA plans to send astronauts to Mars by 2040. If you wanted to be an astronaut on that mission, what special skill do you think would be needed?

LESSON 3

The Hazards for Spacecraft

Quick Write

The New Horizons probe accomplished an impossible mission and provided an immense amount of data for scientists. What challenges do you think a probe would face to travel outside our solar system?

Learn About

- radiation hazards
- the hazard of impact damage
- threats associated with surface landings
- fire hazards in space

On January 19, 2006, NASA launched the New Horizons space probe. The mission of New Horizons was ambitious. The probe would travel past the planets in our solar system and send data back to Earth, taking nine years to reach the last planet, Pluto. It's a long journey, and the New Horizons spacecraft faced many challenges and hazards that could be detrimental to the mission.

Even before launch, the New Horizons probe faced challenges. It survived the torrential rain and winds of Hurricane Wilma, the coordination of many teams and components, and a successful launch schedule. New Horizons had a launch window of January 11 to February 14, 2006. However, if it was launched within the first 23 days of that window, New Horizons would receive a boost from the gravity on Jupiter. If the launch window was missed, the probe would have reached Pluto up to six years later.

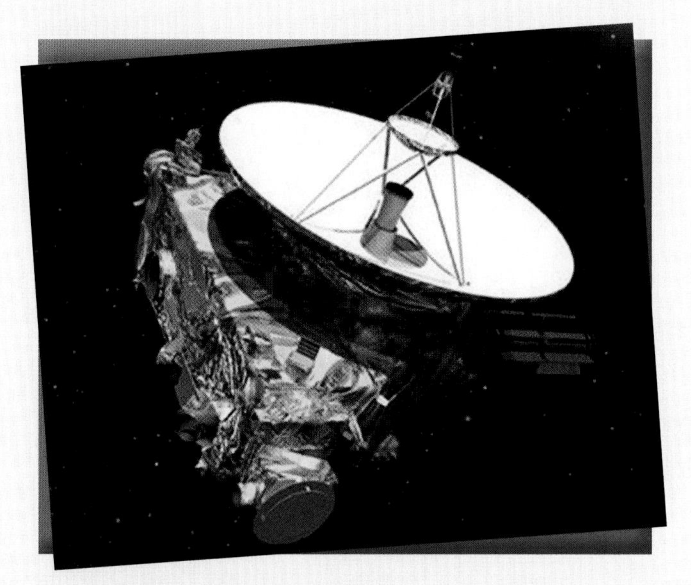

A rendering of the New Horizons probe in space.
Courtesy of NASA

Once airborne, New Horizons had a new challenge: how to power the batteries. Pluto is three billion miles away from Earth, and the Sun's rays are not powerful enough to reach Pluto with enough strength to power the batteries. So, New Horizons is powered by nuclear power that can withstand the radiation hazards and longevity of its mission in space. In addition, multiple backup systems are on board in case of failure or radiation damage to any of the probe's components.

Before New Horizons reached Pluto, several new moons were discovered around the dwarf planet, posing an impact hazard to the spacecraft.

New Horizons successfully reached Pluto in July 2015, sending magnificent images of the dwarf planet's icy surface and its largest moon, Charon. New Horizons did not stop at Pluto, however. Instead, it is continuing into deep space to the Kuiper Belt, which it will reach in July 2019.

Vocabulary

- space radiation
- galactic cosmic rays
- space debris
- orbital debris
- micrometeoroids
- molecular diffusion

Images from New Horizons Long Range Reconnaissance Imager LORRI were combined with data from the spacecraft Ralph instrument to create this enhanced view of Pluto.

Courtesy of NASA/Johns Hopkins University Applied Physics Laboratory/Southwest Research Institute

Your mission is planned, and it's time to travel into space. Unfortunately, many dangers lie ahead. Space travel is a risky business and has many hazards. In this lesson, we will explore the risks that a spacecraft faces during its journey.

Radiation Hazards

What is radiation? Radiation is a form of energy that is emitted in particles or waves. Radiation can occur in many different forms. It is all around us on Earth, but our atmosphere protects us from the harsh radiation that exists in space.

Space radiation *contains atoms that have been stripped of their electrons as they accelerate through space at speeds close to the speed of light.* As the atoms speed up, they will lose all of their electrons and be left with only the nucleus. There are three kinds of space radiation:

- Particles trapped in the magnetic field of Earth
- Particles sent flying into space during a solar flare
- Galactic cosmic rays, which are *high-energy protons and heavy ions that originate outside our solar system*

Radiation can be extremely dangerous to the human body, and astronauts are exposed to a large amount of radiation in space. Spacecraft must be designed to simulate Earth's atmosphere to try to protect astronauts from being exposed to large amounts of radiation. We'll learn more about the hazards of radiation to astronauts in a later chapter.

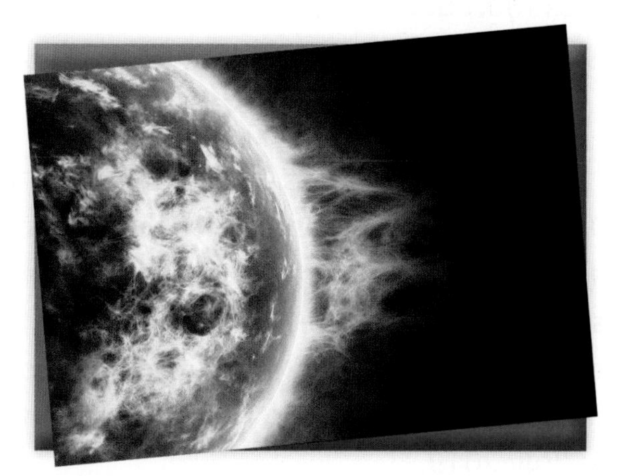

A 3D illustration of the Sun's surface with solar flares.
Pixus/Shutterstock

Radiation's Effect on Electronic Systems

Another concern about radiation in space is the danger to the electronics that protect both the spacecraft and astronauts. Radiation can energize the circuits in computers, which can erase the data on the computers and cause glitches that affect the spacecraft's operability. Most space-based computer failures are a result of single events of high-energized particles traveling through space.

There are two ways radiation affects electronics. Short bursts of high-energy can disrupt electrical charges and penetrate integrated circuits. All programming is a series of bits and bytes–0's and 1's; radiation can actually flip those bits, causing a change in the programming. One small change in programming can have devastating effects on a spacecraft. In addition, radiation can affect microprocessors by causing them to execute a program incorrectly. For example, radiation could cause a program to believe that 2+2 = 5.

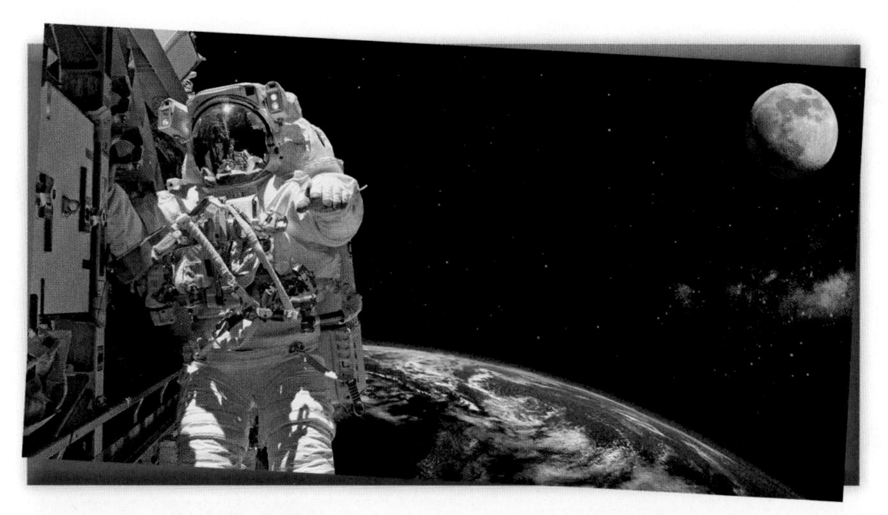

Astronaut on a spacewalk to perform maintenance, with Earth in the background. Elements of this image furnished by NASA.
Castleski/Shutterstock

The second effect of radiation is the long-term threat to circuits. Over time, if circuits are continually exposed to harmful radiation, they become less effective at storing information and making calculations,

The risk of radiation hazards to spacecrafts varies, depending on the location of the craft. For example, radiation around Jupiter is extremely high, and disasters can occur each minute a satellite or spacecraft passes Jupiter. NASA reduces the risk of radiation hazards to some spacecraft, such as the Hubble Space Telescope, by shutting them down entirely when they pass through high radiation zones, such as the Van Allen belt.

So how do we shield electronic equipment from radiation? Radiation shielding is possible, but the shields are heavy and very costly, which does not make them an ideal solution for satellites. Engineers designing the chips and circuits in the electronic systems in space are working to build components that are less susceptible to radiation. Again, this technology is expensive, so it's not an option for many spacecrafts.

The International Space Station (ISS) currently is testing the Radiation Tolerant Computer Mission (RTcMISS), which is a computer system designed to handle the negative effects of space radiation.

The most common option for fighting radiation hazards to protect electronics is Triple Modular Redundancy (TMR). TMR uses readily available electronic components, but in triplicate. Therefore, if one system has an error, the other two can correct it.

NASA astronaut Shannon Walker, Expedition 24 flight engineer, is pictured near a robotic workstation in the Destiny laboratory of the International Space Station.
Courtesy of NASA

Did You Know?

Computer systems are more susceptible to radiation today than when the Apollo mission went to the Moon. As computer systems advance, they become smaller and more complex. This actually allows more interference from radiation.

The Hazard of Impact Damage

Space mission operators may have to maneuver spacecraft to avoid collisions. 1.2 cm is the size of a pea, 18 cm equates to just over 7 inches. In some cases, no maneuvers are possible.

Courtesy of ESA

Space debris refers to *natural and man-made particles in space*. The definition includes orbital debris, which is *man-made particles in space, like trash from spent rockets and broken satellites*. Orbital debris, which orbits the Earth, is also referred to as "space junk."

There currently are more than 500,000 pieces of orbital debris orbiting the Earth. Orbital debris orbits at a speed of 17,500 mph. This causes a collision hazard for spacecraft, satellites, and the ISS. A collision with orbital debris could be catastrophic in space. As a result, NASA has invested in the detection and prevention of collisions with space debris.

Meteoroids, which orbit the Sun, are another hazard to spacecraft and travel at twice the speed of space junk. However, the majority of meteoroids are very small micrometeoroids. Micrometeoroids are *small pieces of asteroids and comets that cannot be tracked*.

Micrometeoroids and orbital debris (MMOD) are the biggest risk for NASA's flight programs. About 20,000 pieces of orbital debris are larger than a softball. About 500,000 pieces of orbital debris are the size of a marble or larger. Larger orbital debris can be tracked and avoided by spacecraft, but there are millions of pieces of small orbital debris that cannot be tracked. NASA uses computer modeling to mitigate the risks of MMOD to spacecraft. The primary tool used by NASA is a computer program named Bumper. Bumper analyzes the probability of spacecraft being damaged by MMOD.

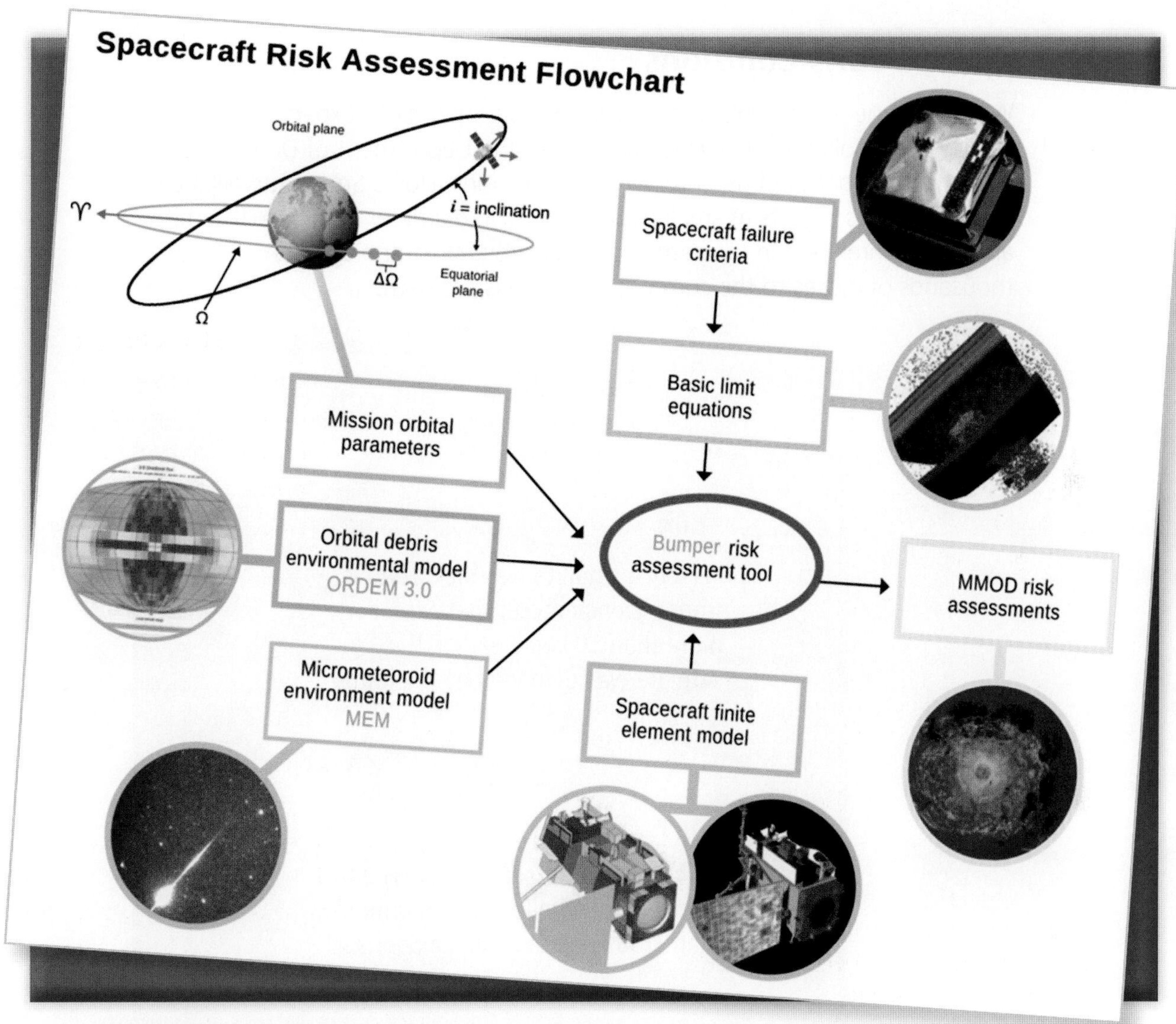

The Bumper risk assessment tool evaluates many areas to determine the risk of impact to a spacecraft.
Courtesy of NASA

Let's think about at a small fleck of paint orbiting Earth at about 17,500 mph. If that small fleck of paint collides with a spacecraft entering space, the velocities of each of these objects can create a surprising amount of damage to a spacecraft. NASA has had to replace many windows in space shuttles due to debris collision with flecks of paint. Thus, the greatest risk to spacecraft is the undetectable space debris.

The Right Stuff

Space Debris Collisions

With the large amount of space debris in orbit, one might expect a high number of disastrous collisions. However, NASA and the Department of Defense (DoD) actively track space debris to try to avoid collisions. DoD's Space Surveillance Network tracks objects as small as two inches in diameter in low earth orbit, and about three feet in diameter in higher orbits. Just one collision could add thousands of pieces to the current space debris problem.

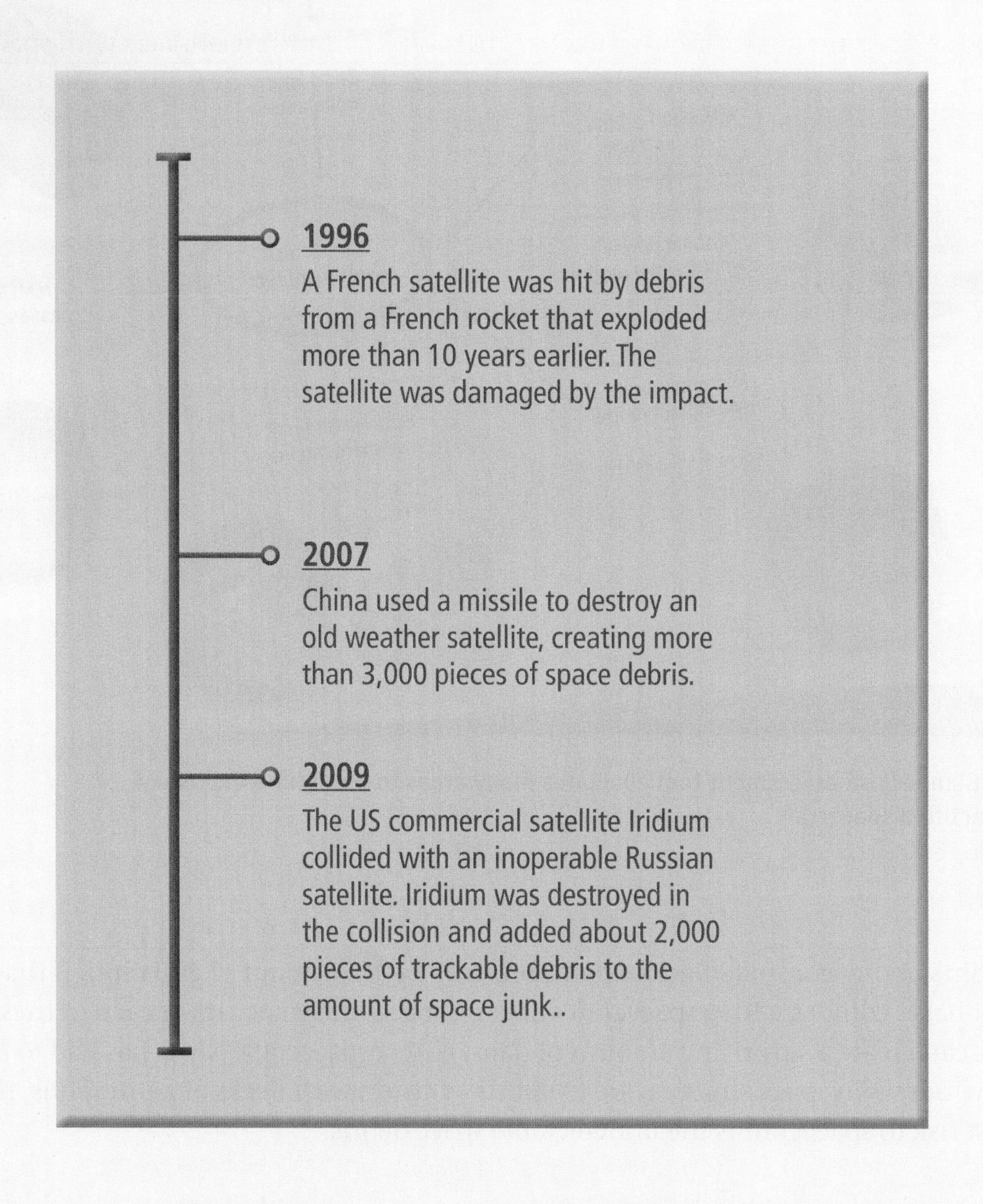

Avoiding a Collision with Space Debris

So how do spacecrafts avoid a collision with so much space debris in orbit? NASA has a standard set of guidelines to assess the risk of impact with space debris. The guidelines create an imaginary box, known as the "pizza box," around a spacecraft. The pizza box is a mile deep and 30 miles across and long. If an assessment determines that debris will pass within the pizza box, Mission Control centers work to determine a course of action. The launch path and/or date may be altered to avoid collision or the spacecraft may need to engage in avoidance maneuvers. These maneuvers are typically small and occur at least an hour before a collision may occur.

Cleaning Up Space

Although NASA and the DoD have effective methods for avoiding impacts with space debris, the ultimate solution is to clean up the space junk. As more junk is created, the risk of collision increases exponentially. International guidelines exist that require the removal of space debris and spacecraft within 25 years of the end of the mission for the spacecraft. That's a long time for, say, an inoperable satellite to be orbiting Earth. And not surprisingly, only 60% of operators actually follow through with the removal of their spacecrafts.

Several plans for space debris cleanup are in the works. In June 2018, astronauts in the ISS deployed the RemoveDebris (RemDeb) spacecraft for its first mission: an experimental phase that began in September 2018. To do its job, RemDeb launches a cube that travels 16 to 23 feet away from the main spacecraft. RemDeb then launches a net to capture the cube and any space debris in the vicinity. Next, it launches a sail to assist with deorbiting, which returns the debris to Earth in about 10 weeks.

Threats Associated with Surface Landings

Your spacecraft has made it to space, but now what? You may be landing back on Earth, the Moon, or Mars! There are hazards to a spacecraft associated with landings.

NASA's shuttle program used Kennedy Space Center's runway as its landing point. The runway is longer and wider than most commercial runways. It was kept clean and free of debris to offer perfect landing conditions to arriving shuttle astronauts. But what about landing on the Moon or Mars? There are no runways available for a smooth landing. The surfaces of the Moon and other planets are rocky, unwelcoming locations. So how do we prevent damage to a spacecraft from landing in these areas?

NASA's Autonomous Landing and Hazard Avoidance Technology (ALHAT) project successfully concluded in 2014. The program's mission was to design and develop advanced technologies that use surface-tracking sensors to identify a landing area. The ALHAT navigation system can successfully navigate a spacecraft to a safe landing area and bring the spacecraft to a landing without human interaction.

The Mars 2020 Rover has a new landing technique that will allow it to land closer to its target. In previous Rover missions, weeks and months were spent travelling to the perfect target location. With the new landing technique, the Rover will take descent photos, compare them to the orbital map, and then divert the landing, if necessary. The landing area will also require 50% less space than previous missions. This is called Terrain-Relative Navigation (TRN). Terrain-Relative Navigation is critical for the exploration of Mars. As discussed in Chapter 2, Mars is a very rocky planet and has a tricky terrain. Of course, these rocky areas are also the most interesting parts of Mars to explore. However, they have been off-limits as potential landing sites because 99% of a landing area has to be free of hazardous slopes and rocks.

The Mars 2020 Rover will use a new landing technique to allow the Rover to land in previously off-limits areas.

Courtesy of NASA

Fire Hazards in Space

Fire is another hazard to spacecraft. Fire burns very differently in space. When a fire burns on Earth, oxygen is pulled into the flames and combustion products are pushed out. Heat and hot gases rise from the flames on Earth. In space, there is no gravity, so heat does not rise. A process called "molecular diffusion" controls a flame's behavior in space. Molecular diffusion is *the thermal motion of all particles at temperatures above absolute zero.* What this means is that, in space, fire draws in oxygen and releases combustion products 100 times slower than on Earth. Flames in space also burn at a lower temperature and need less oxygen to burn. For this reason, the material used to extinguish space flames must be more concentrated.

Fire needs oxygen and airflow to survive. On the ISS, the ventilation system supplies the airflow needed for a fire to survive. This means that if a fire were to ignite on the ISS, it could spread in any direction. On Earth, flames would only burn upward because of the effects of gravity. Because smoke and fire don't rise in the zero gravity of space, the smoke detectors on the ISS are installed in the ventilation system.

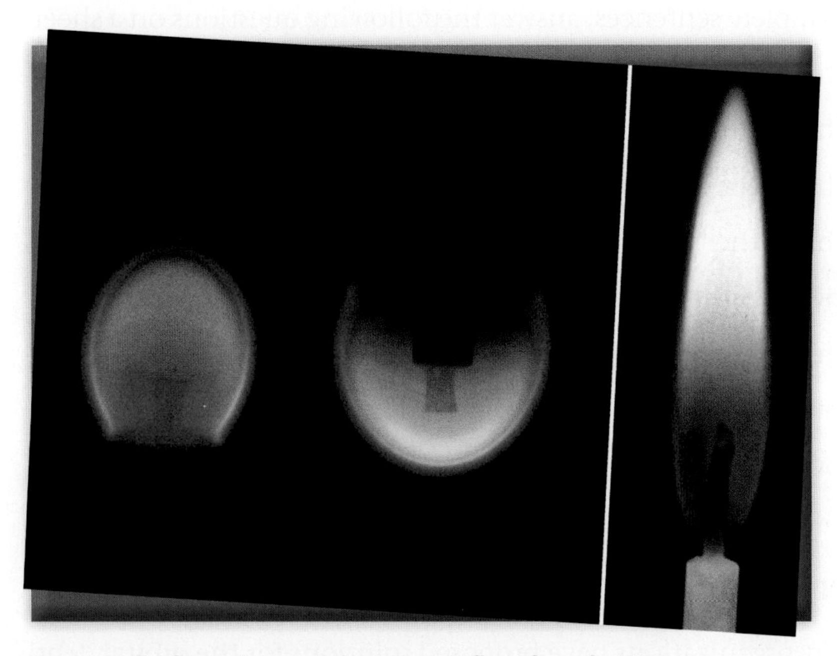

Two candle flame images showing air flow from bottom to top, as compared with how a flame appears on Earth.
Courtesy of NASA

So how do we protect spacecraft from fires in space? The best way to fight a fire in space is prevention. All spacecraft and space components must complete a flammability test before launch to help prevent space fires.

If a fire did occur on the ISS, the astronauts would follow a three-step plan:

1. Turn off the ventilation system.
2. Shut off power to the affected area.
3. Use extinguishers to put out the flames.

As you have read, there are many hazards to both space travel and to a spacecraft itself. NASA and commercial space organizations are constantly researching methods to prevent these hazards from occurring. In later chapters, you'll read about the hazards to astronauts in space.

Did You Know?

The ISS is experimenting with fire in space as part of the Flame Extinguishment Experiment (FLEX). Scientists ignite a fire in a special experiment rack and record the burning process. These experiments have demonstrated that fire is less predictable in space and can be deadlier than on Earth.

Lesson 3 Review

Using complete sentences, answer the following questions on a sheet of paper.

1. What are the three kinds of space radiation?
2. What is the most common option for fighting radiation?
3. What are micrometeoroids?
4. What is the "pizza box" around a spacecraft?
5. How does the RemoveDebris spacecraft work?
6. Why is it so hard to land on the Moon or other planets?
7. How does the Mars 2020 Rover's landing technique work?
8. How is fire different in space?
9. How would astronauts on the ISS handle a fire?

APPLYING YOUR LEARNING

10. Many organizations have proposed solutions for the orbital debris problem. Based what you have read in this lesson, what possible solution would you offer for removing orbital debris from space?

CHAPTER 4

The space shuttle Atlantis and its six-person crew launch into space to rendezvous with the International Space Station.

Courtesy of NASA/Sandra Joseph and Kevin O'Connell.

Space Programs

Chapter Outline

> "We choose to go to the moon in this decade and do the other things, not because they are easy, but because they are hard, because that goal will serve to organize and measure the best of our energies and skills, because that challenge is one that we are willing to accept, one we are unwilling to postpone, and one which we intend to win."
>
> President John F. Kennedy
> 35th US President

LESSON 1

Strategic Significance of Space Programs

Quick Write

How do you think the launch of the Sputnik satellites helped fuel the United States' space exploration endeavors?

Learn About

- US space policy
- Russian space program
- Chinese space program
- space programs around the world

After World War II, the United States and the Soviet Union were locked in a period of competition for global influence. Each was striving to be the first nation to launch a man-made object into space. On October 4, 1957, the Soviets became the first nation to launch a man-made satellite into orbit around Earth with Sputnik 1. The launch came as a surprise to the United States because it occurred much earlier than anticipated.

Sputnik 1 was a simple satellite. Shaped like a sphere, it was about the size of a beach ball. The satellite weighed 183 pounds and carried a radio transmitter that allowed it to be tracked as it orbited the Earth. The launch of Sputnik was announced to the world, and the Soviets provided the radio transmission frequencies so that everyone could track it. This ensured that all would believe the first satellite had been launched. Sputnik 1 orbited Earth once every 90 minutes. Although the transmission details had been provided beforehand, the United States had not expected Sputnik 1 to be launched until months later. The satellite completed two orbits before the US military even realized it was in space.

The success of Sputnik 1 kicked off tremendous support for the space program in the Soviet Union. Political and financial support followed for the space program to continue its efforts.

As a follow-up to Sputnik 1, the Soviets launched Sputnik 2 on November 3, 1957. This was less than a month after the Sputnik 1 launch! Sputnik 2's goal was to put an animal into orbit. A stray dog from the streets of Moscow, Laika, was the sole passenger of Sputnik 2. She was carried into orbit along with equipment to run experiments.

Video and data feeds were sent back to Earth every 15 minutes during orbit. Unfortunately, the mission was not designed to be a round-trip journey for Laika. Although the satellite automatically dispensed food and water for Laika at timed intervals, the Soviets had no intention of returning the satellite to Earth. The satellite eventually ran out of oxygen and Laika died in orbit. The Soviets were not prepared for the enormous backlash they received from around the world following Laika's death.

Despite the death of Laika and the public criticism, Sputnik 2 was considered a huge success. The satellite remained in orbit for 200 days and returned massive amounts of scientific data.

Vocabulary

- space race
- intercontinental ballistic missiles (ICBMs)
- suborbital flight
- orbital flight
- exploitation

An illustration of Sputnik 1 in orbit.
AuntSpray/Shutterstock

US Space Policy

After World War II, the space race began, with the US and Soviet Union in a race to get a human into space. The space race was *the competition between the United States and the Soviet Union to prove their superiority with the technology and man power required to send a man to space*. In an effort to indicate America's commitment to winning the space race, the US Congress passed legislation establishing the National Aeronautics and Space Administration (NASA) on Oct. 1, 1958. Less than a week later, President Dwight D. Eisenhower tasked NASA with putting a man into orbit.

Why is space important to US national interests? During the Cold War, the space race represented not just national pride, but national security, as well. In the 1960s, Vice President Lyndon B. Johnson stated, "Failure to master space means being second best in every aspect, in the crucial arena of our Cold War world. In the eyes of the world, first in space means first period; second in space is second in everything."

The initial push to explore space was heavily influenced by national security. Space was seen as a possible component of war. Reconnaissance from high-altitude orbiting satellites could provide real-time information. And if a satellite could be launched into space, then a weapon could feasibly be launched across the ocean as well. Many saw the need to explore space as a key to US safety.

Who Develops US Space Policy?

The responsibility for developing national space policy ultimately falls onto the President of the United States. Obviously, each president relies on advisors and experts in the field to help develop space policy. In addition, the level of space ambitions varies with each president. New space policies are not developed with each new president; as President Eisenhower recognized, space goals are long-term plans. It takes time, effort, and resources to develop the technology and knowledge to complete a space mission. Therefore, space policies typically have covered several presidential terms.

NASA is a civilian-operated agency that handles both civilian and military space initiatives. The DoD is responsible for directing the military space efforts for the nation. The Army, Navy, and Air Force all operate separate organizations dedicated to space applications. In 2001, the DoD specified that the Air Force was the military's executive agency for space. The Air Force Space Command is responsible for the acquisition and operations of all military space systems.

1950s

History forever changed when the Soviet Union launched Sputnik 1. The US was caught completely off guard, the public feared that if the Soviet Union could launch a satellite into space, they could also launch intercontinental ballistic missiles at the United States. Intercontinental ballistic missiles (ICBMs) are *missiles with a flight capability of over 3,400 miles*.

The US Defense Department responded by providing additional funding for multiple space satellite programs. And, in January 1958, the US successfully launched Explorer 1. This satellite carried a scientific payload that eventually discovered magnetic radiation belts around the Earth. The space race was officially on!

1960s

The Right Stuff

"First, I believe that this nation should commit itself to achieving the goal, before this decade is out, of landing a man on the moon and returning him safely to the Earth."

That proclamation by President John F. Kennedy before a joint session of Congress on May 25, 1961 set the stage for an amazing time in our nation's emerging space program.

NASA created the Space Task Group to set up and manage space flight. The Space Task Group consisted of a small group of NASA engineers tasked with managing manned space flight programs. Robert Gilruth, an expert in rocket testing and development, was assigned to lead the group. Gilruth was a pioneer in aeronautics and is considered the father of the US manned space program.

Robert Gilruth, considered the father of the US manned space program

Courtesy of NASA

Following the Soviet Union's launch of Yuri Gagarin in April 1961, President John F. Kennedy announced a daring plan, challenging NASA to safely send a man to the moon and back by the end of the decade. The US launched its own astronauts in capsules with the mythical names of Mercury, Gemini, and Apollo. From suborbital flights in 1961 to orbital flights in 1962, NASA proved that humans could survive in space. A suborbital flight is *a flight trajectory that does not complete a full orbit of the Earth.* Orbital flight is *when an object is placed on a flight trajectory that keeps it in space for at least one orbit around a planet.* During this period, astronauts became a new breed of national heroes, admired for their courage and dedication to exploring the new frontier of space. The two-man flights of 1965 and 1966 demonstrated that humans could fly in space, undertake complex rendezvous and docking operations, and even leave the spacecraft for extravehicular activity (EVA), or activity done by an astronaut outside the spacecraft, while in orbit. In 1969, NASA achieved President Kennedy's goal of landing a human on the moon and safely returning him to Earth. NASA's space exploration and moon landings will be discussed in a later lesson.

1970s

While NASA continued Apollo missions to the Moon and further space exploration, the US government had turned its attention toward the economy and the Vietnam War. President Nixon had also decided that the US should focus on a new type of space transportation, the space shuttle. By 1975, NASA's budget had been slashed from a high of five billion dollars in the mid-1960s to just over two billion dollars by 1974.

President Richard Nixon with NASA Administrator James C. Fletcher announced the Space Shuttle program in 1972.
Courtesy of NASA

President Nixon also stressed the importance of cooperation with space exploration. In 1972, President Nixon signed a cooperative program agreement between NASA and the Soviet Union that would last for five years. The deal resulted in the 1975 Apollo-Soyuz Test Project between the two nations.

President Carter did not have lofty goals for spaceflight during his term in office, but did direct military policies on space. In 1978, President Carter restated previous resolutions and treaties that addressed the exploration of space for peaceful purposes and for the benefit of all mankind. Carter wanted to restrict the use of space weapons with the policy. This directive also stated that the US shall encourage domestic commercial exploitation of space capabilities for economic benefit under supervision and regulation of the US government. Under this directive, exploitation is *the action of making use of and benefiting from resources such as new technology*.

1980s

President Ronald Reagan was a strong supporter of NASA and the shuttle program. Reagan believed that space should be explored for strategic defense and not only a surveillance platform. Reagan also wanted to streamline space travel and believed in the power of a free market. His policies encouraged commercial organizations to join the space industry. In March 1983, Reagan announced the Strategic Defense Initiative, also known by some as Reagan's "Star Wars." The Strategic Defense Initiative involved the construction of a space-based anti-missile system to defend the nation from attack.

With the announcement of the Strategic Defense Initiative and the Challenger disaster in January 1986 (more on this in a later lesson), the US issued a revised space policy in January 1988 that set new goals for space:

1. Strengthen the security of the United Sates.
2. Obtain scientific, technological, and economic benefits for the general population and improve the quality of life on Earth through space-related activities.
3. Encourage continuing US private-sector investment in space and related activities.
4. Promote international cooperative activities.
5. Maintain freedom of space.
6. Expand human presence and activity beyond Earth orbit into the solar system.

The Reagan administration's shift in policy implied for the first time that space was not a perfect environment, but, like land, sea and air, was another arena for military operations.

1990s

As the Cold War ended between the US and the now-dissolved Soviet Union, presidents were focused on using space as a means for protecting US national interests.

The beginning of the 1990s saw great support for NASA and space initiatives. President George H. W. Bush increased NASA's budget in a slow economy. He also proposed a grand plan for the space program. The Space Exploration Initiative would focus on the construction of a space station, a permanent presence on the moon, and a manned mission to Mars by 2019. The goals were ambitious and estimated to cost over $500 billion. Many did not agree with the extensive plan, and the initiative was never implemented. Space was not a high priority under Bill Clinton's administration and NASA saw decreases in its budget. President Clinton emphasized the subject of international space cooperation. The 1996 US National Space Policy was the first post-Cold War policy on space. It focused on cooperation when it stated, "Access to and use of space is central for preserving peace and protecting US national security as well as civil and commercial interests."

The congressionally chartered "Space Commission" completed an evaluation of US space policy when it reached five unanimous conclusions in its report:

1. Space should be a top national priority. To reach this conclusion, the committee reviewed the current US dependence on space, the pace at which dependence is increasing, and the vulnerability it creates.
2. The US government is not ready to meet the space needs of the twenty-first century.
3. Space programs in the US are essential to peace and stability of the nation.
4. Space will see conflict. As history has proven, conflict occurs everywhere—air, land, and sea. Space will be no different. The US must develop methods to deter and to defend against conflicts in space.
5. It is essential for the US to invest in science and technology resources to remain a leader in space exploration.

Did You Know?

Secretary of Defense William Cohen under President Clinton wrote in a letter to his military service leaders, "Space is a medium like the land, sea, and air within which military activities will be conducted to achieve US national security objectives."

US Department of Defense Secretary William Cohen.
Courtesy of DOD

2000s

Seeing a need to update the 1996 US space policy to reflect both the post-Cold War and post-9/11 situations, on June 28, 2002, President Bush instructed the National Security Council to chair a review of US space policies and report back during 2003. A new National Space Policy was released in 2006 that established a national policy for conduct of US space activities. This new policy replaced the earlier National Space Policy of 1996.

This new document was the first full revision of US space policy in 10 years. The new policy emphasized security, encouraged private investment in space, and the role of US space diplomacy. The policy stated that in this century, those who effectively utilize space will enjoy added prosperity and security and will hold a substantial advantage over those who do not. Freedom of action in space is as important to the United States as air power and sea power. In order to increase knowledge, discovery, and economic prosperity and to enhance national security, the United States must have robust, effective, and efficient space capabilities.

In 2009, the Augustine Commission reviewed America's spaceflight plans at the direction of President Barack Obama. After the commission reported its findings, President Obama directed NASA to focus on manned missions to an asteroid by 2025 and manned missions to Mars by the mid-2030s. Obama's new policy also provided additional NASA funding to help commercial space companies enhance their capabilities.

Present

President Donald Trump's National Space Strategy strives for the United States to establish a leadership role in space. In partnership with private industry and US allies, President Trump decided to base space strategy on four pillars:

1. Accelerate the plans of the US in designing and building inhabited space environments, such as space stations to enhance defenses, and the ability to repair space systems.
2. Strengthen US and allied options to deter potential adversaries from extending conflict into space and, if necessary, counter hostile threats.
3. Ensure effective space operations through improved situational awareness, information gathering, and the process of investment in technologies, programs, and needed support.
4. Streamline regulations, policies, and processes to better support US commercial industry.

On December 11, 2017, President Trump once again set America's sights toward the stars by signing Space Policy Directive -1, which instructed the National Aeronautics and Space Administration (NASA) to return American astronauts to the moon for long-term exploration and utilization, followed by human missions to Mars and other destinations.

Russian Space Program

Before Mercury could launch a man into orbit, Soviet cosmonaut, Yuri Gagarin, became the first human to orbit Earth on April 12, 1961 in the Vostok capsule.

Vostok

The spherical Vostok capsule was a simple design. It used an adapted R-7 rocket to launch into space and, although the craft had manual controls, it was to be controlled by mission control on the ground.

All early Soviet missions were designed to be controlled by mission control for two reasons. If the cosmonaut became incapacitated, the Soviets needed a way to control the capsule and return the cosmonaut to Earth. Second, they worried that if a cosmonaut had control of the craft they may defect, give up allegiance to their country for another country, to the West and cause embarrassment to the country.

The Vostok rocket that delivered Yuri Gagarin to space.
dimbar76/Shutterstock

The Vostok capsule was designed to eject the cosmonaut during re-entry. Since the Soviets could not slow down the capsule and the landing terrain was rough, the safest method of landing was ejection. The cosmonaut would eject from the capsule and parachute back down to Earth.

Yuri Gagarin successfully completed one orbit of Earth during his 108-minute flight. The entire flight was controlled by mission control on the ground.

During Gagarin's flight, the Soviets encountered problems with re-entry. The capsule spun out of control during the initial re-entry when the equipment did not separate completely from the capsule. After about 10 minutes, the connectors burned away, allowing the equipment to separate. This stabilized the capsule, and re-entry continued as planned.

The Right Stuff

Selecting Cosmonauts

The Soviet process for selecting cosmonauts was very private. Sergei Korolev was an aeronautical engineer who designed and developed rockets. He was responsible for the Sputnik program. In September 1959, Korolev created a cosmonaut selection commission that reported to the Scientific Research Institute of the Soviet Air Force. Because NASA's efforts to train and select astronauts were public, the Soviets adapted much of their training from the NASA program.

From a pool of 3,000 military pilots, they whittled down the group to 15 cosmonauts known as the "Air Force Group One." The cosmonauts would train at the new Cosmonaut Training Center outside Moscow. The center would eventually evolve into Star City, which is still in use today.

Fifteen of the cosmonauts trained to go into space and 11 of them successfully made it to space as part of the Vostok project.

The Right Stuff

The First Woman in Space

Sergei Korolev had an idea to send a woman to space. He secretly selected and trained several women cosmonauts for the mission. Valentina Tereshkova soon excelled at the testing and had an ideal background. She had a background as a worker in a textile factory, and her father was killed in the Winter War that was part of World War II; he was regarded as a hero. Korolev felt this made Tereshkova the ideal candidate, as the country would embrace her as a hero. While all other cosmonauts had been pilots, Tereshkova was a parachutist. On June 16, 1963, Tereshkova became the first woman in space. She orbited the Earth 48 times over three days in the Vostok capsule. This was more time in space than all Americans combined, at that point.

However, Tereshkova's time in space was also a publicity stunt. The Soviets wanted to achieve another first in the space war. It would be more than 19 years before the Soviets sent another woman into space. And shortly after Tereshkova's flight, the Soviets disbanded the women's cosmonaut program.

Valentina Tereshkova
Courtesy of ESA

Voskhod

The Voskhod program aimed to explore the effects of space on the human body. The mission had two launches. The Voskhod 1 launched on October 12, 1964 with a three-person crew. On March 18, 1965, the Voskhod 2 launched with a two-person crew. During the 1965 flight, Alexei Leonov completed the first spacewalk in history—another first for the Soviets.

However, in 1964, the Soviet leadership changed. This new leadership had less focus on space. As a result, Korolev cancelled the Voskhod program to focus on the Soyuz capsule.

Soyuz

The Soyuz was designed as a lunar landing program. The capsule would take two cosmonauts and a lander into space. One cosmonaut would then take the lander to the lunar surface, walk around, fly back into lunar orbit to rendezvous with the Soyuz, and then return to Earth. The project was plagued with technical difficulties. And in January 1966, Korolev died and left a major hole in the program. The mission was abandoned in 1974 and the Soviets turned their attention to space stations.

The Soyuz spacecraft orbiting Earth.
Andrey Armyagov/Shutterstock

Chinese Space Program

The Chinese did not begin to engage in space travel until the late 1950s. China's major goal in exploring space travel was the launch of weapons into space. The United States and Soviet Union were well into their space race to put a man on the moon, and the use of rockets to send weapons into space alarmed China.

The Chinese government had an agreement with the Soviets that allowed them to use the R-2 rocket technology. This agreement ended in the 1960s, which forced China to develop its own rockets. China started launching rockets into space in 1960.

The Chinese space program was initially slow-moving. The country was dealing with a major political division and its interest in space was largely due to defense. The Chinese focused on missile development rather than space exploration. The Ministry of Aerospace Industry was established by China in 1988. A few years later, the ministry was split into two organizations: China National Space Administration (CNSA) and China Aerospace Science and Technology Corporation (CASC). The two organizations allow both the government and private companies to participate in space travel.

The Chinese space program has grown immensely over the past 20 years. It developed the Long March rocket, deployed the country's first space station, and launched the Chinese Lunar Exploration Program (CLEP).

The Chinese have both lofty goals for their space program and the money to fund those goals. In 1992, the Chinese set a long-range plan for their space program, and they successfully executed that plan on schedule. In 2016, the Chinese established a new long-range plan. Some of the projects in the plan are as follows:

- A Mars probe by 2020 that will orbit Mars and land a rover on the surface
- A reusable space plane by 2020 that will carry astronauts by 2025
- A nuclear-powered spacecraft by 2040
- Exploration of the asteroid belt
- Solar-powered plants in space
- Mining on the moon and asteroids
- A fully colonized space village

The Right Stuff

Taikonauts

Chinese astronauts are called "taikonauts." The process to become a taikonaut is very long and strenuous. Some taikonauts study for over 15 years before they are ready to go to space. The first generation of taikonauts was selected in 1998. Two of the taikonauts went to the Russian Cosmonaut Training Center to be trained and to learn how to set up their own training program. The first generation of taikonauts included 15 male military pilots.

The second generation of taikonauts was selected in 2010. It included five male fighter pilots and two female military transport pilots between the ages of 27and 34. After being selected, the taikonauts complete a minimum of 2 to 3 years of training before they can train for a mission.

In January 2018, China announced that it will begin the selection for the next generation of taikonauts. Candidates will be selected from space industry companies, research entities, and universities. To become an engineer on the Chinese Space Station, a candidate must have a master's degree. To become a payload specialist on the Chinese Space Station, a candidate must have a doctorate degree.

Shenzhou

The first manned Chinese spacecraft program was the Shenzhou. The Shenzhou spacecraft was similar to the Russian Soyuz, except it was larger and newer. The goal of the program was to perfect manned spacecraft techniques, including extravehicular activity, rendezvous, and docking, and eventually ferry taikonauts to the Chinese space stations.

Yang Liwei's spacesuit on display at the China National Museum
Shan-shan/Shutterstock

The spacecraft itself was comprised of three parts: orbital module, re-entry capsule, and aft service module. The orbital module was able to maintain autonomous flight and had its own propulsion, solar power, and control systems. The re-entry module was similar to the Soyuz, but was 13% larger. The service module housed the electrical power, control, and propulsion for use in orbit.

Work began on the Shenzhou program in 1992, but limited funding slowed its progress. In 1999 and 2000, unmanned flights were sent into orbit. Then, in 2003, the first Chinese taikonaut, Yang Liwei, traveled to space aboard the Shenzhou 5. The flight was relatively short and lasted only 21 hours. Nonetheless, China became the third country to send a human into space.

Shenzhou 6 was the second manned mission and carried two taikonauts to space for five days in October 2005. Then, in 2008, Shenzhou 7 deployed China's first EVA with a spacewalk by crew member Zhai Zhigang. The Shenzhou 7 mission lasted for 2.85 days.

Did You Know?

China currently cooperates with the European Space Agency (ESA) as they partner on space missions. However, the United States and China do not work together. In 2011, Congress passed a bill that forbids NASA from working with China for fear of espionage.

Chinese Space Station

The first Chinese space station to launch was the Tiangong-1 which translates to "Heavenly Palace 1." The space station was about the size of a school bus at 34 feet long and 11 feet wide. It weighed in at nine tons. The space station had two main parts. The experimental module was used to house visiting taikonauts. The resource module housed the solar energy and propulsion systems.

The Tiangong-1 was launched on September 29, 2011 as an unmanned space station. It orbited slightly lower the much bigger ISS. The goal of the Tiangong-1 was to learn the technologies needed to assemble and operate a fully functioning space station.

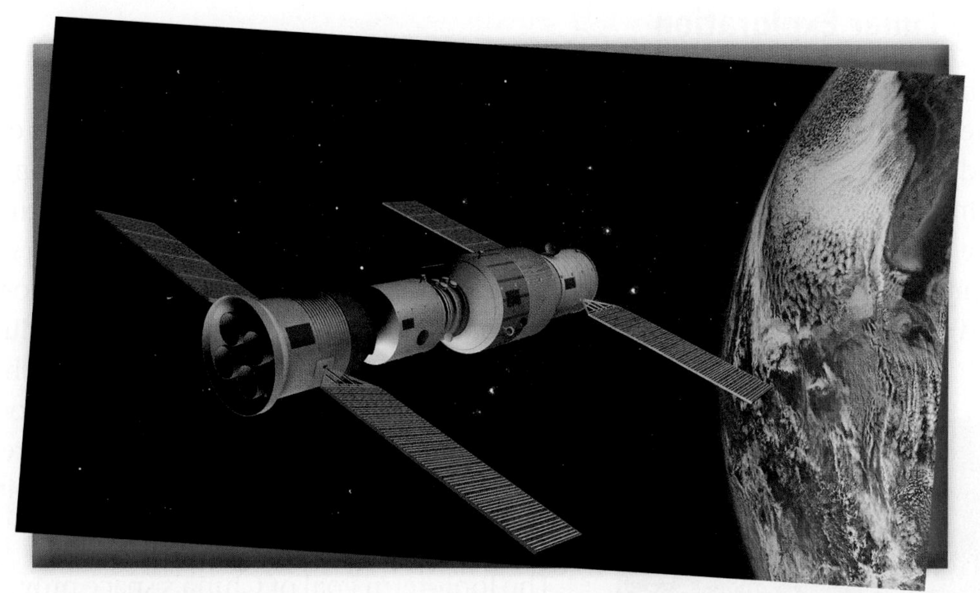

3D module of the Chinese space station Tiangong-1.
Alejo Miranda/Shutterstock

The Shenzhou-8 was the first spacecraft to dock with the Tinagong-1 in November 2011. The first taikonauts to board the space station launched in June 2012 aboard the Shenzhou-9. Shenzhou-10 transported three taikonauts who spent two weeks on Tiangong-1.

The Tiangong-1 was designed to last for two years, and work with the space station ended with Shenzhou-10. Data transmission with the space station ended in 2016. The Chinese have never said if they lost communication or if the data transmission ended intentionally. On April 1, 2018 the Tiangong-1 broke apart and burned up over the southern Pacific Ocean during re-entry. The Chinese claim the re-entry was controlled; however, other agencies disagree with this statement. A controlled re-entry requires the spacecraft handlers to be in communication with the space station.

The stepping stones to a fully operational space station continued with the Tiangong-2. Tiangong-2 was launched in September 2016. The goal of this space station was to test advanced life support, refueling, and resupply. In late 2016, the Tiangong-2 hosted two taikonauts for 30 days. This was the longest human space mission for China. In June 2018, the Tiangong-2's orbit lowered by almost 100 km. It is likely that the Chinese are preparing to de-orbit the Tiangong-2.

The Chinese space program is currently constructing its own 60-ton space station, which will be named Tianhe. The first module will launch in 2020. This is the same year the ISS is scheduled to be decommissioned. Two more modules will be added, with the completion of Tianhe scheduled for 2022.

Chinese Lunar Exploration

The Chinese space program has conducted several missions with the Chang'e project to orbit the Moon. In January 2019, the Chang'e 4 became the first spacecraft to land on the far, or "dark," side of the Moon—the side of the moon that always faces away from Earth. The Chinese have also scheduled Chang'e 5 to land on the Moon in 2019 and return with a sample of lunar regolith, or rocky material that covers bedrock, from two meters deep. The Chang'e 5 mission will require the spacecraft to utilize four modules. Two modules will land on the Moon. One will collect the sample and give it to the second module. The second module will launch and dock with the third module that will be orbiting the Moon. The third module will transfer the lunar regolith sample to the fourth module in lunar orbit, which will return it to Earth.

Illustration of the Chinese flag stuck in the surface of the Moon.
BeeBright/Shutterstock

The long-term goal of China's space program is to send crewed missions to the Moon and perhaps set up a lunar outpost. It is expected that the Chinese will work with the European Space Agency (ESA) to create the outpost, dubbed the "international Moon village." They are already working toward that goal by establishing a simulated lunar base in China. The Yuegong-1, or Lunar Palace, is a laboratory that simulates the lunar environment. Currently, potatoes, wheat, and carrots are being grown in the "lunar" base.

Space Programs Around the World

NASA, Russia, and China are not alone in the world of space exploration. The European Space Agency (ESA) makes several orbital launches each year. ESA's very large launch site in French Guiana specializes in commercial and scientific missions. Israel, Japan, and India each have unique programs for studying space, too. Like NASA, each of these programs exists to support and promote manned and unmanned missions to space. The "Big Six" in space exploration consist of the following:

- National Aeronautics and Space Administration (NASA)
- Roscosmos (Russian space agency)
- Chinese National Space Agency (CNSA)
- European Space Agency (ESA)
- Indian Space Research Organization (ISRO)
- Japanese space agency (JAXA)

European Space Programs

After World War II, the space race was on between the United States and Russia. European scientists were eager to join in, but could not compete on a national level. Their countries were simply too small to invest the research and funds into space exploration individually. In 1958, Pierre Auger and Edoardo Amaldi suggested a joint organization for space research. In 1960, 10 European countries joined forces to set up a commission that would determine the space exploration projects for Europe. In 1961, the European Space Research Organisation (ESRO) was born. The commission decided to set up the European Launch Development Organisation (ELDO) to develop a launch system and the European Space Research Organisation (ESRO to develop the spacecraft.

The ESA's Automated Transfer Vehicle-4 approaches the International Space Station.

NASA Images/Shutterstock

In 1975, the ELDO and ESRO merged to create the European Space Agency (ESA). The members of the ESA include Austria, Belgium, the Czech Republic, Denmark, Estonia, Finland, France, Germany, Greece, Hungary, Ireland, Italy, Luxembourg, the Netherlands, Norway, Poland, Portugal, Romania, Spain, Sweden, Switzerland, and the United Kingdom.

Although the ESA usually is plagued with delays and technical difficulties in getting its rockets completed, it excels at cooperation. The ESA has enabled cooperation between space agencies and was an instrumental part of the cooperation behind the International Space Station.

One achievement of the ESA is the Giotto space probe. The probe was able to examine the core of Halley's Comet in 1986. In addition, the ESA developed the Ulysses spacecraft that launched in 1990 to explore the Sun's polar regions.

The Indian Space Program

In 1969, India established the Indian Space Research Organisation (ISRO). The ISRO strives to provide the country with space services and to develop new technologies. ISRO has one of the largest fleet of communication satellites. These satellites are used all over the world to provide us with reliable communication.

Although ISRO has not sent a human into space, it has achieved 97 spacecraft missions and has 237 satellites in orbit. On February 15, 2017, ISRO had a record-breaking launch when it launched 104 satellites into orbit using one rocket. ISRO is an extremely resourceful organization and has been able to accomplish much on a limited budget. It successfully sent a probe to Mars and began testing a space shuttle that it fully designed and developed.

The Japanese Space Program

Japan was late to the space race, but is making up time! It went from never having launched a satellite, in 1969, to being an emerging space agency by 1994. The Japanese Aerospace Exploration Agency (JAXA) is a merger of the Institute of Space and Astronautical Science (ISAS), National Aerospace Laboratory of Japan (NAL), and the National Space Development Agency of Japan (NASDA). JAXA supports the government's overall effort to conduct space research and development.

JAXA's Earth Observation Center.
Studio 400/Shutterstock

The major accomplishments of JAXA include the Hinotori satellite. This satellite was Japan's first solar observation satellite and studied solar flares in 1981. The satellite was able to return the first ever x-ray images of a solar flare. JAXA continued its studies of solar flares and launched additional satellites in 1991 and 1995.

As you will read in an upcoming chapter, many of these organizations are working together to successfully operate the International Space Station (ISS). Space exploration continues to be an area of interest for many countries, and new developments are continuously being made that will push space exploration farther and farther into deep space.

Lesson 1 Review

Using complete sentences, answer the following questions on a sheet of paper.

1. Why is space important to US national interests?
2. What did the two-man flights of 1965 and 1966 demonstrate?
3. What did the space policy released in 2006 emphasize?
4. Why did the Soviets control their spacecraft from the ground?
5. What problem did Yuri Gagarin have during re-entry?
6. What was the goal of the Voskhod program?
7. What was the main goal of the Chinese space program when it launched?
8. What were the goals for Tiangong-1 and Tiangong-2?
9. Who are the "Big Six" in space exploration?
10. What did the ESA's Giotto space probe accomplish?

APPLYING YOUR LEARNING

11. How do you think space exploration efforts would have been different if international agencies did not work together?

LESSON 2

US Manned Space Program

Quick Write

Read the vignette about the soccer ball survivor. Why do you think it was important for the soccer ball to finally complete its journey to space?

Learn About

- early US manned space program
- why develop a shuttle program
- space shuttle's main components
- lessons learned from Challenger and Columbia

It's another day of practice on the soccer field at Clear Lake High School in Texas in 1986. Young Janelle Onizuka has the thoughtful idea to have her teammates autograph the soccer ball for her father. Passing around a blue ink pen, each teammate autographs the ball and writes "Good Luck, Shuttle Crew." You see, Janelle Onizuka just happens to be the daughter of astronaut Ellison Onizuka. Her father is one of seven crew members who will go on the space shuttle Challenger's 10th mission in the days to come.

Although Ellison is supposed to be in quarantine to avoid sickness before the launch, he sneaks out to catch a soccer game. There, the team proudly presents him with the ball.

On January 28, 1986, Ellison Onizuka boards Challenger with a few personal items, including the soccer ball. Just over a minute into flight, Challenger breaks apart, creating a fiery ball in the sky. Debris is scattered over the area. The seven crew members are lost.

For weeks to follow, the US Coast Guard retrieves tons of debris from the Challenger accident. A most important item found floating in the Atlantic Ocean is Onizuka's soccer ball with blue-inked autographs still visible.

For decades, the soccer ball sits in a trophy case at Clear Lake High School. Thirty-one years after the Challenger accident, Clear Lake High School parent and astronaut, Shane Kimbrough offers to take a memento to space on the school's behalf. Principal Karen Engle has the idea to send the soccer ball into space, as its story was beginning to fade within the school. And so, Kimbrough took the soccer ball survivor into space.

The infamous soccer ball from the Challenger accident that finally made it to space more than 30 years later.

Courtesy of NASA/JSC

Vocabulary

- retrorocket
- space shuttle
- orbiter
- Remote Manipulator System (RMS)
- airlock
- extravehicular activity (EVA)
- aft
- Solid Rocket Boosters (SRBs)
- Space Shuttle Main Engines (SSMEs)
- O-ring
- max-Q

Early US Manned Space Program

After NASA was established, the United States embarked on a mission to send a man into orbit. After each success, NASA pushed the limits and extended its goals for sending humans into space. The US manned space program led to scientific discoveries and technological advances in space travel and paved the way for future space travel.

Project Mercury

NASA answered the call from President Eisenhower to put a man into orbit around Earth. Project Mercury was the initial project of the US manned space program. Mercury was the winged messenger of the gods in ancient Roman and Greek mythology. The name was suggested by NASA's Director of Space Flight Development. Project Mercury consumed most of NASA's resources for the next five years until it succeeded in putting a man into orbit.

Project Mercury was a three-phase project that included six spaceflights. In Phase I, NASA selected and trained astronauts, while also developing safety systems to be used in the launch and re-entry of the space capsule. In Phase II, NASA adapted the Redstone ballistic missile for suborbital flights. A suborbital flight is a spaceflight in which your trajectory takes you in a large arc going into space and then falling back to Earth. Phase III of Project Mercury involved adapting the Atlas rocket to launch a capsule into orbit with a human aboard. An orbital flight is a spaceflight whose trajectory moves around, or orbits, the planet.

NASA created the Space Task Group to set up and manage Project Mercury. Robert Gilruth, an expert in rocket testing and development, was assigned to lead the group. The Space Task Group designed a capsule that would hold one human. The capsule would use retrorockets and parachutes to return to Earth. A retrorocket is *a small auxiliary rocket on a spacecraft that is designed to slow down the craft*. During re-entry, the space capsule would reach extreme speeds, so NASA needed a way to slow down the capsule to safely deliver the astronaut back to Earth's surface. The capsule designed for Project Mercury was extremely small. It had room for only one adult. The cone-shaped capsule was approximately 11 feet long and 6 feet wide. Each launch utilized a different capsule that was given a unique name.

The Mercury capsule during construction.
Courtesy of NASA

The first test flight for Project Mercury launched from Wallops Island, VA on December 4, 1959. This was an impressive feat because the first test flight occurred about a year after the launch of the project. The test flight was a suborbital flight with the capsule occupant being a rhesus monkey named Sam. The test flight was a success, and testing continued. On January 31, 1961, a chimpanzee named Ham flew 157 miles into space during a 16 minute and 39 second suborbital flight. Ham's capsule was launched using a Redstone rocket. NASA was concerned that the Redstone rocket didn't have enough power to launch a heavier capsule, so it began to adapt the Atlas rocket for Project Mercury. A chimpanzee named Enos would complete two orbits around Earth after launches that used the Atlas rocket. Sam, Ham, and Enos successfully and safely returned to Earth, providing NASA the proof that it was ready to begin sending astronauts to space.

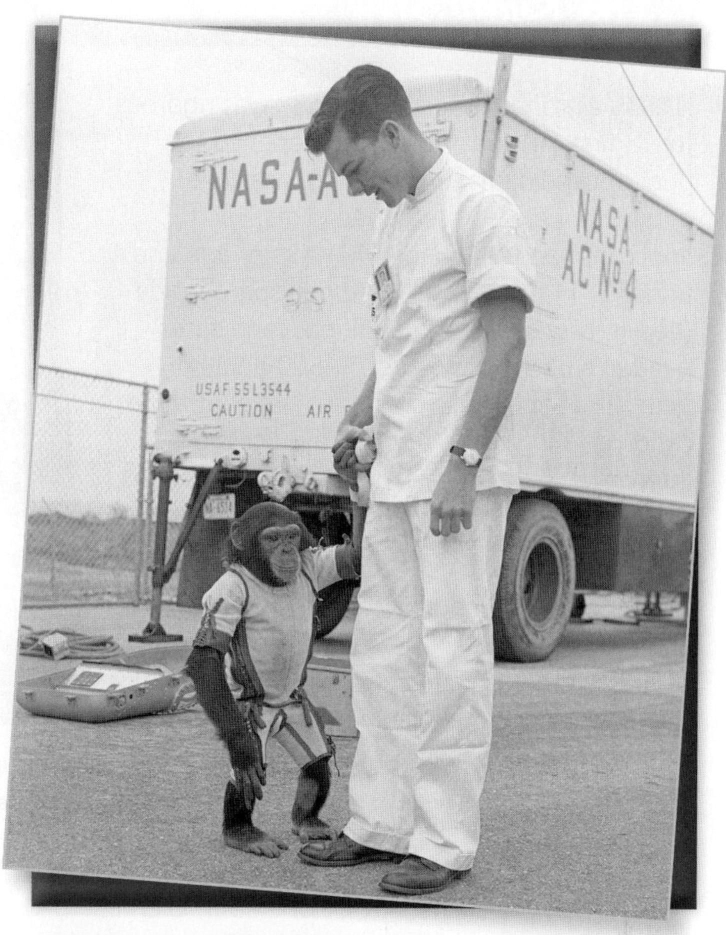

Ham the Space Chimpanzee prepares for flight aboard Mercury Redstone rocket.

Courtesy of NASA

On May 5, 1961 Alan Shepard became the first American to reach space in the Freedom 7 capsule. Shepard completed a 15-minute suborbital flight. The first two manned flights of Mercury used the Redstone rocket, but the remaining Mercury flights used the Atlas rocket. Although this was a huge accomplishment for the United States space program, the Soviets beat the Americans into space by putting a human into orbit in April 1961.

NASA was determined to continue Project Mercury and reach its goal of putting a man in orbit around Earth. On July 21, 1961, Virgil "Gus" Grissom became the second American in space when he completed a suborbital flight aboard the Mercury capsule Liberty Bell 7. The flight had problems after re-entry. After splashdown in the Atlantic Ocean, the capsule's hatch blew off too early. Grissom's capsule was dropped into the sea without a hatch and quickly filled with water. Grissom barely escaped.

The Right Stuff

Group portrait of the "Mercury Seven."
Courtesy of NASA

NASA's astronaut selection process was rigorous for its first mission into space. It opened the pool of candidates to young, physically fit men who were also seasoned pilots. Each applicant had to be under 40 years of age and under 5′ 11″ due to the cramped space inside the Mercury capsule. In addition, candidates needed a bachelor's degree, 1,500 hours of flight time, and qualification as a jet pilot. The candidates would go through a process of interviews, written tests, psychiatric evaluations, and medical history reviews.

In April 1959, NASA selected seven candidates from the original pool of around 3,000 applicants to be the "Mercury Seven." All of the astronauts were active-duty military members who would begin intensive training to prepare for a possible flight to space.

The "Mercury Seven" consisted of the following officers:

- Marine Lieutenant Colonel John Glenn Jr.
- Navy Lieutenant Commander Walter "Wally" Schirra
- Navy Lieutenant Commander Alan Shepard Jr.
- Navy Lieutenant Commander Scott Carpenter
- Air Force Captain Gordon Cooper
- Air Force Captain Virgil "Gus" Grissom
- Air Force Major Donald "Deke" Slayton

On February 20, 1962, John Glenn became the first American to orbit Earth. His capsule, Friendship 7, orbited Earth three times in the 4 hour and 56 minute flight. His flight also had complications. During orbit, the autopilot controls failed and Glenn had to fly the last two orbits manually. John Glenn already had a reputation as one of the best test pilots in the country. After his flight to space, he quickly became an American hero for his accomplishments and boosted the reputation of the NASA program.

Between 1962 and 1963, Project Mercury completed three additional flights. On May 20, 1962, Scott Carpenter completed three orbits of the Earth aboard Aurora 7. And on Oct. 3, 1962, Walter "Wally" Schirra completed six orbits aboard Sigma 7. Finally, on May 15, 1963, Gordon Cooper completed a record 22 orbits in 34 hours aboard Faith 7. Remember, the capsules for Project Mercury were very small and the astronauts were confined to their seat for the entire flight. Project Mercury had exceeded its goals and put a man into orbit.

John Glenn entering the Friendship 7 capsule prior to launch.
Courtesy of NASA

Project Gemini

Project Gemini would be the next major project for NASA. Project Gemini had three main goals:

1. Learn how to maneuver, rendezvous, and dock with another spacecraft.
2. Allow astronauts to practice working outside the spacecraft.
3. Gather physiological (human performance) data on long spaceflights.

Gemini means "twins" in Latin and is the name of the third constellation of the zodiac. It seemed an appropriate name for the new two-man crew and the project's relationship to Mercury. Project Gemini was the stepping stone to reaching the Moon. NASA and its astronauts needed to successfully complete the goals of Project Gemini in order to send a man to the Moon.

This image of NASA's Gemini 7 in space was taken by Gemini 6A as the two capsules meet during a rendezvous.
Courtesy of NASA

The Project Gemini capsule was designed to hold a two-astronaut crew. The program ran from 1965 to 1966. It was a relatively short project for NASA, but was an extremely successful one. Over the course of the program, Project Gemini flew 10 missions.

TABLE 2.1 Project Gemini Timeline

Mission / Date	Event
Gemini 3 March 23, 1965	Virgil "Gus" Grissom and John White were the first Gemini mission crew.
Gemini 4 June 3, 1965	Edward White II was the first American to spacewalk.
Gemini 5 August 21, 1965	The Gemini 5 mission stayed in orbit for more than a week.
Gemini 6A December 15, 1965	Gemini 6A and 7 were in space at the same time.
Gemini 7 December 4, 1965	Gemini 7 stayed in space for two weeks.
Gemini 8 – March 16, 1966	Gemini 8 successfully connected to an unmanned spacecraft in orbit.
Gemini 9A June 3, 1966	Gemini 9A tested methods for flying near another spacecraft.
Gemini 10 July 18, 1966	Gemini 10 successfully connected to an unmanned spacecraft in orbit and then used its engines to move both spacecrafts.
Gemini 11 September 12, 1966	Gemini 11 flew higher than any other NASA mission.
Gemini 12 November 11, 1966	Gemini 12 worked on solving complications with spacewalks.

Did You Know?

On June 3, 1965, during the Gemini 4 mission, Edward White II set another first for the US by becoming the first American to spacewalk. White spent 30 minutes outside of the spacecraft. He used a propulsion gun that released nitrogen to move through space. He was connected to the spacecraft with only an oxygen hose.

Astronaut Edward White during the first American spacewalk.
Courtesy of NASA

Project Apollo

NASA's Apollo project began on January 27, 1967. This was the NASA program that would put a man on the Moon. The proposed name, "Apollo," was the name of the Greek god of archery, prophecy, poetry, and music; most significantly, he also was god of the sun in Greek mythology. In his horse-drawn golden chariot, Apollo pulled the sun on its course across the sky each day. The Apollo missions had four goals:

- Establish the technology to meet other national interests in space.
- Achieve preeminence in space for the United States.
- Carry out a program of scientific exploration of the Moon.
- Develop human capability to work in the lunar environment.

The Apollo spacecraft had three parts. The command module (CM) housed flight control and the crew's quarters. The service module (SM) housed the propulsion and spacecraft support systems. When these two systems were connected, they were referred to as the CSM. The lunar module (LM) was the portion of the craft that would take a two-man crew to the lunar surface.

The project began in disaster. During a preflight test on January 27, 1967, a fire broke out in a pressurized, oxygen-rich area of the CM. The oxygen-rich environment fueled the fire, and it spread rapidly. Because the module of the spacecraft was pressurized, it took NASA longer than usual to open the hatch. By the time they were able to access the spacecraft, three astronauts had perished: Virgil "Gus" Grissom, Edward White, and Roger Chaffee. These men were the first fatalities of the US space program. After the accident, NASA halted the mission for a complete investigation. It made improvements to the design and established safety measures for the Apollo spacecraft. NASA agreed to resume the mission.

Apollo 1 astronauts, left to right, are Virgil "Gus" Grissom, Edward White, and Roger Chaffee.

Courtesy of NASA

On December 21, 1968, Apollo 8 launched with Frank Borman, James Lovell, and William Anders on board. Four days later (on Christmas Eve), Apollo 8 broadcast live from space for all of America to see. They provided the country with spectacular images of the Earth as they successfully orbited the Moon.

Did You Know?

Apollo 8 captured pictures of Earth while orbiting the Moon. The images soon became famous and were known as "Earthrise" because they showed the Earth rising as a small planet above the Moon's surface.

The first Earthrise image taken by the crew of Apollo 8.
Courtesy of NASA

The Apollo 9 and 10 missions were used to further test the lunar module and orbit the Moon. On July 16, 1969, Apollo 11 launched from Kennedy Space Center with astronauts Buzz Aldrin, Neil Armstrong, and Michael Collins aboard. Over the next three days, the astronauts would travel to the Moon to start their orbit. On July 20, 1969, the lunar module dubbed "Eagle" was separated from the command/service module (CSM) of Apollo 11 and began its descent to the Moon.

Neil Armstrong and Buzz Aldrin were in the lunar module (LM), while Michael Collins stayed in the CSM. Armstrong and Aldrin had a stressful lunar landing because the automatic landing controls were directing the LM to land in a boulder field. Armstrong took over the controls to manually pilot the LM to a more suitable landing location. The fuel on the LM quickly dropped. With barely 30 seconds of fuel remaining, the LM touched down on the lunar surface. Shortly after, Neil Armstrong announced to Mission Control, "The Eagle has landed." NASA had successfully put a man on the Moon. Now it was time for the astronauts to have some fun!

An astronaut's footprint on the Moon. The Moon does not have erosion from wind or water so the footprints left by astronauts on the Moon will be there for a long, long time.
Courtesy of NASA

Armstrong and Aldrin were scheduled to take a five-hour rest after they landed on the Moon. They skipped their scheduled rest because they were too excited. They put on their spacesuits and stepped out of the LM. Neil Armstrong was the first to exit the LM. As he stepped on the lunar surface, he exclaimed, "That's one small step for a man, one giant leap for mankind." Buzz Aldrin joined him on the lunar surface for the first moonwalk. While on the moon, they planted an American flag, collected soil and rock samples, and set up experiments for NASA.

This would not be Apollo's last mission to the Moon. In November 1969, Apollo 12 returned to the Moon with Charles "Pete" Conrad and Alan Bean. The astronauts completed two moonwalks, each approximately four hours in length. They set up experiments on the moon to measure seismic activity, solar wind flares, and the Moon's magnetic field.

Apollo 13's famous mission launched on April 11, 1970. It was to be the third trip to the Moon for Americans. Astronauts James Lowell, Fred Haise, and John Swigert were aboard Apollo 13. At 56 hours into their mission, an oxygen tank ruptured, damaging the power, electrical, and life support systems in the command module (CM). As a result, the CM had limited resources to support the astronauts aboard. The lunar module (LM), however, was not affected. The astronauts remained in the LM until Mission Control could figure out how to get them home safely. NASA had to think quickly and work with limited resources to adapt the spacecraft so that it could return to Earth safely. The success of Apollo 13 proved that NASA had the ability to not only send a man to the Moon, but to successfully avoid catastrophes in space.

The original Apollo project was to have 20 missions. Due to budget cuts, the program was scaled back and only completed 17 missions. During the Apollo missions, astronauts deployed over 50 experiments on the moon, 12 astronauts walked on the Moon, and the Lunar Roving Vehicle (LRV) was deployed on the Moon to allow astronauts to travel greater distances on the Moon. To date, the United States is the only nation to have sent humans to the Moon.

Buzz Aldrin walking on the Moon.
Courtesy of NASA

Lunar Roving Vehicle (LRV) photographed against a desolate moon landscape during the Apollo 15 mission.
Courtesy of NASA

Did You Know?

A movie, Apollo 13, was released in 1995 to document the safe return of the spacecraft and astronauts to Earth.

Why Develop a Shuttle Program

In early 1969, President Richard Nixon formed a new Space Task Group, a group of NASA engineers. Their task: identify the future vision of space exploration, beyond the Moon. Later that year, the group presented a plan involving a shuttle, space station, and eventually, manned missions to Mars.

Much like we may take a subway to travel between work and home, the shuttle would take a crew of astronauts, reusable supplies, and food, to a larger space station. The space station would remain there for future shuttles. The shuttle would be used to exchange crews, make any necessary repairs to the space station, and conduct new projects.

NASA proposed a fully reusable space shuttle that would go to and from the space station. This plan was short lived. Nixon thought the plan was much too pricey and cut NASA's budget, focusing on one aspect of the plan: the space shuttle itself. A space shuttle is *a reusable spacecraft designed to transport people and cargo between Earth and space.*

The space shuttle, or STS (Space Transportation System), provided a way to easily, economically, and safely leave the surface of Earth. NASA wanted to move away from a costly spacecraft and utilize a reusable shuttle with a large payload bay to haul equipment.

The space shuttle prototype, Enterprise, flies free of NASA's 747 Shuttle Carrier Aircraft (SCA) during one of five "free flights" carried out at the Dryden Flight Research Facility, Edwards, Calif., in 1977 as part of the shuttle program's Approach and Landing Tests (ALT).

Courtesy of NASA

The space shuttle could haul up to 45,000 tons of cargo and fly up to 250 miles above Earth. In addition, the shuttle could hold up to 10 astronauts, although the typical flight carried a crew of seven.

The first space shuttle, named Enterprise, rolled out in September 1976. Enterprise made some flights, but never reached space. Over the 30 years of the Space Shuttle Program's history, NASA built five other space shuttle orbiters for flight. These include Columbia, Challenger, Discovery, Atlantis, and Endeavour.

With 135 completed missions, the Space Shuttle Program was extremely successful. It ferried 852 astronauts to space.

The Shuttle's Missions

Did You Know?

Columbia was the heaviest space shuttle, weighing in at 178,000 pounds.

The space shuttle Columbia made the first space flight of the Space Shuttle Program, achieving many milestones in space exploration. Columbia's first flight launched on April 12, 1981 from Kennedy Space Center in Florida. It had been six years since an American had been in space. The first Columbia flight made 36 orbits in a two-day span and then landed as an aircraft at Edwards Air Force Base in California. The Columbia missions provided proof that a reusable spacecraft could complete a mission into orbit and return safely to Earth. Columbia finished three more missions, testing its performance at each phase of the flight.

Columbia conducted its first operational mission on November 11, 1982. With a crew of four, Columbia recovered several satellites from orbit and became the first manned spacelab mission designed for human medical research.

The next space shuttle for NASA was the Challenger. For its first mission in 1983, Challenger took off from NASA's Kennedy Space Center and marked the first spacewalk of the Space Shuttle Program. It also deployed the first Tracking and Data Relay System satellite, among many other accomplishments. You will learn more details about Challenger later in this lesson.

Following Challenger, Discovery launched a mission to deploy several communication satellites in 1984. Discovery's claim to fame is the number of missions it flew. Completing over 30 successful missions, Discovery made more flights than any other orbiter in NASA's history.

Discovery carried satellites into space, along with crews of astronauts to perform many experiments on the International Space Station (ISS). Discovery also made two flights to the Russian space station, Mir. Importantly, Discovery took the Hubble Space Telescope into space in 1990.

In 1981, space shuttle Columbia and STS-1 lifted off from NASA's Kennedy Space Center, marking the first flight of the Space Shuttle Program.

Courtesy of NASA

Did You Know?

Like Columbia and Challenger, Discovery was named for an exploration ship from our past: the ship used by Henry Hudson to explore the Hudson Bay.

Lastly, we have Atlantis and Endeavor. Atlantis conducted several important missions for the Department of Defense in 1985. Atlantis also linked with the Russian space station, Mir, to form the largest spacecraft in history.

Endeavor, endorsed by Congress as a replacement for Challenger (Challenger's fate will be covered later in the lesson), was first launched in 1992. Along with satellite repairs, Endeavor is credited with the longest spacewalk in NASA history.

TABLE 2.2 Project Gemini Timeline

Shuttle	Number of Flights	Earth Orbits	Miles Traveled	Total Crew	Time in Space
Columbia	28	4,808	121,696,993	160	300 days
Challenger	10	995	23,661,290	60	62 days
DISCOVERY	39	5,830	148,221,675	252	365 days
ENDEAVOUR	25	4,671	122,883,151	173	299 days
ATLANTIS	33	4, 848	125,935,769	207	307 days
TOTAL	135	21,152	542,398,878	852	1,333 days

Astronaut Paul Richards, STS-102 mission specialists, works in the cargo bay of the space shuttle Discovery.

Courtesy of NASA.

Did You Know?

The longest spacewalk was 8 hours and 56 minutes, performed by Susan J. Helms and James S. Voss during STS-102 on March 11, 2001.

The Right Stuff

Over the 30 years of its history, the Space Shuttle Program accomplished some of the greatest achievements in NASA's history. Not only did it help build the International Space Station (ISS) and launch the Hubble Space Telescope, it also brought nations together. For Americans, it marked many first-in-flights for individuals of all backgrounds.

In 1983, Guy S. Bluford Jr., a mission specialist, became the first African American in space. His first launch, aboard the Challenger on NASA's STS-8, was especially memorable. At lift-off, the audio recording captured someone laughing as they ascended into space. That person was Guy Bluford, clearly delighted to be going into orbit!

Also, in 1983, Dr. Sally K. Ride was the first American woman to go into space. Ride had an impressive education, including a doctorate in astrophysics. She was one of the few women selected for astronaut training.

Training was hard and complicated. Ride trained in parachuting, water survival, microgravity, and g forces. The hard work paid off. Ride made it to space and also became a pilot. She contributed to the design of a robotic arm that would be used to deploy and retrieve satellites.

In 1985, Air Force Colonel Ellison Onizuka, became the first Asian American in space. He was aboard Challenger on NASA's STS-51C Defense Department mission and the STS-51L mission.

Other firsts in flight include Franklin Chiang-Diaz, the first Hispanic American to fly into space aboard Columbia STS-61-C in 1986. In 1992, Dr. Mae Jemison became the first African American woman in space, aboard Endeavor STS-47. In 1993, Dr. Ellen Ochoa was the first Hispanic woman in space, on the STS-56 Discovery mission.

At NASA's Johnson Space Center, astronaut Sally K. Ride takes a break from training as a mission specialist for NASA's STS-7 spaceflight in Earth orbit.
Courtesy of NASA/JSC

Current State of the Shuttle Program

The Space Shuttle Program's achievements are considerable! The shuttle program was used to retrieve, repair, and replace satellites. It was designed as a workhorse and certainly proved itself as such. Although the shuttle program was not designed specifically for science, each mission included a science component. Some of the greatest cutting-edge research was performed during the shuttle missions.

Thanks to the shuttle program, the US successfully assisted in the building of the largest structure in space, the ISS. During all of the flights, the shuttles accumulated over 21,000 orbits and more than 500 million miles. There have been hundreds of shuttle crew members, including men and women from 16 different countries, with a total flight time of almost four years. The program also allowed NASA to expand its astronaut program to include non-pilot astronauts. Guests, mission specialists, and scientists were able to experience the view from space through the shuttle program.

Despite all of these accomplishments, in 1994, President George W. Bush revealed a new vision for space exploration. The new vision did not include the shuttle program. President Bush announced the shuttle program would come to an end, and it did with the final flight of Atlantis on July 21, 2011.

Bush's vision was to develop a new Crew Exploration Vehicle (CEV). The goal would be to go beyond the Moon, to Mars. Unfortunately, Congress did not provide enough funding for Bush's vision.

The launch of Atlantis on the STS-135 mission and the final flight of the shuttle program.

NASA/Bill Ingalls

Since the end of the Space Shuttle Program in 2011, the US has paid Russia to transport American astronauts and cargos to the ISS. NASA has requested additional funding from our government. It has also picked two private companies, SpaceX and Boeing, to develop a spacecraft to replace the shuttles. SpaceX is scheduled to test the Crew Dragon spacecraft in 2019. When testing is complete, NASA will use this spacecraft to ferry US astronauts to and from the ISS, replacing the need for Russian Soyuz spacecraft.

In addition, NASA's Orion spacecraft is being developed; however, it is not expected to carry astronauts until 2023. In this American-European project operated by NASA, the Orion is being built by the US company Lockheed Martin and European partners.

The Space Shuttle's Main Components

Each space shuttle has four main components. These include the spacecraft (orbiter) itself and three components that supply the energy needed to propel the craft into space. The external tanks, rocket boosters, and engines supply the energy.

The space shuttle's main components. All of the components are reused, except for the external fuel tank, which burns after each launch.
Courtesy of NASA

The Orbiter

An orbiter *is what NASA calls a space shuttle. It is a space plane that is one of the four components of what most people think of as the space shuttle.*

The orbiter launches into space like a rocket but looks like a jet plane and returns to Earth by landing on an airfield just as a plane does. It is the crew's home during the flight and is capable of docking with the ISS for studies there.

Each orbiter is about 122 feet long (almost half the length of a football field), 57 feet high (about the height of a six-story building), and has a wingspan of 78 feet (about half the width of a football field).

An orbiter is approximately the size of a DC-9 passenger plane. It weighs about 4.5 million pounds and travels in orbits of 115 to 400 miles high. It travels at speeds greater than 17,000 miles per hour.

The front end of the orbiter is the crew cabin, which includes the cockpit or flight deck. This area is pressurized and can accommodate seven crew members. It also has a nose gear and landing wheel because it lands like a plane. The flight deck is further divided into three regions: the front flight deck, the mid deck, and the airlock.

Did You Know?

The shuttle flies as a glider for reentry and landing, so the crew only gets one attempt at landing. The shuttle must also slow from 17,300 mph to 250 mph for the landing.

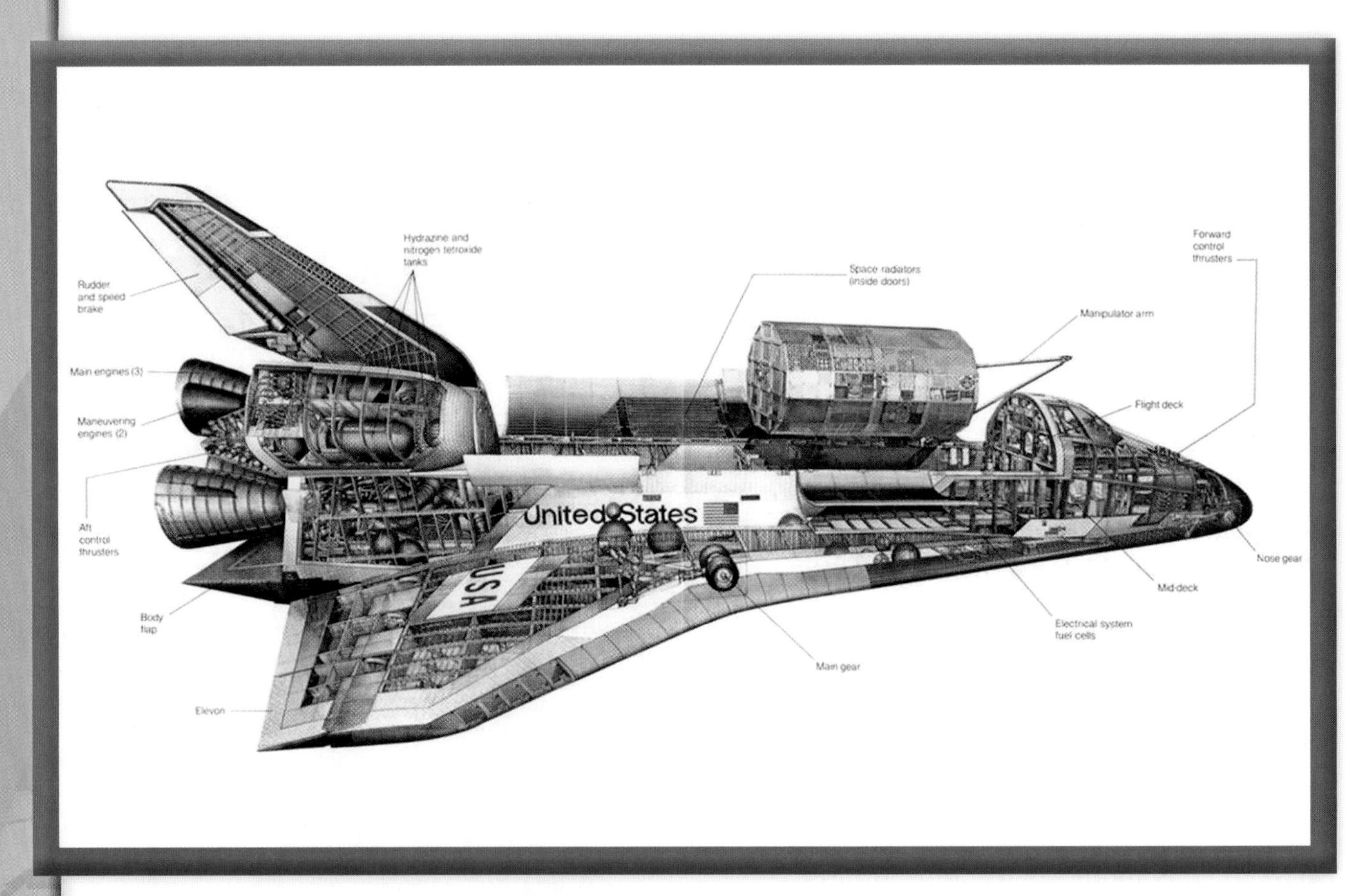

Parts of the orbiter. A detailed cutaway diagram.

Courtesy of NASA

Astronaut conducts spacewalk attached to RMS.

Mike_shots/Shutterstock

The front flight deck is the area for the flight commander and pilot with the commander sitting on the left and the pilot on the right. This is much the same as the seating arrangement in a plane for the pilot and co-pilot. The commander controls the operations of the orbiter, and the pilot is responsible for operating the payloads. The pilot is able to use a Remote Manipulator System (RMS), which is *a 50-foot long mechanical arm that deploys and retrieves things on the outside of the orbiter*.

The mid deck area contains the food and supplies and is the area used for eating, sleeping, and personal hygiene. The *last area of the flight deck* is the airlock. This pressurized area allows access for spacewalks.

The official term for a spacewalk is extravehicular activity (EVA), *when an astronaut leaves the protective environment of a spacecraft and enters outer space*. Behind the flight deck is the middle part of the orbiter, which is the cargo/payload area. This is a storage region for research materials and projects that are to be conducted on the flight and for items that are taken to or returned from the ISS.

The rear part of the orbiter is the *region where the main engines and maneuvering systems are located* to propel and guide the orbiter. This is known as the aft.

External Tank

A very large external tank feeds the fuel to the engines. It also acts as the backbone of the shuttle by supporting the orbiter and the booster rockets until the tank is empty. This is the only part of the shuttle that is not reused. After the first eight and a half minutes of flight, the empty tank is released. Most of the tank disintegrates in the Earth's atmosphere, and the rest falls into the ocean.

The tank weighs about 78,000 pounds empty and about 1.6 million pounds when loaded with fuel. It holds about 670,000 gallons of liquid fuel. The one-inch thick protective thermal coating keeps the fuel at the right temperature. The electrical system distributes the power evenly and protects the tank from lightning. The fuel is released from the tank through a 17-inch diameter tube that branches into three smaller hoses, one for each main engine.

Splashdown of the right hand SRB from the launch of STS-124.

Courtesy of NASA.

Rocket Boosters

Attached to the shuttle when it is launched are two solid rocket boosters (SRBs), *large solid propellant motors that provide 80% of the thrust needed during the first two minutes of launch*. After the initial thrust and at about 24 miles in altitude, the SRBs separate from the shuttle and return to Earth on parachutes by landing in the ocean. There, they are retrieved and can be used again. The fuel from the boosters burns off in just two minutes!

Besides providing the thrust for launch, the SRBs also guide the shuttle along a flight path. Each SRB has more than one million pounds of propellant. The solid fuel is mixed in 600-gallon bowls and poured into a mold. The final product looks and feels like a hard, rubber compound.

Space Shuttle Main Engines

The Space Shuttle Main Engines (SSMEs) *are three large engines located on the rear of the orbiter*. These provide about 20% of the thrust needed to launch the shuttle into space and to continue powering it for another eight and a half minutes after launch.

The SSMEs are powered by fuel from a large external tank during this period of operation. They accelerate the shuttle from about 3,000 miles per hour up to over 17,000 miles per hour within six minutes to reach its orbital velocity.

Did You Know?

The SSMEs burn half a million gallons of liquid hydrogen (the second coldest liquid on Earth) and liquid oxygen. The amount of water in an average family swimming pool is about equal to the amount of liquid fuel burned in the SSMEs in 25 seconds!

Lessons Learned from Challenger and Columbia

On January 28, 1986, millions of Americans, young and old, gathered around the television to watch the much-anticipated Challenger launch. This was a milestone launch, as it included the first civilian crew member in the shuttle program's history. From over 11,000 applicants, teacher Christa McAuliffe was taking the civilian seat as part of the Teacher in Space program. She was going to teach lessons from space by satellite and help restore America's interest in space exploration.

At lift-off, the crowd was full of smiling faces and enthusiastic cheers. At first, everything seemed fine. The Challenger launched into the sky as normal. At 73 seconds into flight however, something went terribly wrong. The cheers were silenced as the Challenger burst into a plume of smoke and fire. The crew was lost.

President Ronald Reagan appointed a commission to investigate the causes of the Challenger accident. Using computer graphics and NASA photos, the Commission was able to identify the origin of the smoke. They determined the first puff of smoke came from a joint on one of the SRBs. They further concluded that there must be a break in the joint's seal.

As Challenger took off, one of the seals broke enough to allow exhaust to leak out. Hot gases filled the external tank causing the first sign of fire. After that, the external tank started leaking liquid hydrogen fuel. The liquid hydrogen met with the flame and the liquid oxygen tank, creating a fiery ball in the sky.

The Commission's final conclusion was failure of the joint between the two lower segments of the right SRB. No other element of the space shuttle system was to blame. But were there more than technical issues that contributed to the Challenger accident?

Did You Know?

Many people describe the Challenger accident as an explosion. It sure looked like an explosion with the huge plumes of smoke and fire. In fact, there was no explosion at all, at least not by scientific standards. It was the flood of liquid oxygen and hydrogen that created the huge fireball in the sky.

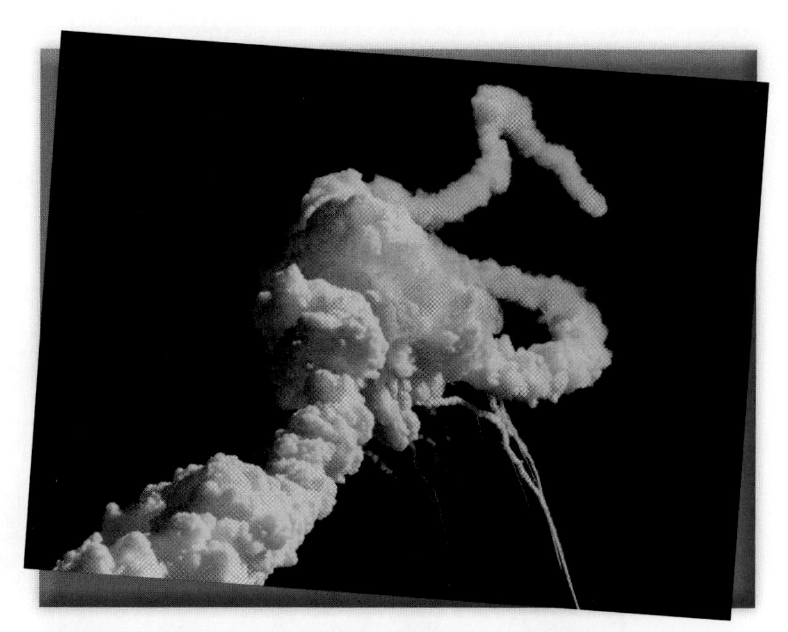

Space shuttle Challenger disaster. Space shuttle exhaust plumes entwined around a ball of gas after a few seconds after the rupture of the O-rings.

Courtesy of Everett Historical/Shutterstock

The Commission concluded that the failure of the solid-fuel rocket booster seal was due to a flawed design. The Commission focused on a part called an O-ring, *a type of gasket, used in the field joints between each fuel segment of the SRBs*. O-rings are designed to spring back into place after being compressed.

The O-ring design used in Challenger did not take into account a number of factors, such as temperature. Challenger launched on the coldest day (32 degrees to be exact) recorded in the history of all launches. The O-ring simply did not provide a tight enough seal in the cooler temperatures.

The Challenger flight was originally planned to launch on January 20, 1986. It was postponed several times during the eight days that followed, for technical issues and bad weather.

Engineers knew of the O-ring problem from previous flights. They realized it would be most susceptible to failure during the launch at max-Q, *when aerodynamic forces reach their maximum*. The engineers notified NASA management personnel about the problem before Challenger could be approved for launch. Management ignored the problem instead of investigating it.

Did You Know?

The coldest temperature during a launch before Challenger was 52 degrees. That is 20 degrees warmer than the day the Challenger launched.

The Right Stuff

The crew of the Challenger shuttle mission in 1986

The Challenger crew. Left to right are Teacher in Space payload specialist S. Christa Corrigan McAuliffe; payload specialist Gregory Jarvis; and astronauts Judith A. Resnik, mission specialist; Francis R. (Dick) Scobee, mission commander; Ronald E. McNair, mission specialist; Mike J. Smith, pilot; and Ellison S. Onizuka, mission specialist.
Courtesy of NASA

The night before the launch, engineers from the company that manufactured the SRBs also told NASA they were concerned about the problem and wanted to postpone the launch. NASA opposed the delay, so the manufacturer reluctantly approved the flight.

The launch went ahead as scheduled on January 28, 1986. The max-Q point was reached 58 seconds into the launch, but the seal had been damaged and could be seen releasing gases from the right SRB. Aluminum oxides from the burned solid fuel sealed the joint temporarily but not long enough.

Francis R. (Dick) Scobee, Challenger's mission commander, received pilot training through the US Air Force and an aerospace engineering degree from the University of Arizona. After receiving his pilot's wings, Scobee flew a combat tour in Vietnam. With NASA, Scobee piloted the STS-41-C, the fifth orbital flight of the Challenger, prior to Challenger STS-51L.

Michael J. Smith was a commander in the US Navy. He received training through the US Naval Academy and the Naval Postgraduate School. Smith worked as a test pilot for the Navy and was selected as a NASA astronaut and shuttle pilot. Challenger was Smith's first space flight.

Judith A. Resnick received a doctorate in electrical engineering from the University of Maryland and worked in the medical field for several years. She was selected by NASA as a mission specialist. Resnik was the second American woman in orbit on the first flight of Discovery STS-41D.

Ronald E. McNair was also a mission specialist aboard Challenger. He received a doctorate in physics from MIT, with a focus on quantum electronics and laser technology. His work with lasers and satellite communications led to his acceptance into NASA. McNair was the second African American in space on Challenger STS-41B.

Ellison Onizuka was also a mission specialist on Challenger. He received a graduate degree in aerospace engineering from the University of Colorado, where he participated in the Air Force ROTC program. Onizuka served as a flight test engineer, officer, and a chief of engineering at Edwards Air Force Base (AFB). With status as an Air Force officer, Onizuka was selected as a mission specialist for a Department of Defense classified mission.

Gregory B. Jarvis, a payload specialist, received a graduate degree in electrical engineering from Northeastern University. Jarvis worked for Hughes Aircraft Corp.'s Space and Communications, where he was awarded a spot in the astronaut program. He was one of two nongovernment employees aboard Challenger.

S. Christa Corrigan McAuliffe, another nongovernment employee, was the first teacher in space. McAuliffe taught English and social studies at Concord High School in New Hampshire. NASA selected McAuliffe out of 11,000 applicants.

How NASA Changed After the Challenger Accident

President Reagan ordered NASA to put in place the recommendations made in the Commission's report to ensure safety of the shuttle. The report included nine recommendations. With just 30 days to develop a plan, NASA responded.

NASA redesigned the solid rocket booster, replacing the O-rings with rings made of more reliable material. They added an orbiter to spread out the flight schedule of the fleet. The shuttle management structure and communication procedures were improved at all levels. The pre-launch processes and review of shuttle components were made much more stringent and thorough.

In addition, President Reagan created a policy that would forbid NASA from holding the monopoly for launching government satellites. Reagan also opened the doors for the public sector in space by taking NASA out of the commercial launch business for satellites.

NASA spent three years diagnosing and correcting the Space Shuttle Program. In September 1988, the program successfully returned to flight with the launch of Discovery. Some questioned whether it was too soon and if all of the issues from Challenger's accident had been successfully addressed.

The Columbia Accident

On February 1, 2003 at Kennedy Space Center in Cape Canaveral, Florida, a couple of hundred viewers gathered to watch the space shuttle Columbia land after a 16-day mission. The landing was expected to be just another normal landing.

Sadly, just 16 minutes before touchdown, space shuttle Columbia broke apart on its way back to Earth. Hydraulic fluid temperatures and tire pressure went abnormally low. Other sensors failed and then communications with Mission Control were lost. Onlookers in the southwestern United States could see falling debris. Another shuttle and its crew were lost.

An aluminum tank from space shuttle Columbia's STS-107 mission was uncovered in Lake Nacogdoches, Texas after being underwater for eight and half years. It was uncovered when the lake's water level receded during an ongoing drought.

Courtesy of NASA

An investigative group concluded that a large piece of foam broke off from the shuttle's external tank and breached the spacecraft wing during lift off. NASA had to conduct one of the largest land searches in American history, as debris was scattered all over east Texas.

After several months of debris retrieval and analysis, the Columbia Accident Investigation Board (CAIB) confirmed that faulty design was the technical cause. Instead of O-rings, this time it was the foam insulation. The damage from the foam striking the wing led to almost immediate breakup of the orbiter upon re-entry. When the crew module separated from the fuselage, the module immediately depressurized.

Just as NASA's management was held largely responsible for ignoring warnings about Challenger's O-rings, the same accusations were made about the Columbia disaster. In previous flights, similar but smaller pieces of foam had detached. They were much more numerous but smaller in size.

NASA did not identify the detached pieces of foam as a flight safety issue, despite engineering requirements stating that nothing should impact the shuttle during launch. NASA simply did not address the possibility of a larger piece hitting a vulnerable area like a wing.

As with Challenger, NASA management did not listen to its engineers, who were concerned that the landing gear could be damaged by such an impact. The engineers had requested the use of imaging satellites to view any damage to the landing gear. This equipment could help them take additional measures to ensure a safe landing.

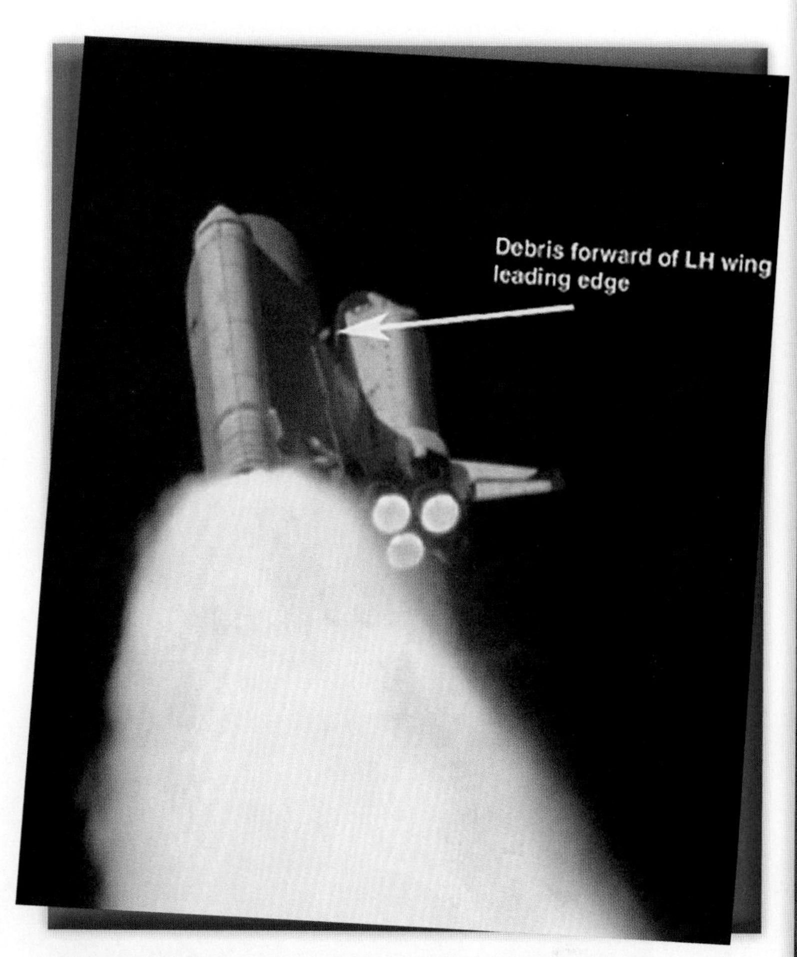

Approximately 80 to 84 seconds after liftoff of space shuttle Columbia, a large piece of debris is observed striking the underside of the left wing.
Courtesy of NASA

Senior managers said there was no need for such information and did not approve the request. If approved, NASA would have had to redesign some of the instruments.

How NASA Changed After the Columbia Accident

As mentioned, the CAIB confirmed that faulty design was the technical cause of the Columbia accident. Once again, NASA responded with several changes.

NASA redesigned the orbiter's heat shield to minimize the loss of foam. Pre-launch inspection procedures were improved as well. NASA also created the Technical Engineering Authority to evaluate control hazards within the shuttle system. Video cameras, radar, and sensors mounted throughout the main systems were added.

As a result of the Challenger and Columbia accidents, NASA has much-improved protocol for rapid responses to serious events. For the future of space missions, NASA and the private sector will continue to improve spacecraft design and reduce the risk of future accidents.

The Right Stuff

The crew of the Columbia STS-107

Seated in front, from left, are astronauts Rick D. Husband, mission commander; Kalpana Chawla, mission specialist; and William C. McCool, pilot. Standing, from left, are David M. Brown, Laurel B. Clark, and Michael P. Anderson, all mission specialists; and Ilan Ramon, payload specialist, representing the Israeli Space Agency.

Courtesy of NASA

Rick D. Husband, Columbia's mission commander, received a graduate degree in mechanical engineering from California State University and pilot training at Vance AFB. After receiving his pilot's wings, Husband served as a test pilot in the US and in England with the Royal Air Force. With NASA, he was a pilot on Discovery STS-96, during which the first docking with the ISS was completed.

William C. McCool was pilot of Columbia. He received a graduate degree in aeronautical engineering from the Naval Postgraduate School, where he also completed Test Pilot School (TPS). After TPS and two years of NASA training, McCool was selected for flight assignment as a pilot. Columbia was McCool's first space flight.

David M. Brown received a doctorate in medicine from Eastern Virginia Medical School and worked as a US Navy flight surgeon. He was the only surgeon in over a decade to be selected for pilot training. He was selected by NASA as a mission specialist. Columbia was Brown's first space flight.

Laurel Blair Salton Clark was also a mission specialist aboard Columbia. She received a doctorate in medicine from University of Wisconsin, with a focus on pediatrics. She also served on the Navy's Diving Medicine Department as an Undersea Medical Officer and later as a Submarine Medical Officer, and Naval Flight Surgeon. Columbia was her only space flight.

Michael P. Anderson received a graduate degree in physics from the Creighton University and was commissioned as a second lieutenant. After pilot training at Vance AFB, he served as an aircraft commander, instructor pilot, and tactics officer. He was selected by NASA as a mission specialist. With NASA, Anderson also flew on the STS-89 Endeavor mission, prior to Columbia.

Kalpana Chawla, a mission specialist, received a doctorate in aerospace engineering from the University of Colorado. Chawla performed complex technical research related to EVA and robotics. She flew on STS-87 Columbia, prior to STS-107 Columbia.

Ilan Ramon, an Israeli Air Force fighter pilot and colonel, served as a payload specialist on Columbia. He received degrees in electronics and computer engineering from the University of Tel Aviv. Columbia was Ramon's only space flight.

Lesson 2 Review

Using complete sentences, answer the following questions on a sheet of paper.

1. What were the three phases of Project Mercury?
2. What were the three main goals for Project Gemini?
3. Which Apollo mission was the first successful mission to the Moon?
4. What did the Apollo 13 mission demonstrate to America?
5. In what ways did Columbia's first mission achieve NASA's vision for the Space Shuttle Program?
6. What were some of the accomplishments of the Discovery space shuttle?
7. What is one major achievement of either shuttle Atlantis or Endeavor?
8. Is there a difference between a space shuttle and an orbiter? Explain.
9. What role do SRBs play in getting a shuttle off the ground?
10. What do the SSMEs do after the initial launch?
11. What were the technical causes of the Challenger accident?
12. What design changes did NASA put in place in response to the Commission's report on the Challenger accident?
13. What were the technical causes of the Columbia accident?

APPLYING YOUR LEARNING

14. How have the Challenger and Columbia tragedies changed NASA operations?

LESSON 3

Making Space People-Friendly

Quick Write

If you had the opportunity to spend a year in space as Scott Kelly did, would you do it? What would be the pros and cons of spending that much time in space?

Learn About

- how microgravity affects the human body
- threat of radiation to astronauts traveling in space
- space biomedicine

Imagine spending an entire year in space without being able to just sit down and eat a meal. Astronaut Scott Kelly has just completed his year in space and returns to a "Welcome Home" dinner with his family, — something he has been dreaming about for a long time. It's a normal occurrence for those of us on Earth, but for Scott it is a very different experience. It is the first meal in over a year that doesn't require him to use Velcro to hold down his utensils or duct tape to hold his plate in place.

There is another reality for Scott at his family dinner. He can feel gravity pressing him down into his chair. It has been 48 hours since his return from space, and he is struggling to make it through dinner. He's exhausted, and getting out of his chair is a struggle more like that of an 85-year-old man, not a healthy 51-year-old.

Now that he's out of his chair, he has to walk to his room. It's only 20 steps, but his body complains with each step. He hasn't walked in over a year and now his body has to get used to walking. Not only are his legs not cooperating, but his sense of gravity is off and the floor seems to tilt under him. As he stumbles into a planter, his twin brother, Mark Kelly, who is also an astronaut, congratulates him on doing so well. This is the first stumble that Mark has seen and he knows exactly what it's like for the body to return to Earth.

Scott's body hurts! His joints and muscles are not used to the pressure of gravity. Scott wakes in the middle of the night with his whole body in pain. As he struggles to get up from bed, he can feel the blood rush to his legs. This is a sensation he hasn't felt in a while. When he gets to the bathroom, he realizes his legs are very swollen. The blood has indeed rushed to his legs, and you can squeeze them and move the liquid around like a stress ball.

That's not all, though! Scott has a rash in every place that the sheets touched his body. His skin feels like it is burning, and he is covered with hives. Scott knew the risks when he went to space, and he knew the return would not be easy. He will be a science experiment for the rest of his life. NASA and scientists will study his body and the long-term effects that space has on the human body in order to reach the next step, Mars.

Vocabulary

- osteoporosis
- muscle atrophy
- potassium perchlorate candle
- gene expression
- radiation sickness
- cataracts
- biomedicine
- in vitro fertilization (IVF)

Expedition 43 NASA Astronaut Scott Kelly gives a thumbs as he has his Russian sokol suit pressure checked ahead of the mission's launch to the ISS.

Courtesy of NASA

How Microgravity Affects the Human Body

The United States met its goals of traveling to space and putting a man on the Moon. But what effects does space have on the human body? The environment in space and on the Moon is very different from Earth. Now that the goal is to return to the Moon and travel to Mars, it's essential we understand the effects of space on the human body to undertake these future missions.

NASA's Human Research Program (HRP) is responsible for discovering the best methods and technologies to support safe, productive human space travel. This may involve such diverse goals as making sure the astronauts have appetizing food to managing the risks of radiation.

Microgravity and the Human Body

The effects of microgravity on the human body are immense. In a zero-gravity environment, the skeleton isn't required to hold up the body. As a result, the body loses bone density. In fact, astronauts lose 1 to 2% of bone density each month on the ISS. In a matter of six months, they will have a 10% loss of bone density, which is the equivalent of 10 years of aging on Earth.

In addition, bones begin to dissolve into the bloodstream from the bone loss. As bones dissolve, calcium levels increase in the blood, putting astronauts at a greater risk of kidney stones. Although kidney stones are easily treatable on Earth, astronauts are at a greater risk because surgery cannot occur in space.

Effects of Microgravity

The effects of microgravity on the human body include the following:

- Increased risk of neurodegenerative disease
- Eye abnormalities
- Puffy face
- Nasal congestion, loss of smell, and diminished taste
- Loss of muscle mass and bone density
- Lower red blood cell count
- Lower blood plasma volume and increased kidney output
- Greater risk of kidney stones
- Lower immune system
- 10 to 30% decrease in leg circumference

Did You Know?

Astronauts are experimenting with the drug bisphosphonate during ISS missions. Bisphosphonate is an osteoporosis drug used to lower calcium levels in the blood. Osteoporosis is *a medical condition in which bones become brittle and fragile from loss of tissue*. The initial results of this experiment were positive, and NASA will continue to test the use of bisphosphonate, combined with exercise, to decrease the loss of bone density and lower the risk of kidney stones

Another skeletal concern for astronauts in space is the spine. Human spines are naturally curved, but without gravity the spine begins to straighten. Approximately 66% of all astronauts who have returned from space report back problems. And astronauts are four times more likely to have herniated disks--a condition in which the rubbery cushions (disks) between the bones (vertebrae) are damaged.

The muscular system can also take a hit from living in a zero-gravity environment. Muscle atrophy is *when muscles waste away from lack of use*. In space, an astronaut's muscle capacity for physical work can drop 40% in just 180 days. This is the equivalent of obtaining the body of an 80-year-old in a matter of six months.

Did You Know?

NASA experiments have shown that astronauts can lose up to 20% of their muscle mass on spaceflights lasting just five to 11 days.

On Earth, gravity pulls the blood in your body away from the heart and to your lower extremities. The heart then pumps the blood in your body and pulls it back to the heart. In space, there is no gravity to pull blood to the lower regions of the body. This results in a buildup of blood in the upper body, which astronauts commonly refer to as "puffy face," and a lack of blood in the lower body, which astronauts refer to as "chicken legs." While the nicknames given to these conditions can be humorous, the effects are quite serious. "Puffy face" can cause astronauts to feel as though they have a stuffy head or blocked sinuses. In addition, their vision, sense of smell, and sense of taste can be affected. Astronauts suffering from "puffy face" and "chicken legs" find that their heart weakens over time and their blood pressure drops because there is no need to pump the blood back to the upper body. In addition, tests have revealed that the shape of astronauts' hearts change in space and become more spherical.

Expedition 10 commander Leroy Chiao on the ISS displays the effects of "puffy face."
Courtesy of NASA/JSC

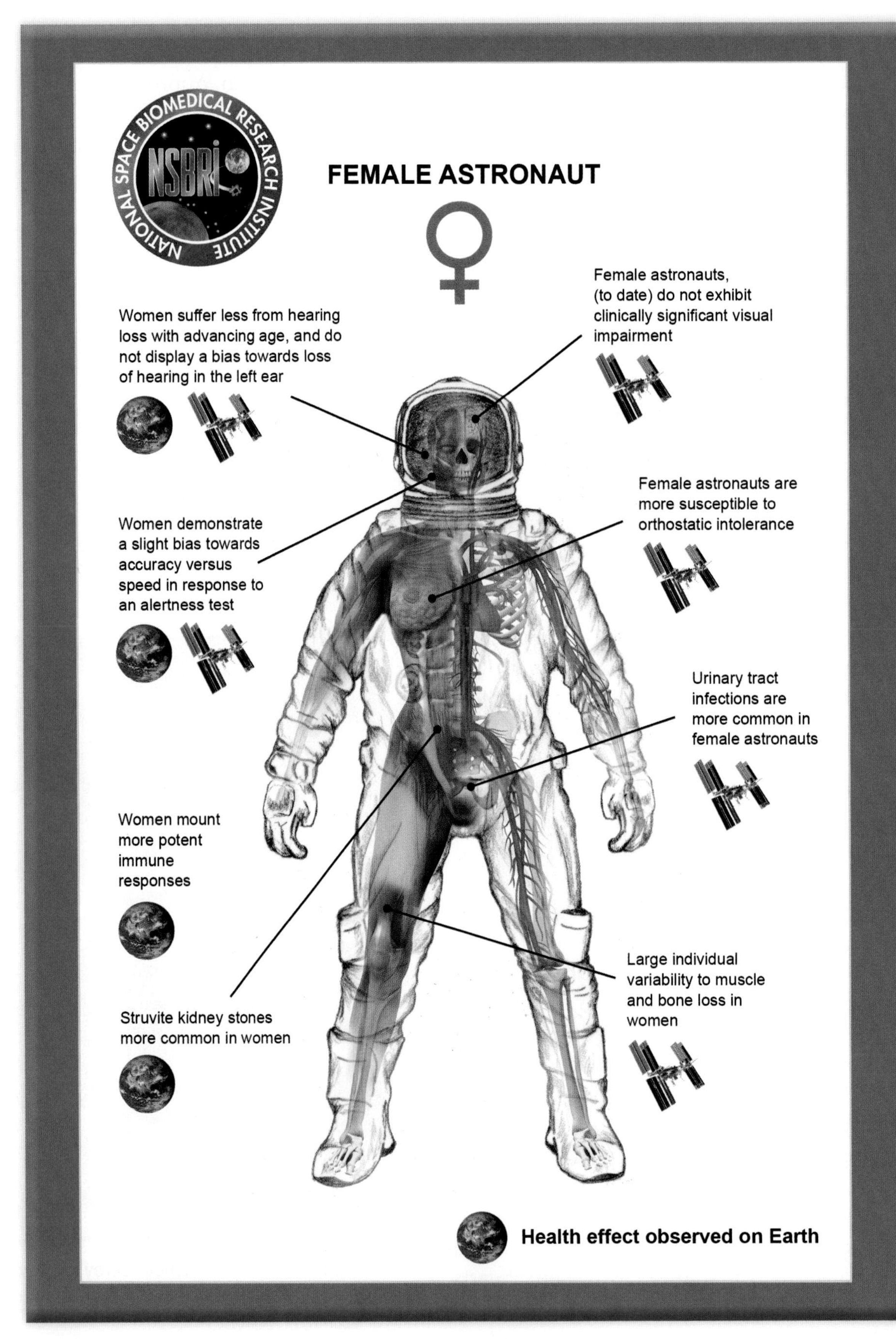

The male and female body are affected in different ways by living in a microgravity environment.

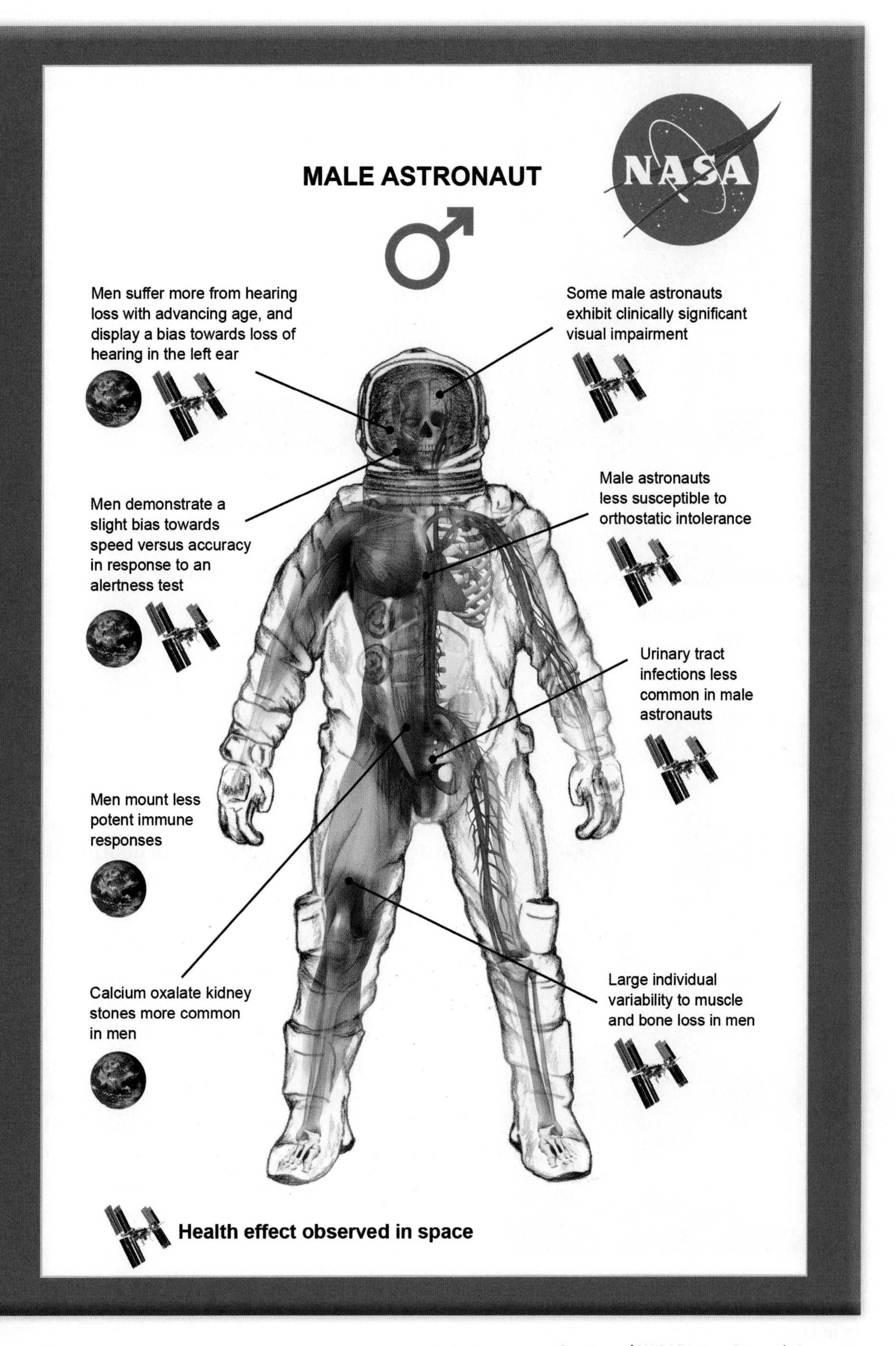

Courtesy of NASA/Human Research Program

NASA astronaut, Dan Burbank, exercising on the ISS.
Courtesy of NASA/JSC

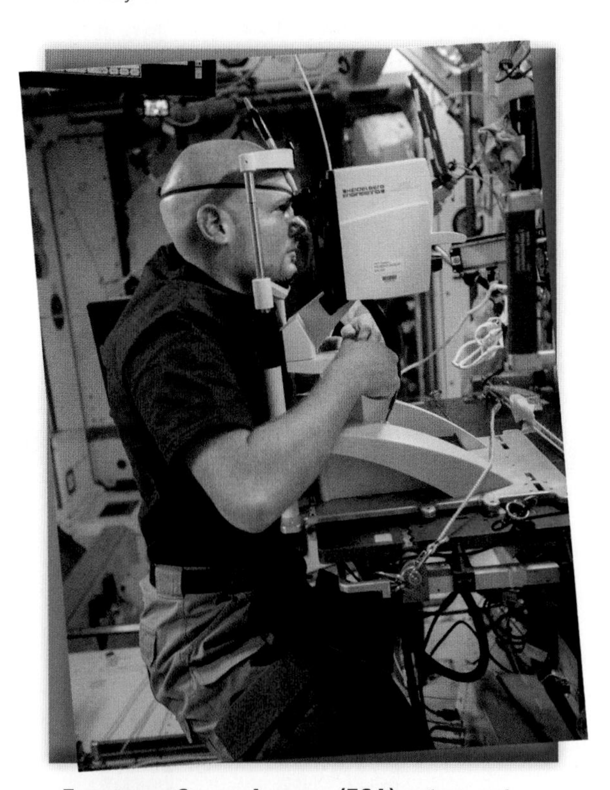

European Space Agency (ESA) astronaut Alexander Gerst conducts a vision test while in space.
Courtesy of NASA/HRP

To combat the muscular and skeletal decay in the human body, astronauts must adhere to a strict exercise program in space. They must spend at least 15 hours a week exercising while on the ISS. In addition, scientists are working on a method to expel kidney stones with ultrasound while in space. Once back on Earth, astronauts can recover from the loss of bone density. It will take 3 to 4 years for the skeleton to fully recover.

Did You Know?

Just as you might get sea sick traveling on a boat, astronauts can get space sickness. After about two weeks, astronauts develop their "space legs" and recover from their space sickness. Unfortunately, when they return to Earth, they typically need to work hard to get their "Earth legs" back.

Astronauts also suffer from eye problems in space. On the ISS, particles are floating about due to the lack of gravity. These particles can get into the eye and cause abrasions to the eye itself. In addition, "puffy face" leads to a buildup of fluids in the face and head. This additional fluid pushes up against the back of the eye and causes permanent changes to the shape of the eye. While in space, astronauts typically require the use of eyeglasses to assist with their vision, due to "puffy face." When they return to Earth, their vision does not necessarily return to normal. Most astronauts deal with permanent vision deterioration.

The Right Stuff

Breathing in Space

Breathing oxygen is a requirement for human life. In a zero-gravity environment, such as space, there is no oxygen. If a human were to enter space without the protection of a spacesuit, the results would be devastating. Human lungs can hold about 15 seconds of oxygen when a person is holding his or her breath. But this would not help astronauts if they decided to hop out of a spacecraft and hold their breath. Space has no pressure, so the air in the lungs would expand, causing the lungs to rupture. Bubbles would form in the blood; the body would be in excruciating pain as it swelled to twice its normal size. Death would shortly follow. This is why astronauts must use protective spacesuits that supply oxygen and provide protection from the harsh environment in space.

On the International Space Station (ISS), it is not feasible to send oxygen tanks into space, so the ISS creates its own oxygen from electrolysis. This process involves passing electricity through wastewater (H2O) on the ISS. The H2 and O atoms are pulled apart during the process. The oxygen (O) is then pumped into the ISS for astronauts to breathe. Using the electrolysis system, the ISS requires six gallons of water per day to produce the 12 pounds of oxygen that would sustain four astronauts. Of course, with any vital system, you need a backup. What if a problem occurs with the electrolysis system in space? The ISS astronauts would use potassium perchlorate candles. A potassium perchlorate candle is *a chemical candle that contains sodium chlorate and iron, which when burned produces oxygen*. These chemical candles produce six and a half hours of oxygen each and take five to 20 minutes to completely burn.

Mental Health in Space

Can you imagine being cramped in a small space for over 100 days? It is bound to affect an astronaut's mental health. Many countries have conducted tests on astronauts to see how they would handle being confined to a small space by having them live in isolation chambers for extended periods of time. NASA kept volunteers in their closed environmental chambers for 91 days as part of their Lunar-Mars Life Support Test Project. The longest experiment of this type was the Mars500 experiment conducted by the European, Russian, and Chinese space agencies, in which a crew of six was confined for 520 days. This is the time it would take to reach Mars and back.

A view of astronauts in the Lunar Module Simulator at the Kennedy Space Center.

Courtesy of NASA

The results have shown that astronauts suffer from sleep disorders, alterations of time sense, homesickness, anxiety, depression, psychosis, and more. Another result that was seen was the buildup of tension. The tension can then be taken out on other crew members, causing behavioral issues in space.

Mental health during space missions is an area that is still quite unknown. With our sights set on Mars, it is important to explore the mental effects of space. Imagine floating in a small confined space for an extended period of time, with only darkness outside. Researchers are examining the use of meditation, positive impact pictures, and even virtual reality.

Did You Know?

In 2007, Lisa Nowak brought a knife, mallet, rubber tubing, and a BB gun to the Orlando airport...900 miles away from her home. Once at the airport, she disguised herself and followed Air Force Captain Colleen Shipman to the parking lot. Nowak tried to lure Shipman into giving her a ride and eventually tried to pepper spray her. According to the arrest record, Nowak was intending to harm Shipman over a love triangle. Months earlier, Nowak was controlling robotic instruments during spacewalks aboard ISS as part of the shuttle Discovery mission. Lisa Nowak was a NASA astronaut who was now diagnosed with a psychotic disorder and depression.

NASA conducted medical research on astronauts between 1981 and 1998 to determine possible mental health issues that may arise due to extended periods of time in space. Over the course of 89 missions, there were 1,800 medical events and only 2% of them related to behavior. The majority of the behavior episodes were related to anxiety. However, there is evidence that some missions may have ended early due to behavioral issues.

The evidence isn't all bad! Researchers also found that space travel can create joy and happiness, which leads to happy astronauts. After all, astronauts are getting a once-in-a-lifetime opportunity to travel to space and view Earth from the outside. It is an awe-inspiring view! Most astronauts who visit the ISS come back with a change in perspective: viewing life as a small part of the universe.

NASA focuses mainly on prevention of mental health issues. Candidates complete hours of psychiatric screening. Once selected for a mission, NASA has psychologists and psychiatrists to support astronauts on their missions. On the ISS, astronauts must complete a psychological conference with medical staff every two weeks.

Did You Know?

The ISS has antipsychotics, antidepressants, and anxiolytic drugs on board to administer to any astronaut who may require them based on mental health issues that arise. In addition, they have a "restraint system" that can be used if necessary.

Kelly Brothers Experiment

Scott and Mark Kelly are identical twins born on February 21, 1964. NASA had a unique opportunity when both Scott and Mark applied for the 1996 astronaut class. They were both test pilots at the time and were both accepted as astronauts into the program. In 2015, NASA began the Twins Study. Scott would spend a year on the ISS, while Mark remained on Earth (but still assigned to duty as an astronaut). Samples would be taken from both men before, during, and after the mission.

Mark and Scott Kelly.
Courtesy of NASA

Scott ended up spending 340 consecutive days in space. Combining that with his previous missions, Scott Kelly has accumulated 520 days in space. He returned to Earth on March 2, 2016.

So, what happened? NASA discovered that Scott's DNA did not fundamentally change and the brothers remained identical twins. However, they found changes in gene expression, which is *how your body reacts to your environment.* Scott had approximately a 7% change in gene expression after being back on Earth for six months. The 7% of gene changes relate to his immune system, DNA repair, bone formation networks, hypoxia, and hypercapnia.

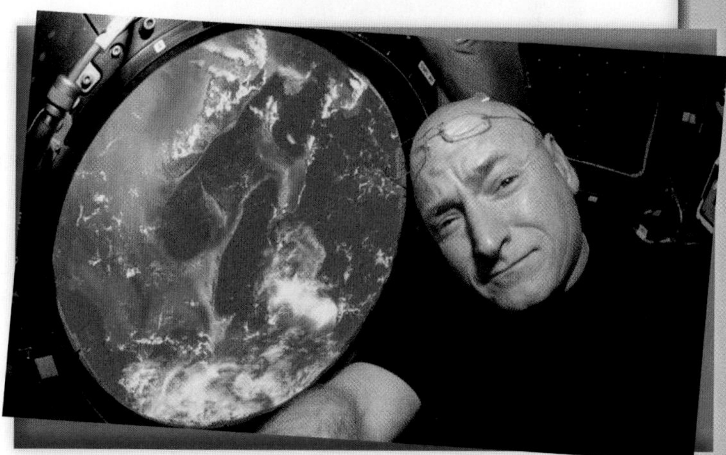

Scott Kelly on the ISS looking toward Earth.
Courtesy of NASA

Most of the effects of space travel, such as "puffy face" and reduced bone density and muscle mass, returned to normal upon Scott's return to Earth. Some changes reverted within days, while others took up to six months to reverse. The twins will continue to be studied as they age to determine how their bodies react to space travel on a long-term basis.

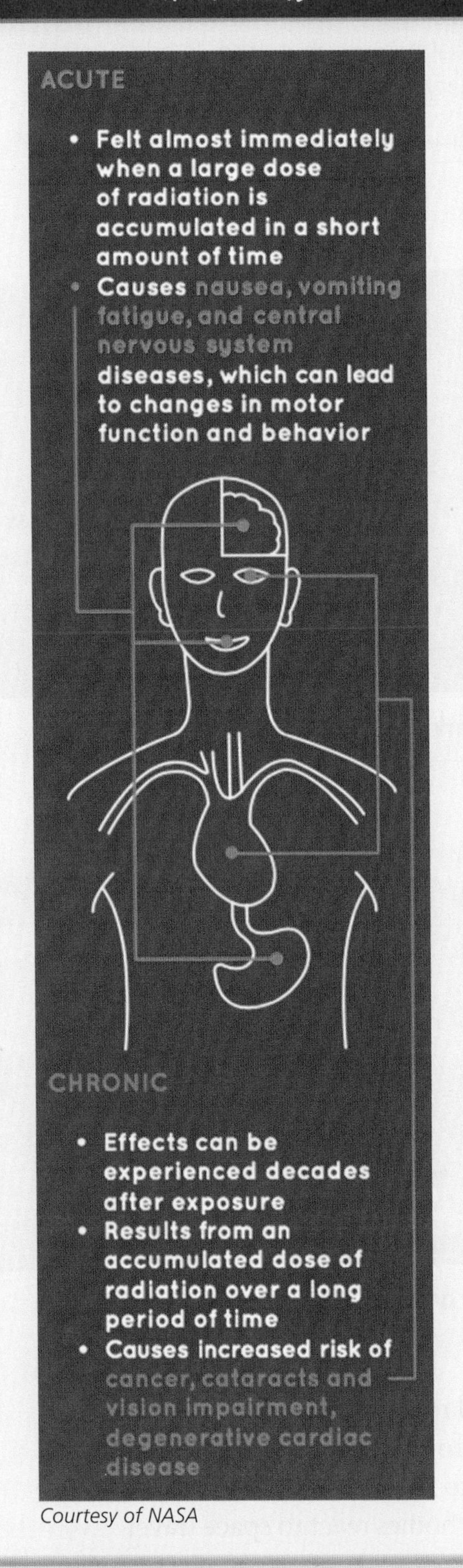

Courtesy of NASA

Threat of Radiation to Astronauts Traveling in Space

In Chapter 3 Lesson 3, we talked about radiation in space and how it affected the spacecraft. But how does this radiation affect humans traveling in space? Remember, radiation is all around us on Earth, but our atmosphere protects us from the harsh radiation that exists in space.

The ISS orbits just within the Earth's magnetic field so that astronauts have some of the protection of Earth's atmosphere. However, on the ISS, astronauts are exposed to more than 10 times the radiation they would receive on Earth. This level of radiation increases the risk of cancer, central nervous system effects, and degenerative diseases for astronauts. In addition, astronauts may suffer from radiation sickness. Radiation sickness is *the damage to the body caused by a large amount of radiation over a short amount of time.* Symptoms of radiation sickness include nausea, vomiting, diarrhea, headache, fever, dizziness, weakness, and fatigue.

There are three factors that affect the amount of radiation astronauts receive in space:

1. Altitude – The farther the astronauts travel from Earth's protective atmosphere, the greater the risk of radiation.

2. Solar cycle – The Sun has an 11-year cycle, and toward the end of the cycle there is a large increase in the number and intensity of solar flares, which produces increased levels of radiation.

3. Susceptibility – Each person is unique, and an individual astronaut may be more susceptible to the effects of space radiation. This is an area that NASA is actively investigating.

Radiation and Cataracts

Astronauts develop an increased risk of cataracts from the radiation they endure in space. Apollo astronauts first described the phenomenon of flashes of light in their eyeballs. What they experienced was space radiation flying through their eyes. When the space radiation hits the retina, it triggers a signal that the brain sees as a flash of light. Obviously, this phenomenon is not good for your eyes.

Many former astronauts have developed cataracts after traveling into space. Cataracts *occur when the lens over the eye becomes cloudy*. Someone with cataracts has blurry or hazy vision. Some astronauts develop cataracts four to five years after returning to Earth. For others, it may take up to 10 years to develop cataracts. Cloudy lenses occur on Earth as a natural progression of aging. Over 50% of people over 65 will develop cataracts. Scientists know that radiation plays a role in cataracts, but they are not yet sure how it makes the eye develop cataracts. If researchers are able to figure out how radiation effects the lens, then they may also be able to prevent cataracts here on Earth.

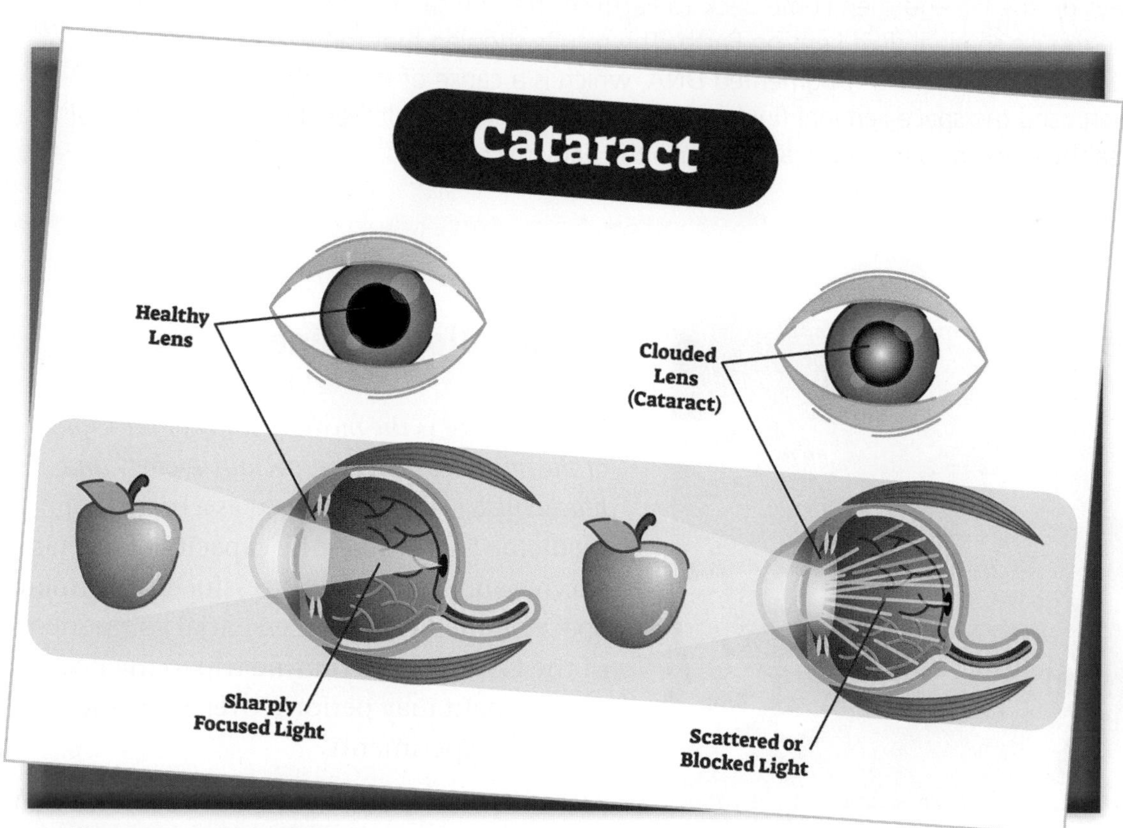

An illustration of a healthy eye and an eye with a cataract.
Courtesy of VectorMine/Shutterstock

Reproduction Risks with Space Radiation

Another concern for astronauts traveling in space is the risk of radiation on reproduction. The particles in space do direct damage to a human's DNA, so how does that affect an astronaut's ability to reproduce? Researchers suspect that astronauts who return to Earth will have no greater chance for infertility based on a short-term spaceflight. This has been proven by several astronauts who returned to Earth and had children.

However, what are the long-term effects of space radiation on reproduction? In order to establish a new colony on the Moon, or even Mars, reproduction must be researched. The only method of sustaining a new colony is for humans to reproduce on that colony. The ISS conducts many experiments to determine the effects on reproduction in space for plants and other animals. Testing has not yet been completed for human reproduction.

Did You Know?

Mouse seminal fluid has been freeze-dried and sent to the ISS so that it could spend 288 days on the ISS and then come back to Earth so that the sample could be compared with the mouse seminal fluid kept on Earth. The results showed that the space seminal fluid had higher amounts of fragmented DNA, which is a cause of male infertility. Scientists then used the space seminal fluid to fertilize an egg. The result was the same number of healthy embryos and they produced healthy baby mice.

NASA astronaut Dan Burbank using the Integrated Cardiovascular Resting Echo Scan on the ISS to conduct research experiments.
Courtesy of NASA

Space Biomedicine

Biomedicine is *the medical study of principles of the natural sciences, especially biology and biochemistry*. It is often seen as a branch of medicine that studies the capacity of humans to survive and function in stressful environments. Space biomedicine is an essential component of the ISS. During a routine trip to the ISS, an astronaut may perform over 50 medical research experiments.

As we discussed earlier, one of the studies that scientists have been exploring is reproduction in space. For a species to survive on a newly colonized planet, it must reproduce. On the ISS, experiments have been conducted on certain animals, such as fruit flies and fish. Scientists have used in vitro fertilization (IVF) where *scientists inject the seminal fluid of the animal into the egg to create an embryo*. The initial studies show that while pregnancies can occur in space, the embryo's development may be slowed down based on lower gravity. Embryos and pregnant animals have been sent to space, but the entire cycle from an animal getting pregnant, the fetus developing, and a healthy baby being born has not been successful in space yet.

The microgravity studies on pregnant rats, geckos, and sea urchins in space have shown that unusual abnormalities appear during fetal development, which is most likely from the exposure to increased radiation. NASA has a strict policy that forbids human pregnancy in space, but as more testing is completed and the goal of colonizing another planet is reached, pregnancy in space may become a reality.

Another biomedical study completed on the ISS looked at the effect of space travel on the human brain. Dr. Donna R. Roberts and Dr. Michael U. Antonucci from the Medical University of South Carolina in Charleston set about studying the MRI scans of astronauts before and after space missions. The study involved 18 astronauts on long-term missions to the ISS. The average stay on the ISS was 164 days. The study also included 16 astronauts who went on short-term space shuttle flights that averaged 13 days. The findings were remarkable. Ninety-four percent of the astronauts on long-term missions experienced narrowing of the central sulcus, which is a groove at the top of the brain. Only 19% of the short-term space shuttle travelers experienced this same condition. In addition, the brain scans of the long-term space travelers showed that all of them suffered from their brains shifting upward in the skull. These studies provide a starting point for researchers. Now that they know what can occur to the human brain in space, they can develop ways to counteract these effects to the human brain and ensure the safety of astronauts on long-term missions.

The study of space biomedicine directly impacts medical research and development on Earth. Some projects that will have direct effects on Earth include the following:

- No-drill dental care
- Noninvasive treatment for skin disorders
- Emergency wound closure
- Biofilm eradication
- Surface and water decontamination
- Waterless cleansing of garments

This lesson may have made space seem like a scary place to visit, but there is good news: NASA has been working on solutions for the health problems associated with space travel for years so that when it is time to travel to Mars, they can ensure the safe return of our astronauts.

Lesson 3 Review

Using complete sentences, answer the following questions on a sheet of paper.

1. What is NASA's Human Research Program (HRP) responsible for?
2. Why do astronauts have an increased risk of kidney stones?
3. What is the cause of "puffy face" and "chicken legs" in astronauts?
4. How long does it take the skeletal system to fully recover from its time in space?
5. What were the results of the confinement studies conducted by NASA and other space programs?
6. What does NASA focus on with regard to astronauts' mental health?
7. What was the initial result of the Twins Study?
8. What are the three factors that affect the amount of radiation an astronaut receives in space?
9. What causes the phenomenon of flashing lights that astronauts see in their eyes?
10. Why is reproduction an experiment on ISS?
11. What were the results of the ISS study on the human brain in space?
12. What are some of the space projects that will have a direct effect on Earth?

APPLYING YOUR LEARNING

13. What do you think would be the biggest drawback to space travel as far as the effects on the human body? Would this deter you from volunteering for future space missions?

CHAPTER 5

A depiction of the International Space Station orbiting over Earth as it passes by Florida.
Courtesy of NASA

Space Stations and Beyond

Chapter Outline

"When I look at what we're doing on the ISS and what we did on Skylab, we are learning how to use the new environment with expertise. We are extending the frontier. We are extending the boundaries."

Ed Gibson
Skylab astronaut

LESSON 1 From Salyut to the International Space Station

Quick Write

Do you think a rotating space station with artificial gravity will ever be developed? Why would something like this be needed for space exploration?

Learn About

- Salyut Space Program
- Skylab Space Program
- Mir Space Station
- International Space Station

Humans have dreamed of living in space for centuries. Movies, comics, and science fiction books written about space stations and living in space helped fuel the desire to explore the unknown. However, early designs for space stations were much different than the space stations that were actually sent into space.

Dr. Wernher von Braun was the first director of NASA's Marshall Space Flight Center in Huntsville, Alabama. Von Braun was a rocket scientist who worked for the US government to develop ballistic missiles after World War II. Von Braun was a huge advocate for space exploration in the 1950s. In a 1952 series of articles written in Collier's weekly magazine, von Braun, then Technical Director of the Army Ordnance Guided Missiles Development Group at Redstone Arsenal, wrote of a large wheel-like space station in a 1,075-mile orbit. This station, made of flexible nylon, would be carried into space by a fully reusable three-stage launch vehicle. Once in space, the station's collapsible nylon body would be inflated much like an automobile tire. The 250-foot-wide wheel would rotate to provide artificial gravity, an important consideration at the time because little was known about the effects of prolonged zero-gravity on humans. Von Braun's wheel was slated for a number of important missions: a way station for space exploration, a meteorological observatory, and a navigation aid.

The rotation would produce centrifugal force, which would provide artificial gravity in space. Centrifugal force is *a force that makes objects move outwards when they are spinning around something*. The wheel would be rotated in space by thrusters so that the astronauts would experience a pull similar to gravity. The size of the wheel and the speed of rotation would dictate the amount of gravity they felt.

NASA's research on the future of space stations also concluded that artificial gravity was essential for the program. So, what happened? As space stations became a reality, this initial design was abandoned for something simpler. The new design had no option for artificial gravity. This allowed researchers to conduct research on the effects of microgravity. After all, wasn't the entire purpose of a space station to conduct research on the effects of microgravity? If artificial gravity were produced, humankind would be no further along in determining the effects created by microgravity.

Vocabulary

- centrifugal force

This is a von Braun 1952 space station concept. This concept was illustrated by artist Chesley Bonestell.

Courtesy of NASA

Salyut Space Program

The first Russian design of a space station was the Almaz. The Almaz was designed as a Soviet military station to hold three crew members. The Almaz would be used for reconnaissance and radar imaging. The Almaz never made it past the design stages, but the design was adapted to use the Almaz space station in combination with the Soyuz capsule to create the Salyut 1. The Salyut 1 was the first space station to enter orbit. The Soviet Union launched the non-military space station on April 19, 1971. Once launched, the Salyut 1 had many technical issues to work out. It was the first space station of its kind, after all. The Soyuz 10 capsule was to be the first capsule to dock with the Salyut 1, but the docking failed, and the mission was aborted. However, Soyuz 11 successfully docked with Salyut 1, and its three cosmonauts lived on the station for about three weeks. The cosmonauts transmitted images back to Earth each evening, and the images were broadcast to the nation. Unfortunately, there was a malfunction on re-entry and an air valve opened. The cosmonauts were not able to close the air valve and perished on their return trip to Earth. Less than six months later, Salyut 1 was intentionally destroyed by deorbiting, when it ran out of fuel. A new crew could not inhabit later versions of the Salyut station until a redesigned Soyuz spacecraft was complete.

Orbital docking of the Soyuz 11 spaceship and the Salyut orbital station.

Courtesy of SPUTNIK/Alamy Stock Photo

The Salyut 2 was the Soviet Union's second attempt at a space station. Unfortunately, after the Salyut 2 reached orbit, it lost altitude and depressurized. The Soviets continued their research and successfully launched a third, fourth, and fifth Salyut that supported five crews of cosmonauts. The most successful of the Salyut space stations was the Salyut 6. Modifications were made to the station to allow crews to stay in space longer. A second docking port was added to allow for refueling and resupply by autonomous spacecraft named "Progress" freighters. Autonomous Progress freighters would dock automatically and, once sealed, the cosmonauts would unload supplies. The new design also allowed for visiting cosmonauts from the Soviet Union and their allies. Cosmonaut Vladimir Remeck of Czechoslovakia was the first person from another nation (other than the United States or Soviet Union) to go to space. The Salyut 6 supported 16 crews and hosted cosmonauts from Hungary, Poland, Romania, Cuba, Mongolia, Vietnam, and East Germany.

The Salyut 7 was the final space station in the Salyut series. It launched in 1982 and continued to conduct experiments and host international visitors. The last crew left the Salyut 7 in 1986. The station hosted a total of 11 crews and was occupied for 800 days. The Salyut 7 remained in orbit until 1991 when the space station, out of fuel, made an uncontrolled re-entry over the country of Argentina.

At the Gagarin Cosmonaut Training Center in Star City, Russia, Expedition 40/41 Flight Engineer Reid Wiseman of NASA admires a model of the old Russian Salyut space station.

Courtesy of NASA

Skylab Space Program

In the summer of 1969, the United States announced plans for the first space station, Skylab. The space station was sent into orbit in 1973. Skylab was the United States' first space station and, although it was announced before Salyut, it entered orbit two years after Salyut. Skylab launched on May 14, 1973 and was designed as a scientific laboratory. Skylab housed three flights of astronauts over the course of 18-months. The astronauts lived in Skylab during their time in space and conducted experiments.

This diagram shows the wings and the five major assemblies of Skylab.

Courtesy of NASA

To achieve Skylab, NASA had to be resourceful and innovative. Due to budget limitations and the Vietnam War, NASA did not have the funds to construct a new space station. Instead, they used ingenuity to repurpose the hardware from the Apollo and Saturn capsules to construct Skylab.

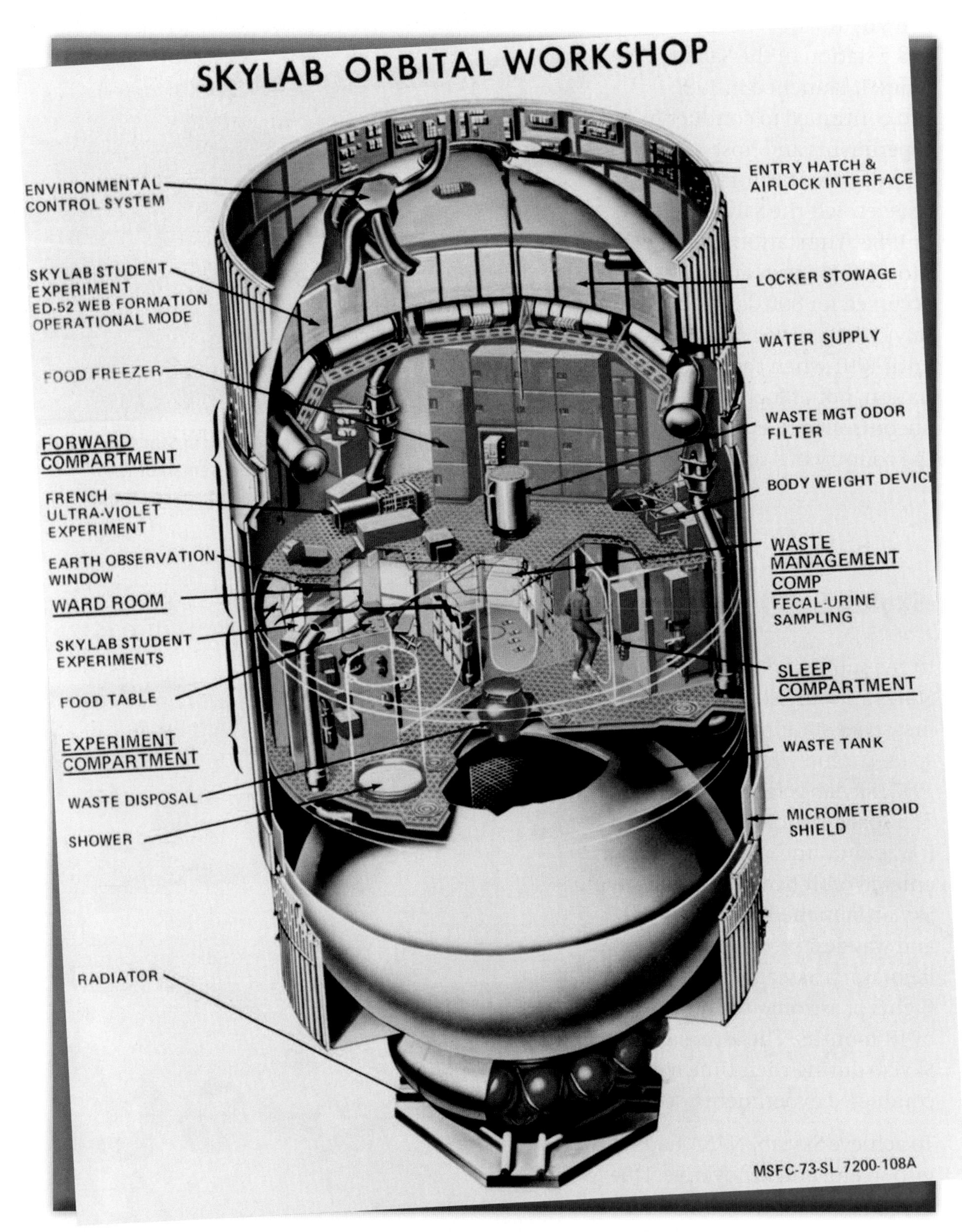

An artist's concept illustrating a cutaway view of the Skylab 1 Orbital Workshop (OWS). The OWS is one of the five major components of the Skylab 1 space station cluster, which was launched by a Saturn V rocket on May 14, 1973 into Earth orbit.

Courtesy of NASA

The Skylab program consisted of four missions. Skylab 1 was an unmanned launch of the space station, followed by three manned missions. However, during launch of Skylab 1, the meteoroid shield was torn off. The meteoroid shield was designed to protect Skylab from damage in orbit and shade it from the sun's rays. Without the shield, the internal temperature would quickly reach 325°F. In addition, as the meteoroid shield fell off, it took one of the station's solar panels with it. Power to the station would be limited without the solar panel. However, the space station was still successfully orbiting. The first Skylab crew was scheduled to launch the next day. NASA postponed the crew launch for 10 days so it could devise a rescue mission for Skylab 1.

The first manned mission to Skylab 1 was dubbed Skylab 2. It launched on May 25, 1973 with three astronauts aboard. Upon rendezvous with Skylab, the crew conducted extensive extravehicular repair to Skylab 1, while living on Skylab 2. NASA had developed a space parasol to maintain the temperature in Skylab at a comfortable 75°F and was able to re-route power to the space station. The crew was successful in its repairs and completed 404 orbits over 28 days. Prior to leaving, the crew prepared Skylab 1 for unmanned experiments until the next manned crew visit, planned in 60 days.

On July 28, 1973, Skylab 3 sent another crew of three for a 60-day mission to the Skylab station. A fourth and final Skylab mission, Skylab 4, launched in November 1973 with three more astronauts. The crew spent 84 days in orbit, conducting experiments and photographing a passing comet.

The Right Stuff

One of the most important projects of its time was the Apollo-Soyuz Test Project. In July 1975, the United States and Soviet Union both launched capsules so that they could test the capability to dock with each other's systems. The project was a diplomatic mission to improve relations and open the door for future projects. NASA developed a universal docking module that could be used between the Apollo Command Module and the Soyuz capsule.

Launch occurred on July 16, 1975, and 45 hours after launch the two capsules completed their rendezvous. Three Americans and two Soviets were aboard. They swapped gifts and conducted experiments together. This project led the way to the International Space Station and space collaboration between the two countries.

Over the length of the project, the Skylab space station completed approximately 3,900 orbits of Earth and flew over 75% of the Earth's surface. It was a flying scientific laboratory with sophisticated equipment. Skylab was a fruitful project that exceeded expectations in many areas. Astronauts performed extensive repairs to Skylab after the initial launch and gained an immense amount of experience working outside the space station.

A key project for Skylab was the solar experiment program. The Skylab program captured remarkable solar images of the surface of the Sun. It recorded solar activity and its effects on weather and communications on Earth. Skylab also looked beyond the solar system to observe stellar objects. Skylab collected and studied interstellar dust, observed the Comet Kohoutek, and recorded energetic particles trapped in the Van Allen belts.

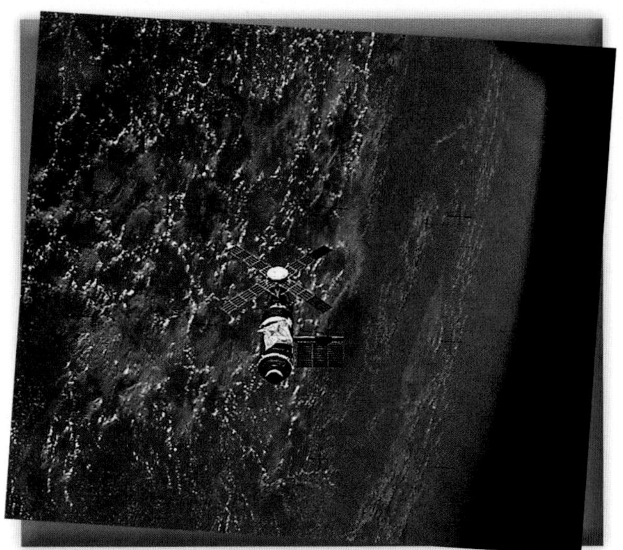

This image of Skylab in orbit was taken by the Skylab 2 crew before departing for Earth.
Courtesy of NASA

The amount of data collected from the Skylab program took years to analyze. The project successfully exceeded every goal it was given. Skylab, however, was designed to enter Earth's orbit, but never to come back from that orbit. Skylab remained in orbit uninhabited until late 1978 when it was discovered that the space station was losing altitude. On July 11, 1979, NASA used the Skylab's booster rockets to send the station back to Earth. It was hoped that the station would land in the Indian Ocean. However, the space station remnants were littered across Western Australia and the Indian Ocean; luckily no one was injured as the space station de-orbited.

Did You Know?

High school students from across the country submitted experiments for the Skylab space station. Seventeen experiments were chosen from those submitted and were actually performed on Skylab.

Mir Space Station

After the success of the Salyut, the Soviet Union continued its development of the space station program with Mir. Mir was launched as several pieces and connected in space to create a larger space station. The goal for Mir was to create a permanent Soviet presence in space. This space station became a research facility to help prepare for long duration flights. The core of Mir launched on February 20, 1986. It was 22.4 tons and had 3,000 cubic feet of habitable space. Another three modules were launched over the next four years.

Mir Modules

Module	Launch Date	Purpose
Kvant 1	3/31/1987	Astrophysics instruments, life support, and altitude control
Kvant 2	11/26/1989	EVA airlock, solar arrays, and life support equipment
Kvant 3	5/31/1990	Science equipment, solar arrays, and docking node

In total, Mir weighed 121 tons and had 13,000 cubic feet of internal space. The space station was 107 feet long and 90 feet wide. This is the equivalent of six full-size cars parked close together. That isn't a very large space for cosmonauts to live in, but that is exactly what they did. Mir was a huge success and was occupied by cosmonauts for the majority of its lifespan.

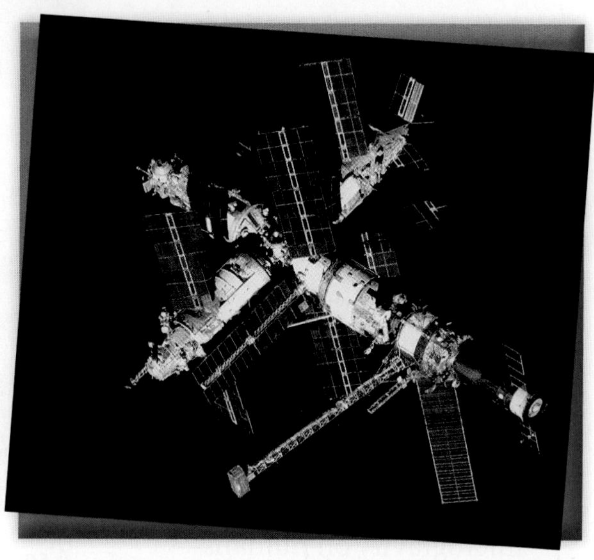

Russia's Mir Space Station is backdropped against the darkness of space, as photographed from the approaching space shuttle Atlantis on June 29, 1995.
Courtesy of NASA

In the longest Mir mission. Vladimir Titov and Musa Manarov spent 366 days living on Mir. Cosmonaut Valery Polyakov broke the record for the longest time in space during a single trip when he spent 439 days living on Mir.

Even with the Soviet Union collapse in 1991, Mir continued under Russian control and orbited with a crew until 1999. By this time, construction of the International Space Station (ISS) was in progress. The Russian Space Agency could no longer afford to keep Mir occupied and in orbit when the ISS would render it outdated. Mir deorbited in 2001 and landed in the Pacific Ocean as part of a scheduled deorbit.

Did You Know?

The US and Russia collaborated for the Shuttle-Mir program. In 1994, in preparation for the ISS, NASA astronauts spent time onboard Mir but not without incident. Mir was damaged by an onboard fire and a Russian resupply ship crashed into Mir. By 1995, NASA's space shuttle had docked with Mir on nine missions. This program was seen as Phase 1 for the International Space Station.

International Space Station

In 1984, a group of space agencies began negotiating an international space station. These agencies determined the scope, design, manufacturing responsibility, and organization of the operation. It was an enormous feat and could only be accomplished if the space agencies worked together. Five space agencies–NASA, Roscosmos (Russia), European Space Agency (ESA), Japanese Aerospace Exploration Agency, and Canadian Space Agency–worked together to develop ISS with the support of 15 countries.

The ISS project has provided many political and scientific achievements. Many diverse nations worked together to build the largest space station ever attempted. The project also revived space aspirations across the world. A new generation now dreams of exploring the unknown and going farther into space. Technical achievements in space engineering thrived from the ISS development. In addition, it allowed the world to study the effects of microgravity on humans—an experiment that is not possible on Earth.

A space shuttle docked to the International Space Station (ISS) in this computer-generated representation of the ISS.
Courtesy of NASA

This impressive undertaking required $150 billion dollars and more than 12 years of collaboration between the space agencies. The rewards from the space station have been immense and have proved invaluable in the study of humankind.

The ISS orbits 250 miles above Earth. It takes approximately 90 minutes for the space station to complete an orbit around Earth at a speed of 17,500 mph. During the course of one day, the ISS travels the same distance as if it were going to the moon and back. Astronauts witness 16 sunrises and sunsets in one 24-hour Earth day. In addition, the ISS supports experiments and research in biology, biotechnology, technology development, and bioscience. The ISS has been directly responsible for developing a vaccine, searched for dark matter, discovered how fire behaves in space, and advanced new technologies.

Mission control for the ISS is located in Houston, TX and Moscow, Russia. There is also a payload control center in Huntsville, AL.

Did You Know?

You can spot the ISS from your own backyard. Visit spotthestation.nasa.com to see when the ISS will pass over your location. The ISS will be the third brightest object in the sky and is visible to the naked eye. It will appear as a very fast-moving plane—only much higher and much faster! You'd better look quickly; the ISS will only be visible at a specific location for approximately two minutes!

Components of the ISS

The ISS is four times as large as Mir and five times as large as Skylab. It's about the length of a football field but has the space of a typical five-bedroom house with two bathrooms and a gym. Astronauts commonly compare the living quarters to the inside of a Boeing 747.

Due to the size and requirements of the ISS, it was built in 34 components and launched into space through 100 space launches. The first component, the Russian Zarya, was launched on November 20, 1998. Zarya, meaning "Dawn" in English, was assembled and launched by Russia. However, Zarya was financed and owned by the United States. Zarya provided guidance, propulsion, power, and storage for early ISS assembly. It was about 41 feet long and 13.5 feet wide and included three docking ports. Zarya is the oldest component of the ISS. It was designed to orbit for 15 years; however, it will almost double its lifespan with the plans to keep ISS orbiting until 2024.

Two weeks after Zarya's launch, the US Unity module was launched and carried to space in the space shuttle. Endeavour shuttle astronauts then connected the two parts together via several spacewalks on December 6, 1998.

Completing the assembly of the ISS required physical, hands-on labor in space. This provided many opportunities for astronauts to complete spacewalks. It took over 200 astronauts and cosmonauts to fully construct the ISS, and new space robots were created to assist in the assembly process.

Astronaut Robert L. Curbeam, Jr., STS-116 mission specialist, smiles for the camera in the Quest Airlock of the International Space Station (ISS). Curbeam had just completed the mission's first spacewalk in which the P6 truss installation was conducted.
Courtesy of NASA

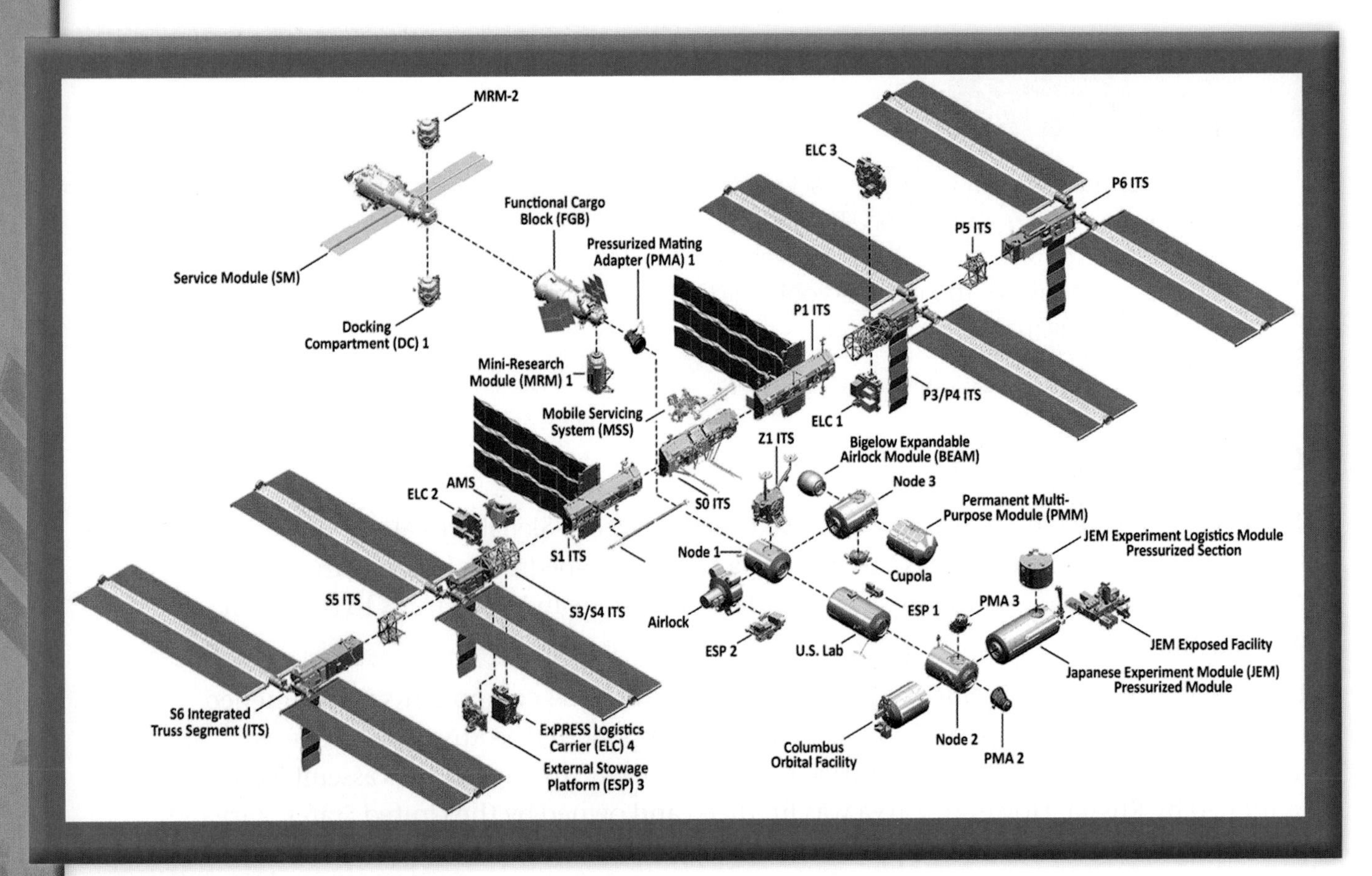

This diagram shows the components and assembly of the ISS.

Courtesy of NASA

Over the next 12 years, additional components were launched and added to the ISS. There was an 11-piece backbone truss, the four solar array wings that provide power, and support components. Canada supplied the Canadarm2 robotic arm, and Japan supplied the Japanese Experiment Module-Exposed Facility (JEM-EF). The EF allows up to nine experiments to be continuously exposed to the space environment at one time.

In short, the ISS became a marvel of technological and political abilities as the largest spacecraft ever built in space. Even after the space station was complete, additional modules were added to enhance the abilities of the ISS

The Crew of the ISS

The first crew went to the ISS on October 31, 2000. American William Shepherd, Russian Yuri Gidzenko, and Russian Sergei Krikalev were the selected astronauts. Since then, the ISS has been occupied at all times. The ISS is designed to be permanently occupied by a crew of three to six people. On several occasions, the ISS has briefly housed up to 13 people during crew changeovers or shuttle visits. During the first occurrence, the ISS had crew from five different countries: the United States, Russia, Canada, Belgium, and Japan.

Over 230 people from 18 countries have visited the ISS. Most occupants are from the US (145 people) and Russia (46 people). So, what determines which countries get to send astronauts to the space station? The amount of money or resources a country contributes to the space station directly affects the astronaut and research time they get on ISS. The major contributors are NASA, Roscosmos (Russia), and the European Space Agency (ESA). In addition, the Japanese Aerospace Exploration Agency and Canadian Space Agency also partner on experiments at the ISS.

On the second day of joint activities between the STS-127 astronauts and ISS Expedition 20 crew members, Canadian Space Agency astronaut Julie Payette and Japanese Aerospace Exploration Agency astronaut Koichi Wakata are pictured onboard the ISS.
Courtesy of NASA

Did You Know?

A one-way ticket to the ISS costs about $70 million. Each crew member usually spends six months on the station to make the trip economical.

The NASA shuttle program and Russian Soyuz were used to shuttle astronauts to the ISS. Since the retirement of the shuttle program, NASA purchases seats to the ISS on the Russian Soyuz. In 2019 or 2020, commercial crew vehicles are planned to be available to shuttle astronauts to the space station. The SpaceX Dragon and the Boeing CST-100 can hold more astronauts than the Soyuz, so they are expected to increase the number of residents aboard the ISS.

Did You Know?

In an emergency, the crew aboard the ISS can return to Earth using two Russian Soyuz capsules that remain docked to ISS.

The Right Stuff

Living on the ISS

Even though astronauts experience 16 sunrises and sunsets a day, they still operate on a normal Earth day. When they wake in the morning, they wash their hair, brush their teeth, and go to the bathroom. Of course, each of these things is a bit more challenging in space. To shower, astronauts squirt a bit of water on their skin and hair and use liquid soap and rinseless shampoo to clean themselves.

Do you know the most common question asked of astronauts who return from the ISS? It's "How do you go to the bathroom?" When going "#1," astronauts use a tube that sucks the urine into a water recycling system. Yes, that's right, ...the urine gets recycled into drinking water. You may be surprised to learn that after the recycling process the water is cleaner than most drinking water on Earth. Now for "#2," things get a bit more complicated. Astronauts have a special toilet that collects the waste in a bag. The bag then gets suctioned away.

NASA astronaut Michael Hopkins, Expedition 37 flight engineer, holds a spoon containing a piece of food in the Node 2 - Harmony - module of the International Space Station.

Courtesy of NASA

After getting cleaned up, it's time for astronauts to eat breakfast. Some foods can be eaten in their Earth form, such as brownies and fruit. But most foods are dehydrated, and water needs to be added before they can be consumed. The ISS does not have refrigerators, so perishable foods must be eaten rather quickly. However, the ISS does have an oven, so foods can be heated up. Astronauts have a full array of condiments, including salt and pepper. However, salt and pepper are in liquid form because the tiny particles of regular salt and pepper could get wedged into the equipment and cause damage. Astronauts eat three meals a day: breakfast, lunch, and dinner. Their preferred meal plan is decided before they launch into space so that the proper items are sent with them or replenished with resupply missions.

Did You Know?

The ISS has a Sunday "Come to Supper" during which astronauts of each country bring cuisines from their nations to share with other astronauts

Exercise is an important part of an astronaut's day. Astronauts must complete a minimum of two hours of exercise per day to minimize the loss of muscle mass. However, think about being in a swimming pool and trying to pick up a weight. It's much easier in a swimming pool because of the weightlessness provided by the water. This is the same in space. An astronaut can easily lift hundreds of pounds in space because there is no gravity. Exercise equipment in space is adapted to be used in space and give the astronauts a workout.

Did You Know?

Sweat does not evaporate in space. Astronauts have to constantly wipe the sweat off their skin in space

The main objective for the ISS is to conduct experiments to help lead to long-term flights in space. NASA is trying to learn how microgravity affects the body long-term so that humans can explore farther into the solar system. An astronaut's day aboard the ISS consists mostly of conducting research experiments and maintenance to the ISS. Astronauts also complete spacewalks to help maintain the space station and conduct school events from the space station.

Astronauts do have free time in space. But what do they do during that free time? One of the most popular pastimes on the ISS is spent looking out the 360-degree window of the ISS at the magnificent views of Earth. The "Cupola" is a panoramic control room and observation deck on the ISS. It is most often described as astronauts' favorite room on the ISS. The Cupola has seven windows in a hexagon layout to provide optimal viewing of Earth. It is also used by astronauts controlling the Canadarm2 robotic arm to grab visiting vehicles and visitors.

Tracy Caldwell Dyson looking out the 360-window of the ISS at Earth.
Courtesy of NASA

Crew members also have fun with microgravity. They play with their food or with equipment that was sent for testing. They watch movies, play music, read books, and talk to their families. All astronauts get weekends off, just as if they were working a job here on Earth.

Finally, it's time to go to sleep. How do astronauts stay in one place to sleep in space? They all have their own area with a work station and a sleeping bag strapped to the wall. At the end of the day, the astronauts strap themselves into the sleeping bag for sleep. It's quite hard to get used to sleeping in space because they miss the feel of laying down in a bed. Each astronaut is scheduled for eight hours of sleep a day, although that often doesn't happen just like on Earth.

Astronaut Scott Kelly's work space and sleeping quarters on the ISS.
Courtesy of NASA

Records set with the ISS

- Most consecutive days in space: 340 days (NASA astronaut Scott Kelly)
- Longest spaceflight by a woman: 289 days (NASA astronaut Peggy Whitson)
- Most time spent in space by a woman: 665 days (NASA astronaut Peggy Whitson)
- Biggest space gathering: 13 people in 2009
- Longest single spacewalk: eight hours and 56 minutes (NASA astronauts Jim Voss and Susan Helms

Future of the ISS

The ISS was developed and designed as a collaboration between multiple countries with the intent to orbit with a crew through 2024. So, what happens in 2024? Is there a possible extension to the life of the ISS? Many possible scenarios are being discussed.

NASA's current plan is to transition its current ISS responsibilities to commercial space organizations after 2024. Under this plan, NASA would become one of many customers to the ISS. NASA is assessing its long-term strategies to determine if a lower-cost alternative to the ISS is available or if its research needs to continue past 2024 on the ISS. This strategy is a bit controversial. Is there enough public interest in space travel to truly privatize the space station?

Another possible alternative is to extend the life of the ISS. Initially, the US determined that the life of the ISS would end in 2015, but since then the operations have been extended twice. NASA is currently evaluating the feasibility of extending the ISS operations through 2028. The ISS currently costs the US $3-4 billion annually. The cost of an extension, compared to the benefits of the research, needs to be thoroughly examined. The cost to maintain the ISS is half of NASA's annual budget, leaving limited funding for other space exploration initiatives.

Only time will tell what the future of the ISS is, but the station has provided 20 years of remarkable research and space exploration initiatives for the world. The collaboration between various countries is a remarkable feat, and the future of space travel has expanded due to the existence of the space station.

Lesson 1 Review

Using complete sentences, answer the following questions on a sheet of paper.

1. What happened to the crew of the Soyuz 11?
2. Why was the Salyut 6 the most successful?
3. What happened during the launch of Skylab?
4. How did NASA rescue the Skylab project?
5. What did Skylab capture during the solar experiment program?
6. What was the longest Mir mission?
7. Who all worked together to design and develop the ISS?
8. How long does it take the ISS to orbit Earth?
9. What was the first component of the ISS and what did it provide to the space station?
10. What determines which countries get to send astronauts to the ISS?
11. What is the main objective of the ISS?

APPLYING YOUR LEARNING

12. What do you think about the possibility of extending the life of the ISS? Is there a benefit to maintaining the presence of the ISS or is it time to focus on new projects?

LESSON 2

The Future in Space

Quick Write

If you were designing a new spacesuit for NASA, what do you think would be the most important features to include?

Learn About

- the potential of mining asteroids
- lunar exploration
- the exploration of Mars
- the possibility of space tourism

Front view of NASA's Z-2 spacesuit.

Courtesy of NASA/Bill Stafford

A trip to Mars would be a dream come true for many astronauts. In pursuit of that dream, NASA already has started the planning process for many details of the highly anticipated mission to Mars. One key detail that needs to be developed is what astronauts visiting Mars will wear. NASA has been using the information gleaned from the past 50 years of space travel to develop a new spacesuit.

The Z-2 spacesuit was delivered to NASA in 2016. It is a modern spacesuit with the sole purpose of exploring foreign planets. Astronauts will still use the current spacesuit while on board a spacecraft and during spacewalks. The Z-2 spacesuit will be used by astronauts after they reach Mars. Future iterations of the Z-2 spacesuit could be designed for microgravity and spacewalks.

The Z-2 will allow astronauts to easily maneuver on the Red Planet. They will be exploring, getting in and out of rovers, and collecting samples. The Z-2 has a hard upper torso, but the waist and shoulder straps are adjustable so that one suit will fit all astronauts.

The suit will feature modern electronics with updated life support and an Integrated Communication System (ICS). The ICS will eliminate the need for astronauts to wear earpieces and headsets inside the suit. The ICS will be equipped with three microphones and four speakers.

In addition, the Z-2 is a lightweight suit that allows astronauts to easily move about the new planet. The suit will feature a rear-entry hatch so astronauts climb into the suit as they would a spacecraft. The rear-entry hatch will also allow astronauts to directly dock with a rover. Once docked, astronauts could exit the suit directly into the rover.

The Potential of Mining Asteroids

Vocabulary

- near-Earth asteroids
- monopolies

Imagine being a miner 200 years in the future. Except you aren't mining coal, you are living in the asteroid belt aboard a space station, mining precious metals from an asteroid. Although the concept may sound like a far-fetched science fiction story it may become a reality someday. Consider the asteroid 2011 UW158. In 2015, the asteroid flew by Earth. It wasn't close to colliding with Earth, in fact, it was five times farther away from Earth than the Moon. So why is this an important event? Well, that asteroid contains $5 trillion of precious metals, including platinum, gold, and silver. Asteroids have the potential to be a big business. They could potentially be the future home of cosmic mines and "gas stations".

As discussed in a previous lesson, asteroids are simple space rocks made mostly made of rock and metals. Asteroids are typically irregular in shape, with pits and craters across their surface. Asteroids come in all different sizes. The smallest asteroid ever studied is six feet wide. The largest is 583 miles wide. Asteroids revolve around the sun, and their orbiting pattern tends to be erratic.

Asteroids are found within three areas of the solar system.

- The "main asteroid belt" is considered to be the area between Mars and Jupiter. This region has more than 200 asteroids larger than 60 miles in diameter. In addition, it is estimated that this area contains over 1.1 million asteroids larger than 3,000 feet.
- Trojan asteroids are asteroids that orbit a planet. Jupiter, Neptune, Mars, and Earth all have Trojan asteroids.
- Near-Earth asteroids (NEAs) are *asteroids that orbit closer to Earth than to the Sun*. There are several types of NEAs.
  - Amor asteroids – These asteroids have orbits close to Earth, but do not cross Earth's path.
  - Aten asteroids – These asteroids spend most of their time outside Earth's orbit but do cross Earth's orbit.
  - Atira asteroids – These asteroids orbit within Earth's orbit.

Did You Know?

Over 150 asteroids have their own moon. Some even have multiple moons!

A swarm of asteroids in front of the Milky Way galaxy.
Courtesy of Dotted Yeti/Shutterstock

Composition of Asteroids

The composition of an asteroid depends on what type it is. Many classes of asteroids exist, but we will focus on the most common classes.

- C-type – C-type asteroids are black or gray in color depending on their distance from the sun. This is the most common type of asteroid and includes about 75% of the known asteroids. They are composed of clay and stony silicate rock.
- S-type – S-type asteroids are green to red in color. Approximately 17% of the known asteroids fall into this category. They are composed of silicate materials and nickel-iron.
- M-type – M-type asteroids are reddish in color and consist mostly of nickel-iron.
- X-type – X-type asteroids are made completely of metal. They are thought to be the remnant cores of larger asteroids. These are the most valuable types of asteroids and are also considered rare.

The specific composition of an asteroid depends on its distance from the sun. Asteroids closer to the sun are composed of mostly carbon, with some also containing nitrogen, hydrogen, and water. Asteroids farther from the sun are composed of silicates. Silicates are very common on Earth and contain silicon and oxygen. Examples of silicate minerals include quartz, mica, and olivine. Some asteroids are metallic asteroids that are made of 80% iron and 20% other metals, such as nickel, platinum, gold, and other precious metals. Platinum is one of the most rare and useful elements on Earth, thus making it a very expensive metal. A single asteroid can contain more platinum than ever mined on Earth.

The Right Stuff

There are millions of asteroids in space. They can be dangerous because they pose a collision threat to Earth. An asteroid the size of a car typically collides with Earth on a yearly basis. The good news is that asteroids of this size disintegrate in the atmosphere before they reach the surface. Asteroids the size of a football field have collided with Earth, but this is a rare occurrence. For example, in 1908 an asteroid flattened 772 square miles of forest in Siberia.

NASA and space agencies around the world are constantly studying and tracking known asteroids so that they can determine if a large asteroid will potentially collide with Earth

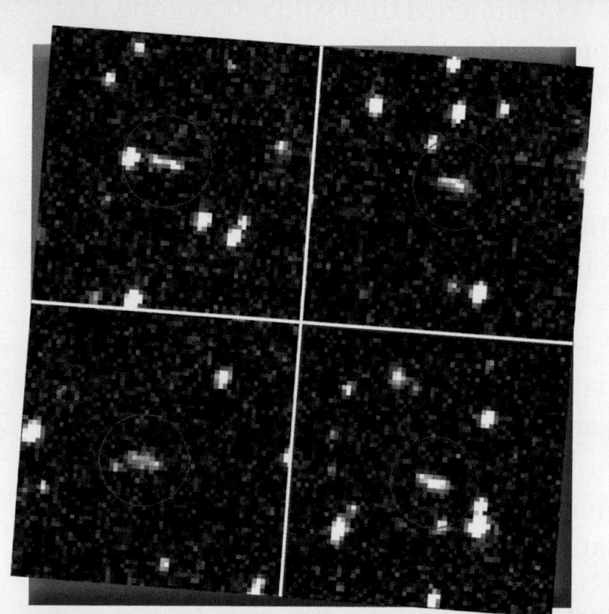

Observations of asteroid 2018 LA from the Catalina Sky Survey, taken June 2, 2018. About eight hours after these images were taken, the asteroid entered Earth's atmosphere and disintegrated in the upper atmosphere near Botswana, Africa.

Courtesy of NASA/JPL-Caltech/CSS-Univ. of Arizona

Space Mining

Plans for commercial space mining already are in the works. Many companies are researching and attempting to raise funding for asteroid mining missions. For example, Planetary Resources Inc. launched the Arkyd-6 satellite in January 2018. The satellite is the size of a home printer and contains infrared cameras. The mission for Arkyd-6 is to survey asteroids near Earth for mining potential. So, what were they looking for when surveying asteroids? They are looking for water! When you have two hydrogen atoms and a single oxygen atom, you can produce hydrogen fuel and liquid oxygen propellant. And don't forget that oxygen provides air to breathe. These elements are all essential for colonization in other regions of space. Planetary Resources Inc. also has plans to launch several satellites in 2020 that will collect and test samples of asteroids.

Did You Know?

James Cameron, the famous movie director, is an advisor to Planetary Resources Inc. In addition, some of its first investors were Google founder Larry Page and Virgin Group Ltd. founder Richard Branson.

NASA has been exploring and researching asteroids for years. In September 2016, NASA launched the OSIRIS-REx probe. The probe will travel for seven years to reach the asteroid Bennu. When it arrives in 2023, it will take a sample of the asteroid and then return to Earth.

In January 2017, NASA's Discovery Program chose to move forward with the Lucy Project and Psyche Project. Lucy's mission is to visit the asteroid belt and then continue to visit six Trojan asteroids. Lucy is planned to launch in October 2021. Psyche's mission is to visit asteroid 16 Psyche, which is a huge metallic asteroid thought to be the core of an ancient planet.

OSIRIS-REx inside the Payload Hazardous Servicing Facility at NASA's Kennedy Space Center in Florida.

Courtesy of NASA

Mineral Wealth of Asteroids

Because of the wealth of minerals found in an asteroid, and the commercial companies diving into the business of asteroid mining, asteroid mining could be the next gold rush. The value of the metals in one asteroid is measured in quintillions of dollars. For comparison, the total yearly mining yield on Earth is $660 billion. NASA currently estimates that the "main asteroid belt" contains $700 quintillion in minerals (or $700,000,000,000,000,000,000). If you divided that among each person on Earth, each person would receive about $100 billion.

Commercial organizations that have entered the space mining race are targeting NEAs because they are the easiest asteroids to reach. In reality, these asteroids are just the tip of the iceberg when it comes to the potential for asteroid mining.

Although asteroid mining is potentially an extremely profitable business, there are issues that need to be overcome. The type of robotics and equipment needed to successfully mine an asteroid do not yet exist. The cost of developing the technology needed could amount to billions of dollars. Consider NASA's OSIRIS-REx probe, for example. This single mission will cost $1 billion.

Supply and demand must also be evaluated. If we suddenly have an influx of supply for gold and other precious metals, the demand would drop and thus the price would also drop. Due to the nature of mining in space, not many organizations could afford to undertake it, so monopolies are sure to arise, as well. Monopolies are *the exclusive possession or control of something.*

Lunar Exploration

What if you could go to the Moon on your summer vacation? It might be possible in the future. In 2019, Vice President Mike Pence called for NASA to return a human to the surface of the Moon by 2024. NASA's Exploration Campaign is a national effort using commercial and international partners to go further with space exploration. The Exploration Campaign will focus on three areas:

- Low earth orbit
- Lunar orbit and surface
- Mars and other deep space objects

Initial missions will be short in length and will consist of science experiments and technology demonstrations on the surface of the Moon. These missions will begin searching for ideal locations for a lunar outpost. After a lunar outpost is set up, astronauts may spend anywhere from 28 days to 6 months living on the Moon. NASA will partner with commercial space organizations to develop and launch rockets and astronauts to the Moon.

In September 2017, NASA and Roscosmos released a joint statement of their intention to build a space station, or deep space gateway, that will orbit the Moon. Additional discussions are being held with Japan, Europe, and Canada to make the space station an international lunar outpost. Construction on the Lunar Orbital Platform-Gateway (LOP-G), or Gateway for short, would begin in 2020 with an anticipated launch of initial components in 2022. Habitation components would be launched in 2024. The close proximity of the Gateway to the lunar surface would ease the missions to the lunar outpost.

Artist's rendering of the Lunar Orbital Platform-Gateway (LOP-G).
Courtesy of NASA

Commercialization of the Moon

In November 2018, the company Lunar Outpost demonstrated its new lunar rover concept. The Lunar Resource Prospector is a small, car-sized rover that will explore the Moon to help determine possible locations for a permanent human presence on the Moon. These rovers will map the surface of the Moon, survey for resources such as water and precious metals, and identify hazards on the Moon's surface. In the past, individual rovers have accompanied spacecraft at the landing sites and have been operated by humans but haven't traveled far. Lunar Outpost plans to send "swarms" of these mini rovers to the surface of the Moon to autonomously map the surface. Details about the first mission by Lunar Outpost are expected in 2019.

SpaceX has two private passengers who have paid to orbit the Moon. The BFR Lunar Mission was scheduled to launch by the end of 2018 but was delayed by the development of the BFR launch rocket. Currently, the flight is anticipated to occur in 2023; the first passenger has been announced as Japanese entrepreneur Yusaku Maezawa. Maezawa is an art collector and plans to take several artists with him on the trip around the Moon in an art project he calls #dearMoon. The goal of the project is to inspire new art and promote peace around the world.

Bigelow Aerospace has developed and tested several space station modules. They currently have an inflatable module being used on the ISS. This company also has announced plans to create a lunar outpost. They would like to establish a space station around the Moon by 2023.

The Right Stuff

In September 2007, Google sponsored the Lunar XPRIZE project to offer $200 million to any private company that could land a rover on the Moon and drive it 1,640 feet over the surface to send back images and video. The second-place prize was set at $5 million. Five organizations secured contracts and were contenders in the race:

- Space IL (Israel)
- Moon Express (US)
- Synergy Moen (International)
- TeamIndus (India)
- HAKUTO (Japan)

In January 2018, the contest ended with no winner due to the loss of funding from Google. However, since each organization secured launch contracts and had plans to land on the surface of the Moon within two years, Lunar XPRIZE will re-launch the competition with new parameters. They are currently searching for a new sponsor for the competition.

The US is not the only country interested in setting up a lunar outpost. As discussed in Chapter 4, Lesson 1, China has already established a mock lunar village so that they can research the possibility of a complete lunar village on the Moon's surface. The European Space Agency (ESA), Roscosmos, China, and Indian space agencies have already announced plans for a joint mission to the Moon. The Chinese space agency and the ESA plan to work together to create a "Moon Village" in the 2020s.

Artistic rendering of a lunar outpost.
Courtesy of Naeblys/Shutterstock

In January 2019, China landed the first-ever probe on the far side of the Moon. The Chang'e 4 landed in the South Pole-Aitken basin of the Moon. The far side of the Moon is an untouched area of the Moon; no lander had ever touched down there. China achieved a remarkable feat in being able to reach the far side of the Moon and send back images and video of the Moon's surface. The far side of the Moon, which is not visible from Earth, has a rough terrain with more craters than the near side of the Moon. On the far side of the Moon, radio communication is cut off from Earth. To allow the probe to communicate back to Earth, China sent a communications satellite into orbit months earlier. The Chang'e 4 probe sends communications to the satellite, which can then relay the communications to Earth.

The Exploration of Mars

The efforts to create a space station in lunar orbit and a possible lunar outpost serve as stepping stones to Mars. The lunar missions provide needed research into the effect of microgravity on the human body. In addition, the Moon serves as a test case for colonization of Mars.

Mars has been the subject of extensive research for scientists on Earth. NASA has taken images of Mars from Earth's orbit, sent probes to the Red Planet, and sent rovers to explore the planet. The next step is to send humans to explore Mars. Within the next 10 years, NASA plans to send humans to Mars. The risks for an astronaut traveling to Mars are immense, so NASA and its international partners have launched several missions to help prepare for a human trip to Mars.

InSight

This is an illustration showing a simulated view of NASA's InSight lander about to land on the surface of Mars.
Courtesy of NASA/JPL-Caltech

InSight is a NASA robot that landed on Mars in November 2018. InSight traveled over 300 million miles in a seven-month span to land on the Red Planet. InSight's objectives are to study the interior of Mars over the next two years. The InSight mission marked the eighth landing of a NASA probe on the planet. Several partners from the ESA are also supporting the InSight mission.

Mars Rover 2020

NASA's Mars Rover 2020 is set to launch in July 2020. The Mars Rover will search for past life and demonstrate oxygen production. In addition, the Mars Rover will collect rock and soil samples from the Red Planet and store them for a future mission. The Mars Rover 2020 mission will be the stepping stone to a round-trip robotic mission that will be a historic first launch from another planet.

An artist's rendition of NASA's Mars 2020 rover studying rocks with its robotic arm. The mission will seek signs of past microbial life.
Courtesy of NASA/JPL-Caltech

Mars500 Project

From 2007-2011, Russia, China and the ESA conducted the Mars500 project to simulate the psychological effects of a trip to Mars. A crew of six spent 520 days in a test facility in Moscow simulating a trip to the Red Planet and back. The six people lived in a habitat of approximately 19,000 cubic feet. In addition, their communication with the outside world was limited, to simulate a trip to Mars. Communications were provided on a 20-minute delay just as they would be on a real trip to Mars. The results of the project showed that, without sunlight, astronauts' sleep patterns became out of rhythm. Participants also suffered from sleep deprivation, low energy levels, and boredom. Most of these symptoms were heightened on the "return" trip to Earth after the excitement of landing on Mars passed.

Did You Know?

The Mars Desert Research Station was created by the Mars Society in 2001 and is located in the Utah desert. The goal of the facility is to educate researchers, students, and the public on how humans can survive on Mars. Visitors can spend one to two weeks living at the facility under simulated conditions to experience what it would be like living on Mars.

ExoMars

ExoMars is a joint robotic exploration mission to explore Mars by the ESA and Roscosmos. The first mission of the program launched in March 2016. The Trace Gas Orbiter began orbiting Mars in October 2016. The ExoMars's first mission aims to study the radiation levels that astronauts would be exposed to on a mission to Mars. It is anticipated that on one mission to Mars, astronauts will be exposed to 60% more radiation than is recommended over their entire career.

The ExoMars 2020 rover was officially named "Rosalind Franklin" after accepting suggestions from residents of ESA countries. Rosalind Franklin was a British chemist/researcher who is known for contributing to the discovered of DNA. Rosalind Franklin will launch in the summer of 2020 and deliver a European rover and a Russian surface platform. The journey will take nine months. The mission will collect samples and analyze them with on-board instruments and travel across the surface looking for signs of life. This will be the first Mars mission that both travels across the surface and studies the planet in depth.

Commercial Space Organizations on Mars

What a Martian colony might look like. Elements of this image furnished by NASA.

Courtesy of u3d/Shutterstock

Of all the commercial space organizations, SpaceX may have the most thought-out plan for Mars. SpaceX has established ideal launch windows for a trip to Mars. Mars's orbit is elliptical in shape, so its distance from Earth varies based on the current location of orbit. The shortest flight to Mars would be 200 days in length. The launch window for this type of flight occurs every 780 Earth days. During the 2022 window, SpaceX plans to send a cargo mission to Mars, followed by a subsequent human and crew mission during the 2024 launch window. The primary missions of these trips will be to build a propellant depot and prepare for future crew flights. SpaceX plans to begin a Mars base that people can visit or live in.

Did You Know?

The efforts that SpaceX is putting into a trip to Mars will provide benefits to commercial travel here on Earth. Using SpaceX's Starship and Super Heavy rocket, the typical flight from Los Angeles to London would go from 10.5 hours to 32 minutes. Imagine visiting London for the day!

The Possibility of Space Tourism

Do you want to be a space tourist? Commercial space organizations are already selling seats to travel to space. Soon they may be offering flights to the Moon or even Mars.

The first space tourist was 60-year-old American engineer and millionaire Dennis Tito. Tito paid $20 million to spend a week aboard the International Space Station in April 2001. He became the first space tourist. During his trip, Tito orbited Earth 128 times. In 2002, Mark Shuttleworth, a South African computer entrepreneur, spent 10 days on the ISS. And in September 2009, Canadian billionaire and founder of Cirque de Soleil, Guy Laliberté, paid $35 million to spend 12 days in space. Laliberté was the last space tourist to visit the ISS due to the space station's increased crew size. With the limited space and the retirement of the shuttle program, NASA needed the extra seats on the Soyuz. The space tourist program with the ISS ended with a total of seven visitors.

Soyuz 2 crewmembers pose for a photo in the Zvezda Service Module on the ISS. From the left are American businessman Dennis Tito; and cosmonauts Talgat Musabayev and Yuri Baturin, Soyuz 2 commander and flight engineer, respectively, representing Russia's space agency.
Courtesy of NASA

It won't be long before private organizations start offering space tourism. In fact, space trips already are for sale, and several companies are almost ready to launch. Virgin Galactic has resumed testing of the SpaceShipTwo. In December 2018, SpaceShipTwo launched pilots Mark Stucky and C. J. Sturckow into space during its suborbital test flight. The craft reached its highest altitude to date and exceeded the height required to be considered in the boundary of space. This accomplishment was the first human spaceflight by an American vehicle since the retirement of the shuttle program. Although a schedule for commercial flight has not been announced, the company is laying the groundwork and has two new SpaceShipTwo vehicles under construction. So far, over 650 people have bought tickets for a suborbital flight aboard SpaceShipTwo at a cost of $250,000 each.

Blue Origin, Jeff Bezos's company, is planning to send its first crewed missions on suborbital space trips in early 2019; its first orbital launches are scheduled for 2021 after the development of the company's New Glenn rocket. Blue Origin is set to announce ticket prices for suborbital space trips once the crewed test flights begin. Suborbital flights are estimated to last 11 minutes, and passengers will be required to complete a day of training prior to flight. So, stay tuned and start saving!

Roscosmos recently selected the private organization KosmoKurs to build a reusable rocket to send tourists to space. The designs have been approved, but testing has not yet begun, so it may be a few years before KosmoKurs is ready to launch. Tourists would be required to complete a three-day training program and pay $200,000 to $250,000 for their 15-minute flight.

Space Hotels

It may sound like science fiction, but at least one company has plans for a space hotel! In April 2018, Orion Span announced plans for a luxury space hotel, the Aurora Station. The company plans to begin construction in 2021. The hotel will have two suites and be able to accommodate four guests and two crew during each stay. Guests wishing to stay in the hotel must complete a three-month training program. The voyage to the space hotel will be contracted with a commercial partner; Orion Span is hoping to partner with SpaceX.

So how much would it cost for a 12-day trip to the Aurora Station? Considering that ISS tourists pay $20 to 50 million, the $9.5 million trip would actually be the most affordable way to experience space life. And just in case you love living in space so much, the Aurora Station has plans to sell space condos!

Orion Span is not the only company diving into space hotels. Roscosmos has announced plans for building a luxury space module for the ISS. Tourists would have private living quarters with four bedrooms and a lounge. The cost for the Roscosmos space hotel would range from $40 to 60 million for a month-long stay. Guests would not only get to stay on the ISS, but they would participate in a guided spacewalk with a cosmonaut. Roscosmos would like to launch the space hotel in 2022.

Space Elevators

Imagine boarding an elevator on the Earth's surface and instead of taking the elevator to the top floor of a building, you take it to space. Is it really possible? NASA claims the concept is sound, and experiments are being conducted on the ISS to determine if, in fact, it is possible. The elevator would extend 22,000 miles above the Earth's surface. Obayashi Corporation in Tokyo has vowed to build a space elevator by 2050, and China intends to build one by 2045.

The concept for a space elevator was developed in 1895 by Russian scientist Konstantin Tsiolkovsky. The elevator would have motorized pods that move up and down a tether. One end of the tether would be secured to Earth's surface and the other secured to a space station in orbit.

High-quality Earth image with space elevators. Elements of this image furnished by NASA.

Courtesy of Vadim Sadovski/Shutterstock

The Japanese Space Tethered Autonomous Robotic Satellite-Mini elevator (STARS-Me) is currently being tested with the ISS. It simulates a 10-meter elevator in a weightless environment to determine if a space elevator can work. If successful, space elevators could provide an affordable method for transporting humans and goods to space. The current cost of sending goods to space is about $3,500 per pound. Using a space elevator, the cost could drop to as little as $25 per pound. Of course, a space elevator carries a price tag of about $10 billion and would be the largest engineering project ever attempted.

As we look toward the future of space travel, it is obvious that the human presence in space will only grow. It is only a matter of time before space may be on the short list for your summer vacation destination.

Lesson 2 Review

Using complete sentences, answer the following questions on a sheet of paper.

1. What are the three areas in the solar system where asteroids are found?
2. What are the four most common classes of asteroids?
3. Why is Planetary Resources Inc. looking for water with the Arkyd-6 satellite?
4. When will the habitable components of Gateway be launched?
5. What is the Lunar Resource Prospector?
6. Who landed the first probe on the far side of the Moon?
7. When does NASA plan to send the first humans to Mars?
8. What will the Mars Rover 2020 be searching for?
9. What is SpaceX's primary mission when it launches cargo and crew to Mars in 2022 and 2024?
10. What was the first spacecraft to achieve human spaceflight since the retirement of the shuttle program?
11. What is the Aurora Station?
12. How would a space elevator work?

APPLYING YOUR LEARNING

13. Based on what you have read and discussed in this lesson, what do you think will be the biggest challenge for getting humans to Mars?

LESSON 3

Space in Your Daily Life

Quick Write

Read the vignette about AFJROTC cadets' real-life experiences with the Lunar Crater Observation and Sensing Satellite (LCROSS). Consider the ways you might put to use your experiences in the JROTC program. What areas of aerospace science are you most interested in pursuing?

Learn About

- history of satellites
- how satellites are used every day
- space technology in everyday use

Combine your knowledge and understanding of science, technology, engineering, math (STEM), and now aerospace science, and you just might find yourself above the clouds. Or perhaps, you will land yourself on the moon!

Air Force JROTC cadets at the Lewis Center for Educational Research's Academy of Academic Excellence (AAE) in Apple Valley, CA have also put their aerospace knowledge and skills to use. For 110 days, cadets tracked NASA's Lunar Crater Observation and Sensing Satellite (LCROSS). LCROSS's mission was to study ice water on the south pole of the moon. LCROSS impacted the moon to create a dust cloud that could be studied.

LCROSS was commanded from the Goldstone Apple Valley Radio Telescope (GAVRT) in the Mojave Desert, CA. The cadets photographed the satellite's impact with the moon and sent the first viewed images to NASA.

This was an exercise in commitment and dedication for the cadets. While NASA controlled the spacecraft for only a few of the 110 days, the cadets controlled and gathered data from the spacecraft for 97% of the mission!

As you read on to learn more about the development of satellites and the growing interest in private commercial satellite launches, consider how we use satellites every day and the possibility of soaring beyond the clouds yourself!

More than 400 AAE students, faculty, and family members gathered in classrooms October 9, 2009, to watch the live viewing of the impact.

Courtesy of Air Force

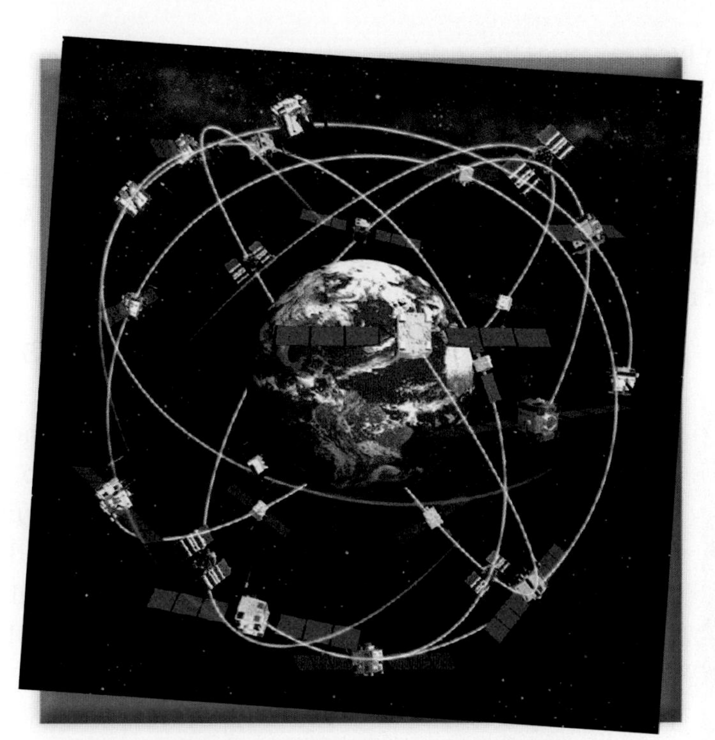

Over 30 navigation satellites are orbiting Earth. These satellites, along with ground stations and receivers, tell us our location at any given time.
Courtesy of NOAA

Vocabulary

- satellite
- geosynchronous orbit
- transponder
- uplink
- downlink
- geostationary orbit
- constellation satellites
- bus
- radiometers
- polar-orbiting satellites
- cellular communications
- satellite cellular phones
- trilateration
- reconnaissance
- spinoffs

History of Satellites

Today, thousands of satellites are orbiting the Earth. By definition, a satellite is *an object that orbits around a bigger object*. Planet Earth itself is a satellite because it orbits around a larger object, the Sun. The Earth and the Moon are natural satellites. In this lesson, our focus is artificial or man-made satellites.

Many of the thousands of artificial satellites orbiting Earth today are fully operational. Fully operational satellites provide us with television signals, communication abilities, GPS, and so much more! With the many current uses and emerging applications, the trend of satellite development and launch is here to stay.

Let's take a look back at the evolution of satellite development.

Satellite Development

NASA has worked with private companies since the 1950s and 1960s to develop satellites for communications and for advancements in meteorology. Companies like Radio Corporation of America (RCA), American Telephone and Telegraph Company (AT&T), and Hughes Aircraft Company partnered with NASA in these efforts over the years.

In 1961, NASA partnered with RCA on a satellite relay program to test across-ocean communications. Think of a satellite relay like a relay race in track and field. A runner goes a certain distance and then passes a baton to the next runner on his or her team. The baton is passed from runner to runner until the finish line. With a satellite relay, data passes from satellite to satellite until it reaches the ground on Earth. For NASA and RCA, a satellite relay was used to measure radiation in space for the purpose of determining the damage that radiation might do to the satellites.

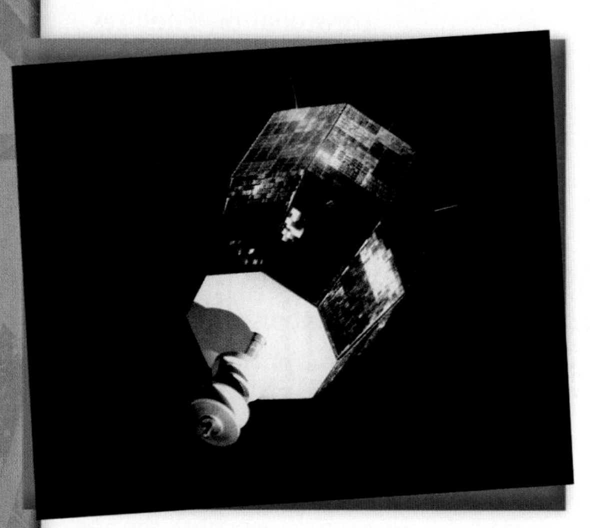

This 1964 image shows the 172-pound Relay satellite.
Courtesy of NASA

Relay satellites were launched beginning in 1963. Each satellite in the RCA program was designed with two sets of circuits. If the first circuit became damaged, the second one would provide backup. With the launch of Relay 1 in 1963, the first circuit did fail, but the second circuit functioned properly. Relay 2 was launched in 1964 using improvements gained from the first flight, and the second circuit was not used. Because of advanced planning and design changes, a planned Relay 3 flight was not needed!

About the same time as the RCA program, AT&T developed its own satellite as part of the Telstar program. AT&T paid NASA $3 million to launch it. Telstar I was the first privately sponsored communications satellite. It was launched into orbit on July 10, 1962. The Telstar I satellite continued to operate successfully into the early part of 1963. On the Telstar I project, the world's first transatlantic television signal was broadcast and the first phone call was transmitted by satellite through space.

The first phone call transmitted through space went something like this:

> *"Good evening Mr. Vice President, this is Fred Kappel calling from the Earth Station at Andover, Maine. The call is being relayed through our Telstar satellite as I'm sure you know. How do you hear me?"*
>
> *"You're coming through nicely Mr. Kappel," said Vice President Lyndon Johnson.*
>
> (Colin Schultz/Smithsonian, 2012)

The Right Stuff

Today, there are 10 Tracking and Data Relay Satellites (TDRs) located about 22,000 miles above Earth, receiving information from the ground telling the satellites what to do (take a picture, turn a sensor on or off, send stored data back, or change its orbit). TDRs allow NASA to have global coverage of all the satellites, all day every day, without having to build additional ground stations on Earth.

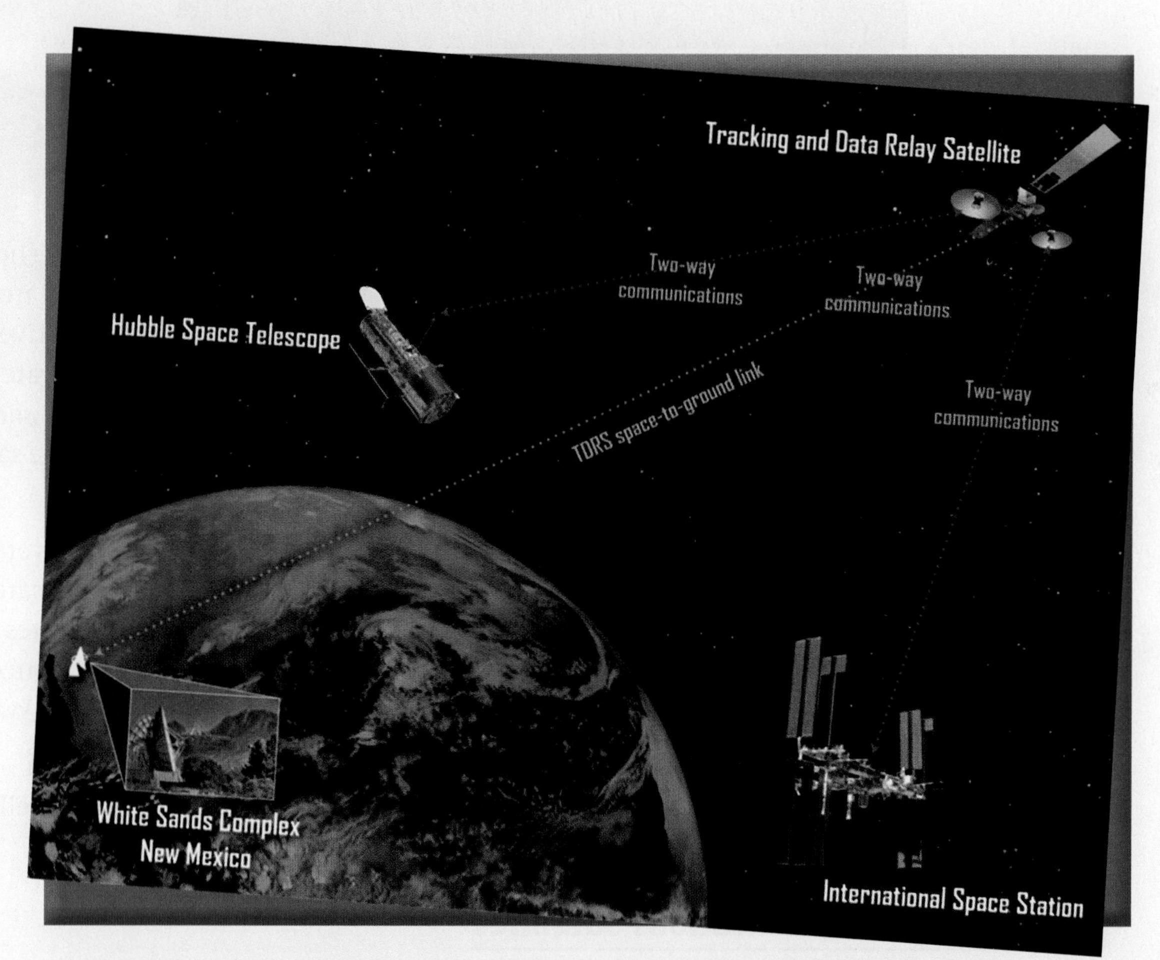

The Tracking and Data Relay Satellites (TDRs) sit about 22,000 miles above the Earth and pass data from a satellite until it reaches a ground station in view..
Courtesy of NASA

The broadcast satellite, Telstar 1, launched into Earth's orbit in 1962. It was 34.5 inches long and weighed 170 pounds.

Courtesy of gfdunt/Shutterstock

The RCA and AT&T projects had something in common: . . . they were expensive! They required large, rotating antennas on the ground to track the orbits of the satellites around Earth. To lower costs, NASA contracted with Hughes Aircraft Company for alternatives. Hughes built a 24-hour communications satellite, called Syncom, that would operate in geosynchronous orbit. Geosynchronous orbit is *an orbit that puts the satellite in the same place in the sky over a specific point on the surface each day*. It looks like it hovers in the same spot, but it actually travels at the same speed as the Earth's rotation.

The Synchronous Communications (SYNCOM) satellite program was in use for several years. Syncom helped receiving dishes on the ground retrieve information from a satellite by pointing at just one point in the sky. Thanks to Syncom, folks in the United States were able to watch live coverage of the 1964 Olympic Games from Tokyo, Japan. Syncom also provided useful Department of Defense communications during the Vietnam War.

NASA also launched several weather satellites during the period from 1960 through 1966. They were called TIROS (Television Infrared Observation Satellites). These satellites contributed to weather and climate studies and predictions.

Commercial Communications Satellites

In 1964, the International Telecommunications Satellite Organization (INTELSAT) was formed to develop a global network and improve access to communications in developing nations. The greatest achievement of INTELSAT at the time, was the management of the first commercial communications satellite. Intelsat I, also known as Early Bird, was the result of a partnership between Hughes and the private company Communications Satellite Corporation (COMSAT). Early Bird was developed to send back the first transmissions from space from over 20,000 miles above Earth. This included telephone and television signals. Expected to work for about a year and a half, it continued to operate for four years.

Did You Know?

Early Bird's name came from the popular saying "the early bird catches the worm."

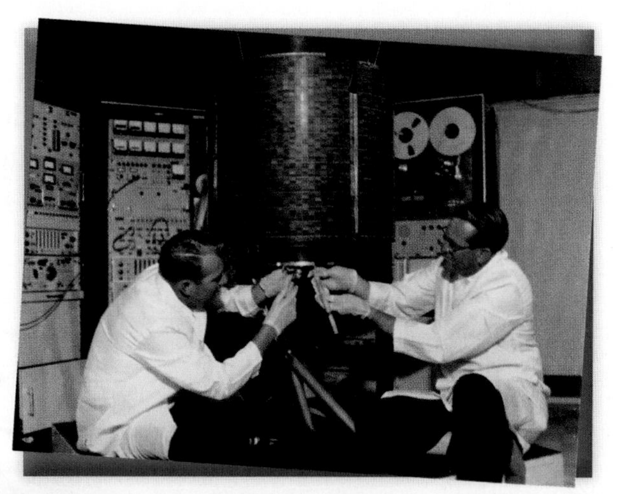

Early Bird was the first commercial communications satellite placed in a geosynchronous orbit.
Courtesy of NASA

By the end of 1965, the Intelsat II system delivered circuits for the NASA Communications Network (NASCOM). By 1969, the Intelsat III system completed its global network with a satellite over the Indian Ocean. This happened just days before Apollo 11 landed on the moon. The INTELSAT satellites made it possible for millions around the world to watch this important moment in history!

Satellites at the time were used mainly to communicate voice, video, and data from one large antenna to another distant satellite. The distant satellite would then relay communication to land-based networks. These communications satellites receive signals from TV or radio. Then, they amplify the signals via a transponder. The transponder *receives a radio signal and immediately transmits a different signal. The signal transmitted from Earth* is known as an uplink. *The signal transmitted from the satellite to Earth* is the downlink.

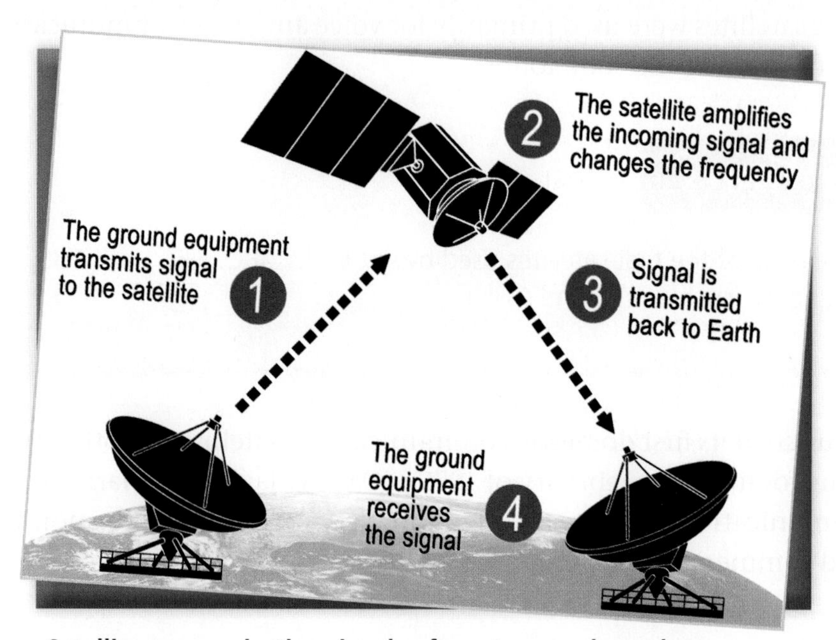

Satellite communications involve four steps as shown here.

In 1972, TELESAT Canada launched Anik, the first domestic communications satellite. It was the first system established to improve the telecommunications and broadcasting services within a specific country. This made Canada the first country with a domestic communications satellite system using a satellite in a geostationary orbit. Unlike a geosynchronous orbit, which can have any inclination, a geostationary orbit *puts the satellite in the same plane as the equator.*

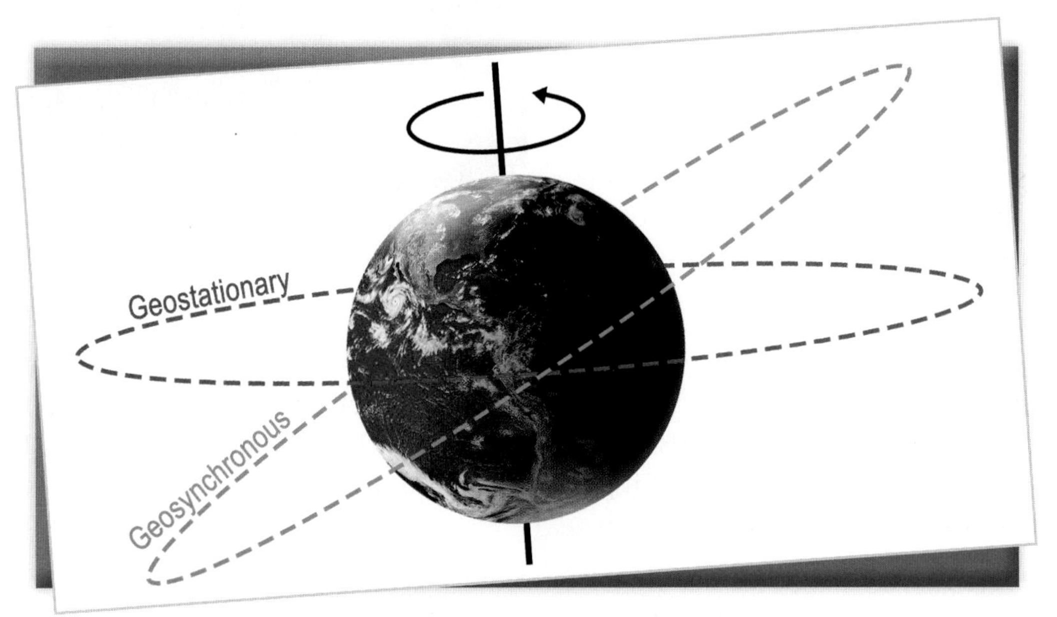

Difference between geosynchronous and geostationary orbits.

RCA leased the use of transponders on Anik until launching its own satellite, Satcom 1, in 1974. Meanwhile, Western Union launched the first US domestic communications satellite in 1974. Westar I was used for voice, video, and fax transmissions throughout the US. These satellites were used primarily for voice and data communications. The distribution of video would soon follow.

Did You Know?

Satcom 1 was one of the first satellites used by networks ABC, NBC, and CBS, as well as cable TV channels, (such as TBS and CNN).

As the US launched its first domestic communications satellites, another type of satellite was preparing for liftoff. In February of 1976, COMSAT launched Maristat, our first mobile communications satellite for telephone service and for traffic monitoring by the US Navy and commercial shipping.

In 1979, the International Maritime Satellite Organization (INMARSAT) was created. It would serve the same purpose as INTELSAT but for sea support communications instead of land. It was hailed as an important milestone in preventing the dangers of collision, loss of life, and pollution.

Satellites were our primary means of communication during the 1980s. With the first fiber-optic cable put in place across the Atlantic in the late 1980s, we established another means for communication. The norm today is to use clusters of *smaller satellites working together*, also known as constellation satellites.

Technology and satellite design continue to improve and advance. Equipment becomes smaller, lighter, and more affordable for launching low Earth orbit (LEO) satellites. LEO satellites orbit the Earth at lower altitudes (about 99 to 1,200 miles) above the Earth's surface, and they take a longitudinal orbit path. They provide faster transmission and coverage of data.

The Right Stuff

The Components of a Satellite

Although satellites perform different functions, they all have common components.

- The bus is *the metal or composite frame and body of a satellite*. The bus must hold all components of the satellite safely inside the frame. In addition, it must be strong enough to survive launch.
- All satellites have a power source. The typical power source for satellites is solar cells with rechargeable batteries.
- All satellites have onboard computers to monitor their systems.
- A radio system and antenna are common to all satellites. This allows ground-control to request information from the satellite or reprogram the satellite.
- All satellites also have altitude control systems. These systems maintain the direction of the satellite.

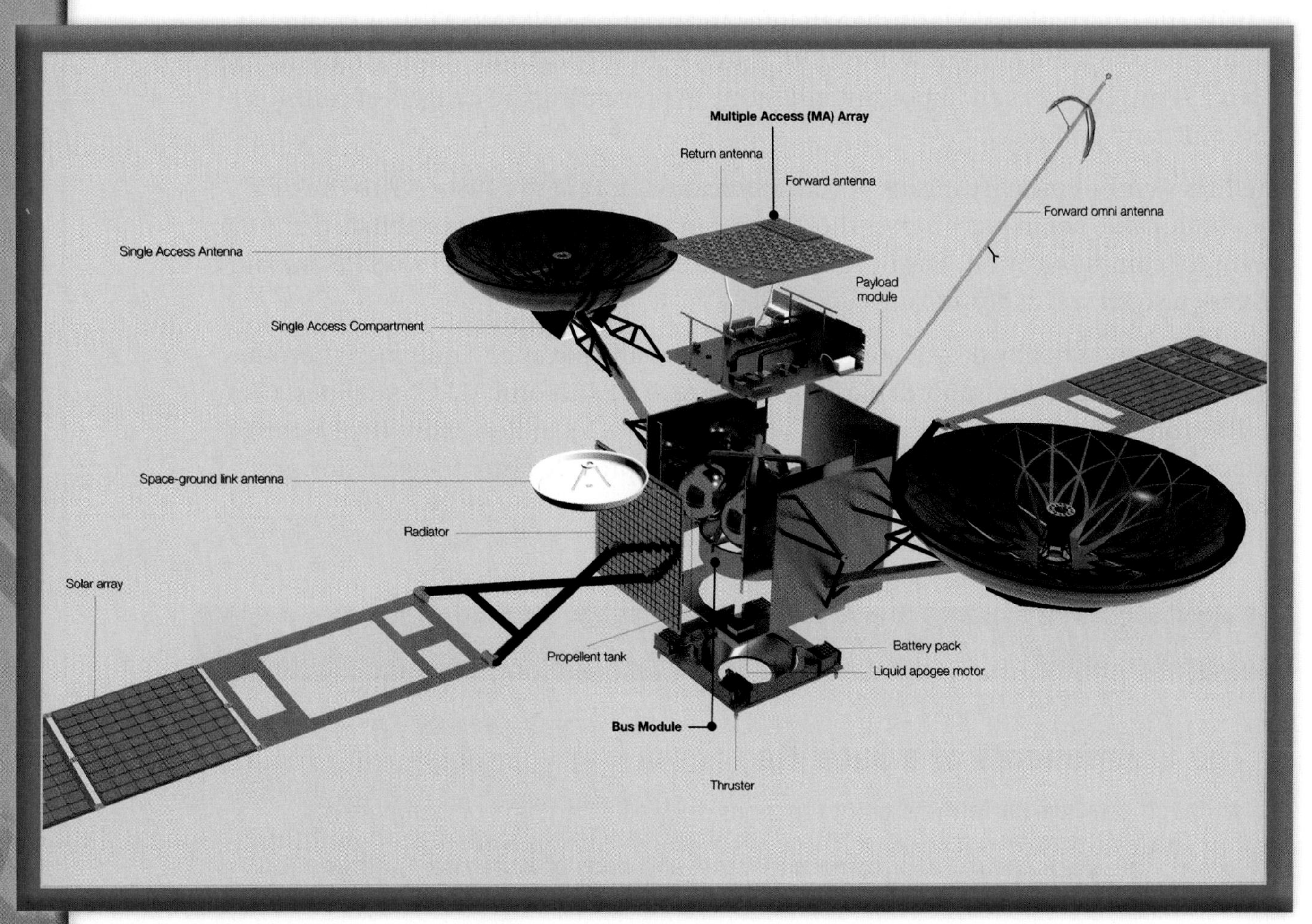

TDRS (Tracking and Data Relay) satellite with the vision of internal parts. 3D rendering - Illustration.
Courtesy of Robysot/Shutterstock

Commercial Launches

In the early 1980s, NASA was the only provider and operator of launch vehicles for the US. The number of launches was more than NASA could manage with limited federal funding. We then saw the commercial development and private operation of space launch vehicles take off. The first private rocket launch took place on September 9, 1982. The Conestoga was the prototype for the Space Services Inc. of America (SSIA).

Federal legislation further propelled the private sector into space exploration. President Ronald Reagan signed the Commercial Space Launch Act of 1984. He approved other legislation that encouraged the development of commercial expendable launch vehicles, orbital satellites, and operation of private launch services.

Through the 1980s and early 1990s, more policies were put in place. They required NASA and other government agencies to buy launch services from commercial companies. Through the 1990s and early into 2000, commercial launches surpassed the number of NASA launches.

Today, the private sector has successfully launched several small satellites without the assistance of any governmental agencies. This shift represents a new era in space exploration and a new version of the space race among private companies.

How Satellites are Used Every Day

Without us knowing it, satellites impact our lives almost every minute of every day. We use them for communications, weather, navigation, computing and more. There are a variety of satellites currently in orbit:

- Weather satellites
- Communications satellites
- Broadcast satellites
- Scientific satellites
- Navigation satellites
- Rescue satellites
- Earth observation satellites
- Military satellites

Weather Satellites

If you check the weather before heading out, you rely on satellites. From hundreds—sometimes thousands—of miles above Earth, these satellites use special tools and sensors to gather information about our weather.

Weather satellites collect data on temperature, precipitation, clouds, and wind. They also help us monitor pollution, smoke, snow, ice, and other environmental data. Weather satellites use radiometers to scan the Earth. Radiometers are *instruments for detecting or measuring the intensity or force of radiation*. Radiometers record the visible, infrared, or microwave radiation in the form of electrical voltages. The satellites convert the electrical voltages to digital images, which are then sent to weather forecast centers.

We use two types of weather satellites to collect data and interpret the weather: geostationary operational environmental satellites (GOES) and polar operational environmental satellites (POES). Remember, geostationary-orbiting satellites are located over the equator. The GOES typically orbit at a very high altitude (22,500 miles) and are used for most of the satellite weather images you see. However, there are disadvantages to GOES. The high altitude requires elaborate telescopes with detailed imaging capabilities to capture Earth at high resolution. Another disadvantage is that GOES only monitors a portion of the Earth. For example, GOES-East and GOES-West cover most of the western hemisphere, but the ESA's Meteosat satellite provides coverage of Europe and Africa.

Polar-orbiting satellites are *satellites that orbit across the Earth's poles*. Polar-orbiting satellites are often called sun-synchronous orbits as well. The POES complements the GOES system and orbits the North and South poles approximately every 100 minutes. The POES allows for complete coverage of the Earth at a closer altitude (520 miles). The disadvantage of the POES is that it only allows imaging of a location every 12 hours.

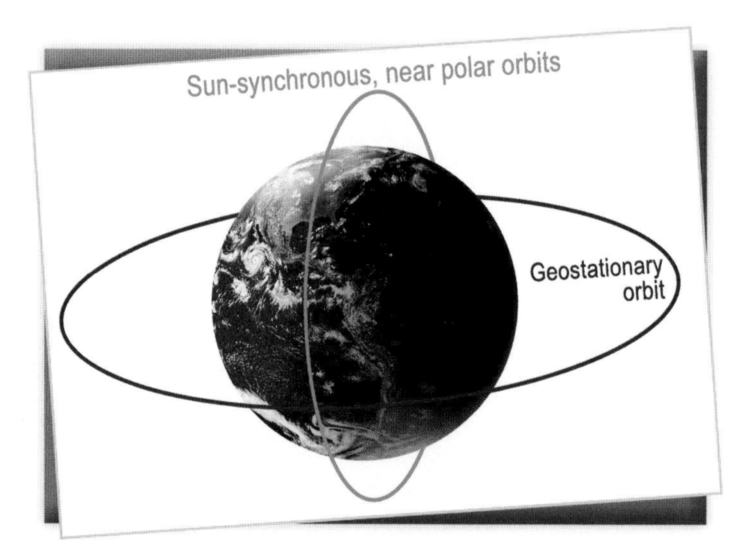

Difference between geostationary and Sun-synchronous orbits.

There are currently two POES in Sun-synchronous orbit so that images are updated every six hours. Because POES provides a full view of the weather on Earth, we often use these images to forecast the weather.

After weather satellites are launched in the US, the National Oceanic and Atmospheric Administration (NOAA) controls them. With the satellites' transmission of data, NOAA analyzes the information to forecast and predict our weather.

Image taken from NOAA's GOES-16 weather satellite. The image shows North and South America and the surrounding oceans.
Courtesy of NOAA/NASA

Did You Know?

The National Weather Service (NWS) uses satellite data to identify quickly cooling cloud tops. These events suggest the presence of snow squalls, or sudden bursts of heavy snowfall that result in near-zero visibilities and gusty winds. With the use of satellite data, forecasters can make life-saving warning decisions.

Cellular and Satellite Communications

We use our cellular phones every day for communication. We take pictures, store information, access the internet, watch movies, and the list goes on. These functions all depend on cellular communications. Does cellular communication use space satellites? The answer is surprisingly no. So, how do they work? Cellular communications *use on-land towers to send and receive signals*. A group of towers, or cells, make up a cellular network, which can be used for voice, data, and other types of content. When we use cellular phones to call friends or to navigate using GPS, our phones transmit and receive data from the nearest cell tower. Some phones, however, contain an actual GPS chip that can locate GPS satellites.

In comparison, satellite cellular phones *use satellites orbiting the Earth to send and receive signals*. This allows much greater coverage than a cellular phone. Why? Unlike our typical mobile phones, satellite phones send signals directly to a satellite orbiting the Earth rather than using on-land towers.

Because satellite communication systems can receive a signal over a much wider area, they are particularly useful in remote locations where cellular communications are not available. Because of the coverage capabilities, satellite phones are valuable for military and disaster relief efforts.

The satellite phone comes with a much greater price tag, though. The makers of satellite phones spend millions of dollars to support the technology with each satellite, making the cost to the consumer much great than that of a cellular phone.

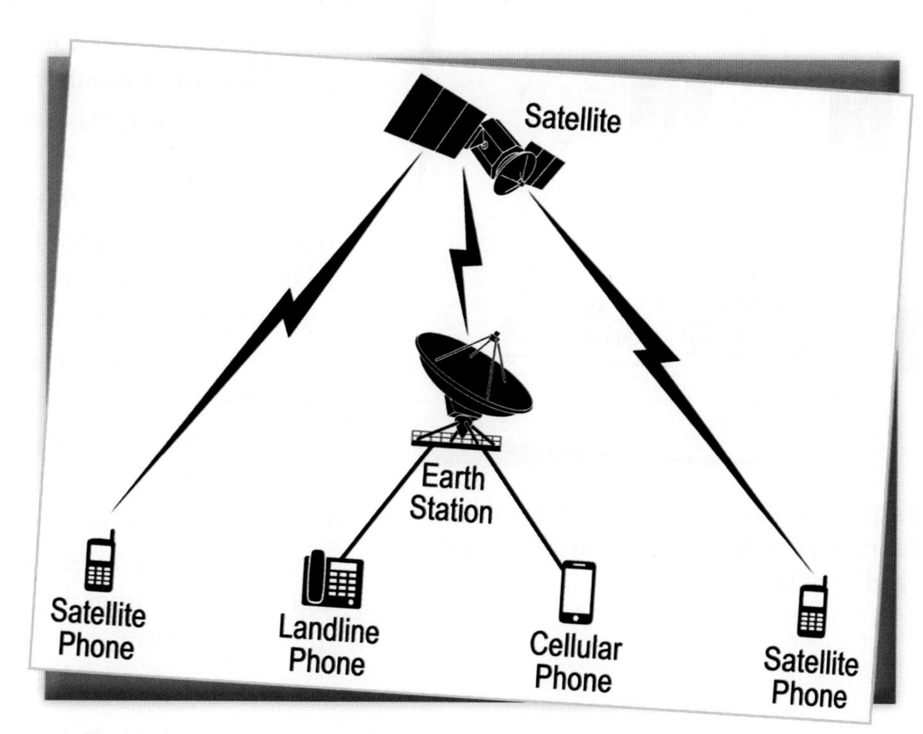

Cellular phones versus satellite cellular phones transmissions.

Did You Know?

Cellular phones may not send and receive signals directly to satellites, but they can act as satellites themselves! In a NASA-sponsored project, PhoneSat , three phones were launched into orbit and performed successfully as very tiny satellites.

Broadcast Satellites (TV and Radio)

Satellite broadcasting is the distribution of multimedia content over or through a satellite network. Television and radio companies use satellites to transmit signals worldwide. From anywhere in the world, they can send a signal across satellite dishes on the ground. The signal is then boosted across the world using geostationary satellites. Satellite broadcasting was first initiated in the 1960s, but by the 1980s a newer, more efficient method was developed. Direct-broadcast satellites (DBS) allow for direct-to-home broadcast programming using a dedicated satellite. Think of companies such as DirecTV®, DISH Network®, or SiriusXM® radio. They deliver programming by relaying a signal off a communications satellite from the broadcaster's location. This signal is received by an antenna or satellite dish, then the signal is decoded through receivers, allowing television viewing or radio listening.

Thin line flat design of satellite communication system, global network service provider, transmitting high speed internet, radio, and TV data.

Courtesy of Bloomicon/Shutterstock

Digital television is an advanced broadcasting method that has changed the viewing experience for many television customers. In fact, you may not remember a time when digital television was not available. In the past, television companies would pick up the broadcast signal from the satellite and then disperse that signal through a series of ground satellites. The signal and picture quality would suffer using these methods. Now the signal is so strong that you can pick up the satellite signal directly from your residence rather than having the television companies broadcast the signal. Most homes are set up for digital television.

The Right Stuff

Today, you probably receive internet access in your home via fiber optic or cable services. The internet signal is sent to your home using fiber optic or cable wiring and then transferred into wireless internet by a router. If you live in a rural area, you may have satellite internet instead. Satellite internet is received directly from space using a satellite installed on your home.

Elon Musk, CEO of SpaceX , suggests a new method of accessing the internet called "space internet." The SpaceX project, known as Project Starlink, has permission from the Federal Communications Commission (FCC) to set up a series of low earth orbit (LEO) satellites. The lower orbit was selected to reduce the risk of space junk collision. The satellites will provide direct Wi-Fi internet access to Starlink customers. Musk aims to have half of all internet traffic go through the Starlink satellites; however, others feel Starlink will most likely appeal to high-internet users who will pay big for faster connections.

In the initial phases, Starlink will launch 4,425 satellites. The satellites will use lasers to send signals between one another and then beam the signals to ground stations using radio waves. The internet speed will be about twice as fast as current fiber optic networks. However, launching a single satellite costs tens of millions of dollars, so the cost for the service will be pricey. And because each satellite only lasts a few years, new satellites launches would need to occur every few weeks to keep up with the satellites needing replaced. But think of the possibilities! Global coverage would provide internet in locations that currently cannot receive it. Internet would be available in remote areas and at sea. Starlink satellites begin launching in 2019.

Navigation Satellites

Mass transportation systems like trains, buses, and airplanes use Global Positioning Systems (GPS) to ensure safe and efficient travel. First responders use GPS to quickly identify the location of emergency situations. Search and rescue teams also depend on the life-saving technology. It helps them find their way to disaster-relief sites and the individuals in need of care. Today in modern automobiles and smart phones GPS can be activated to provide navigation assistance to any location in the world.

How does it work? GPS, of course, is another function of satellites. A GPS receiver looks for one of at least four visible satellites among a constellation of 30 or more satellites orbiting above Earth at about 12,000 miles. Each satellite provides data about its position and time. Satellites transmit data to a GPS receiver at the speed of light. The receiver determines its distance from the satellites based on how long it took to receive the signals.

More specifically, GPS uses a process called trilateration. Trilateration is *when a GPS receiver calculates the distances to satellites by measuring angles*, as shown in the illustration. This gives us data accurate within five meters!

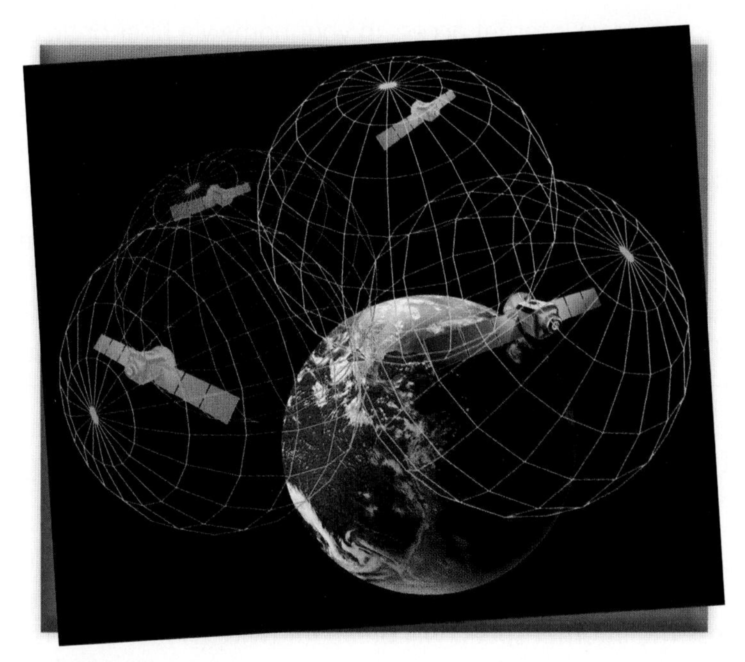

GPS trilateration.

GPS Uses

GPS was developed by the US Department of Defense during the Cold War as a means to fire ballistic missiles accurately from a submarine. In the early 1990s, the US Air Force launched a navigational satellite into orbit that completed development of the Navstar constellation of satellites, or GPS as we now know it. GPS remains owned and operated by the US Department of Defense. It gives us essential positioning capabilities for the military, and it is free to users around the world.

Today, we use GPS receivers most commonly for pinpointing our locations on Earth and for navigating from point A to point B. Simply enter an address into Google Maps, or similar navigational program, and you're there!

As powerful as GPS is for navigating our world, there are many other uses. Aviation and boating industries use GPS for positioning and tracking. The technology allows them to determine the best and safest routes possible. Farmers use GPS zone maps for accurate planting and harvesting.

Other uses include surveying, tracking for law enforcement, animal tracking, emergency services, public safety and disaster relief. Personal uses of GPS continue to expand with technology development. Imagine, in the near future, cars that drive themselves. With GPS and some innovative auto manufacturing, robotic cars can use GPS to navigate safely to their destinations.

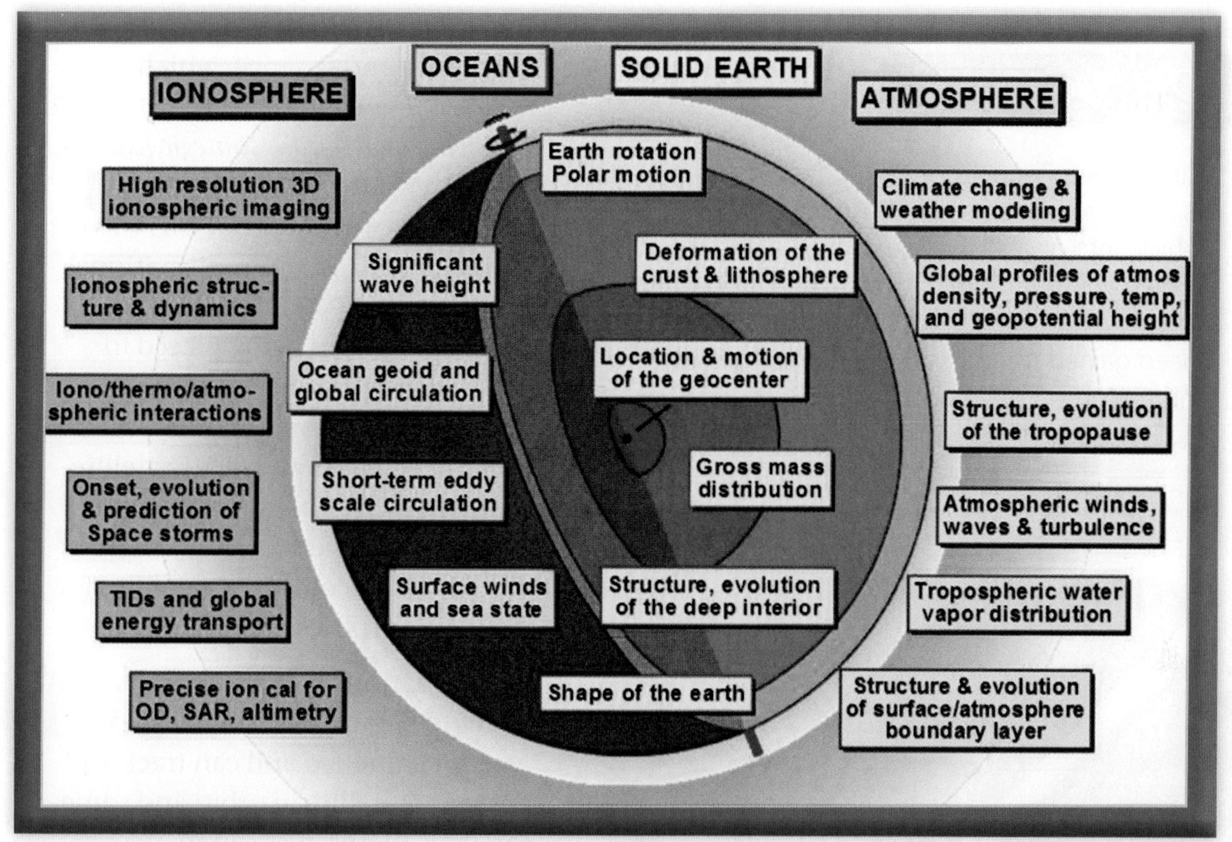

NASA GPS applications enable greater spacecraft autonomy and more advanced space science and Earth monitoring applications.

Courtesy of NASA

Did You Know?

In the 1939 film, The Wizard of Oz, Dorothy finds her way back home thanks to her ruby slippers. Turns out that was not such a far-fetched idea! Thanks to GPS-enabled shoes like GPS SmartSole®, wearers can navigate to any place on Earth. For young children and seniors with memory issues, these special shoes can be lifesaving!

Military Satellites

Hundreds of military satellites orbit the Earth. The most common missions are intelligence gathering, navigation, and military communications. The exact number of military satellites is not known due to secrecy, and some serve a dual purpose serving military and civilian organizations.

Did You Know?

The People's Republic of China demonstrated its ability to launch a land-based kinetic kill vehicle into a satellite when it fired a missile into an aging satellite and destroyed it on January 11, 2007. Kinetic kill vehicles use the speed gained through acceleration to destroy another object on impact.

1983 military artist's concept of an anti-satellite missile after being launched from an F-15 Eagle aircraft.

Courtesy of Everett Historical/Shutterstock

The first use of military satellites was for reconnaissance, which is *the military observation of a region to locate an enemy or ascertain strategic features.* There were attempts to develop satellites as weapons, but this was halted in 1967 with a multinational treaty banning weapons of mass destruction from being placed in space. However, this has not fully prevented the development of weaponized satellites. Anti-satellite weapons (ASAT) are types of space weapons designed to incapacitate or destroy other countries' satellites for strategic purposes.

Several nations have operational ASAT systems; some weapon systems are ground launched and can track and destroy a satellite in orbit; and some satellites themselves can maneuver in space to destroy or incapacitate another satellite. Intercepting a satellite of another country could seriously hinder that country's military operations. There are many ways to destroy or incapacitate a satellite, something every country takes seriously.

How Much Does a Satellite Cost?

Satellites are complex pieces of equipment that contain expensive technology and must last in space for long periods of time. Satellites come with a high price tag! A typical weather satellite costs around $290 million, while a spy satellite could cost around $390 million. A company that has a satellite in space must also pay for satellite bandwidth, which is the cost to maintain satellites. Think of this like your cell phone bill. You purchase your phone and then pay a monthly fee to maintain services on the phone. The bandwidth costs for a satellite are approximately $1.5 million per year.

The development and maintenance costs are not the only cost considerations for a satellite. Launching a satellite into space is quite expensive, as well. Launch of a satellite can cost between $10 to $400 million.

Space Technology in Everyday Use

Since its beginning, NASA has provided leading-edge innovations for space exploration and discovery. New information about scientific principles in our solar system can be used to create new products here on Earth.

Consider the fields of transportation, communications, construction, and food processing. Many of the new products we enjoy today come from NASA research. Stronger, longer-lasting radial tires for cars, GPS technology for navigation, highway safety grooving, and improved baby formulas are just some examples of NASA technology transfer. Additional examples include life-saving insulin pumps, water purification systems, memory foam, and athletic shoes. All of these products come from NASA research!

Not only do these products benefit us individually, they also help to bring us all together as one large human society.

So how do NASA innovations make their way to the private sector? The Space Technology Mission Directorate (STMD) manages NASA's Technology Transfer (T2) program. This program shares the innovations developed for exploration and discovery with the public. The program opens up its patent portfolio , containing nearly 2,000 NASA technologies, its software catalog (with over 1,000 programming codes), all for free.

The program also offers almost 2,000 examples of technology transfer through its Spinoff website and publication. Spinoffs *are the technologies, products, and processes based on NASA innovations.*

Aside from the patent portfolio, software catalog, and Spinoff website, there's more! The T2 University program brings NASA technologies to the classroom for aspiring businesspeople. Interested students build business proposals using T2 research and content to plan commercial applications. The program has already generated success. Building off of the NASA-patented sensor designed for space shuttle orbiter window inspection, a team of graduate business students successfully licensed the sensor as a new commercial product. Royalties from the product are collected by NASA to fund additional technology research and development.

Lesson 3 Review

Using complete sentences, answer the following questions on a sheet of paper.

1. What are satellite relays and how do they work?
2. Why was the Early Bird satellite developed?
3. How did INTELSAT satellites work?
4. What is the difference between a geostationary-orbit and a polar-orbit?
5. How do satellite cellular phones compare to typical mobile phones?
6. How does the process of trilateration work?
7. Who owns and operates GPS in the United States?
8. What are the most common missions for military satellites?
9. How does NASA's Technology Transfer (T2) program share technology innovations?

APPLYING YOUR LEARNING

10. Based on what you have discussed in this lesson, how do you think satellites can be used to improve everyday life?

CHAPTER 6

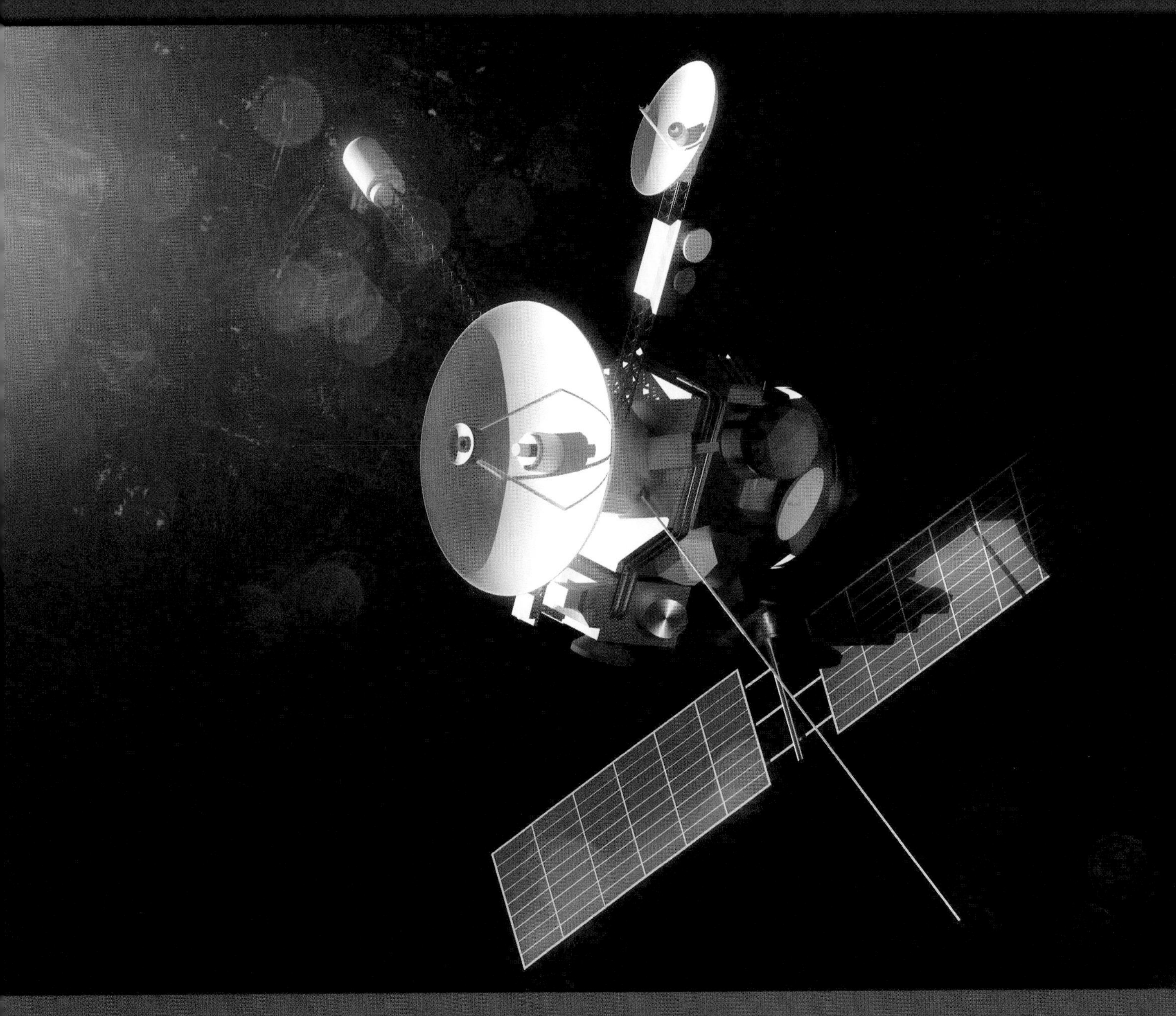

An illustration of a space probe exploring the solar system.
Courtesy of solarseven/Shutterstock

Space Probes and Robotics

Chapter Outline

LESSON 1

Space Probe Missions

LESSON 2

Robotics in Space

LESSON 3

The Mars Rover Expedition

"The people who created this amazing mission of exploration chased their new horizons hard; they never let go of their dream; they put everything they had into it; and eventually they chased it down and accomplished what they set out to do."

Alan Stern

Author of Chasing New Horizons: Inside the Epic First Mission to Pluto

LESSON 1 Space Probe Missions

Quick Write

What do you think we can hope to learn from New Horizons as it continues its journey into space?

Learn About

- missions to the Sun
- missions to the Moon
- missions to Venus
- missions to Mars
- interstellar missions

In the wee hours of New Year's Day 2019, scientists at NASA were celebrating more than a new year. The New Horizons spacecraft reached the most distant target ever explored in history. NASA had successfully gone farther into space than anyone ever before.

The target was Ultima Thule, which is in the Kuiper Belt. This region contains objects that will help unlock the origins of the solar system. Four billion miles from the Sun, the New Horizons spacecraft took images of Ultima Thule. The 19-mile long object looked like a snowman!

It takes about six hours for a radio signal to travel back to Earth, and it is estimated that it will take approximately 20 months to send all the data retrieved from the flyby of Ultima Thule. New Horizons entered a pause in data transmission shortly after the flyby while it passed behind the Sun. On January 10, 2019, the data began to flow back to Earth, and scientists are learning new information on a daily basis. With the data collected from the flyby, NASA researchers have found water and organic molecules on the surface of Ultima Thule. Organic molecules are the kind normally found in living systems, providing evidence of life, this discovery is advancing theories of how the solar system was formed.

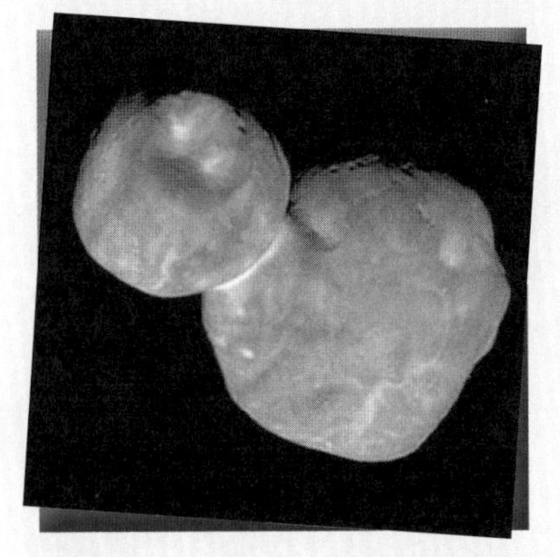

This image of 2014 MU69, also known as Ultima Thule, was captured by New Horizons' MVIC camera on January 1, 2019, and stored in the probe's memory banks.

Courtesy of NASA / JHUAPL/SwRI Photo

Since the discovery of the cosmic snowman, New Horizons has traveled 18.5 million miles past Ultima Thule. Scientists have their eyes set on a more distant object in the Kuiper Belt for a possible flyby in the 2020s. The spacecraft has enough fuel left to make it well into the 2030s, so many more firsts may be in store for New Horizons.

Missions to the Sun

We have discussed manned missions into space, as well as satellites in space. Now we will change gears to discuss space probes. A space probe is *a spacecraft that travels through space to collect science information*. Space probes do not have astronauts, and they transmit data back to Earth for scientists to study.

Sending a space probe to the Sun is complicated. The Sun is a red-hot ball of fire, so space probes can't land on the Sun or get too close or they will burn up. Space probes for missions regarding the Sun are typically launched into space to orbit the Sun and/or Earth and measure solar radiation and solar activity.

NASA's Pioneer probes were the first probes sent into orbit to study the sun. Pioneer 6 was launched on December 16, 1965 into solar orbit. Pioneer 6 was designed to study the positive ions and electrons in solar wind, solar cosmic rays, and the interplanetary magnetic field. Pioneer 6 is the oldest NASA spacecraft still in existence. That's right . . . Pioneer 6 is still orbiting the Sun! The primary transmitter on Pioneer 6 sent data back to Earth before failing in December 1995. In July 1996, the spacecraft was commanded to switch to the backup transmitter. The backup transmitter successfully became active along with some of the scientific instruments on board. Although contact with Pioneer 6 is not maintained anymore, scientists were successfully able to communicate with the probe for two hours on December 8, 2000 to commemorate its 35th anniversary.

Pioneer 7 was launched into solar orbit on August 17, 1966 and was successfully tracked through March 1995. Pioneer 8 and Pioneer 9 were the last probes in the Pioneer mission to be launched into solar orbit. Pioneer 8 is still functioning, but Pioneer 9 failed in 1983. The information sent back from these probes provided information on how solar activity affects the Earth.

In 1974, NASA and West Germany joined forces to study solar wind, magnetic and electric fields, cosmic rays, and dust in interplanetary space. Helios 1 was launched on December 10, 1974 and Helios 2 was launched on January 15, 1976. The probes were given an elliptical orbit around the Sun that at times took them very close to its surface. During the mission, Helios 1 discovered that the Sun has 15 times more micrometeorites than Earth. If you recall from chapter 3, micrometeorites are small pieces of asteroids and comets that cannot be tracked.

Vocabulary

- space probe
- solar maximum
- galactic radiation
- gamma-ray burst
- flyby
- Earth-Moon libration orbit
- magnetosphere
- aerobraking
- rover
- heliopause

Ulysses

Ulysses was a five-year mission by NASA and the ESA to study the north and south poles of the Sun. The probe launched in 1990 so that it would be in solar orbit during the 2001 solar maximum. A solar maximum is *the point in the 11-year solar cycle when the sun is most active*. The mission produced enormous amounts of data on solar activity. And it outlived its expected lifetime by over 13 years.

During its 18-year lifespan, Ulysses orbited the Sun three times. Ulysses was the first spacecraft to study the north and south poles of the Sun. Because the mission lasted so long, scientists were able to study the poles during both calm and turbulent periods of solar activity.

An artist's illustration of Ulysses.

Courtesy of NASA

Ulysses revealed the dimensions of galactic radiation, which are e*nergetic particles produced in solar storms and the solar winds*. Ulysses also allowed scientists to gather the first ever measurements of interstellar dust and helium atoms in the solar system. The research led to the discovery that the magnetic field leaving the Sun is balanced across latitudes.

And in May 1996, Ulysses had an unexpected encounter with the Hyakutake comet proving that the tail of the comet was much longer than scientists originally thought.

Ulysses is part of a network of satellites and probes that are coordinated to detect gamma-ray bursts A gamma-ray burst is *a short-lived burst of energetic explosives in distance galaxies*. They are the brightest electromagnetic events in the galaxy. Ulysses has detected thousands of gamma-ray bursts in its lifetime and on January 31, 2000, it detected the most distant gamma-ray burst ever recorded. Scientists believe that gamma-ray bursts occur quite frequently in the universe; in fact they believe that one occurs every minute in the universe. Some of the bursts occur when rapidly moving stars fall into black holes. The bursts can be used to study the universe during its infancy.

On November 21, 2001, Ulysses was credited with detecting another gamma-ray burst. But why was this detection so special? The Mars Odyssey on Mars's surface also detected the gamma-ray burst. Gamma-ray bursts are typically very short and only last a few milliseconds so it is very difficult to locate the source. Because both Ulysses and the Mars Odyssey detected the burst, scientists could use the data to help pinpoint the location of the gamma-ray burst. If NASA has three satellites or probes that are detecting gamma-ray bursts they can use triangulation to identify the source of the burst. Spacecraft operating close to Earth that have taken on the role of detecting gamma-ray bursts include NASA's WIND and HETE-II satellites, the Italian-Dutch BeppoSAX, the Indian satellite SROSS-C2, the Japanese X-ray observatory Yohkoh, the Chinese SZ-2 mission, and ESA's gamma-ray observatory Integral.

Hinode

An artist's illustration of the polar orbit of Hinode.

Courtesy of ESA - AOES Medial

Hinode is led by JAXA with collaboration between Japan, NASA, the United Kingdom, and Europe. NASA helped in the development, funding, and assembly of the spacecraft's three science instruments. Hinode is part of the Solar Terrestrial Probes (STP) program within the Heliophysics Division of NASA's Science Mission Directorate in Washington. The Solar Terrestrial Probes program is managed at NASA's Goddard Space Flight Center in Greenbelt, MD.

Hinode was launched in 2006 into a Sun-synchronous orbit around Earth. Remember, a Sun-synchronous orbit is an orbit in which the probe passes over the same part of the Earth at roughly the same time each day.

Hinode is equipped with three telescopes that are used to study eruptive phenomena on the Sun and to study space weather. The goal of Hinode is to learn to predict the Sun's influence on Earth. The three telescopes are:

- The Solar Optical Telescope - a specialized suite of instruments that will precisely measure small changes in the Sun's magnetic field.
- The X-ray Telescope – captures X-ray images of the Sun's outer atmosphere to help scientists' study how changes in the Sun's magnetic field trigger solar flares.
- The Extreme Ultraviolet Imaging Spectrometer – measures the speed of solar particles and diagnoses the temperature and density of solar plasma.

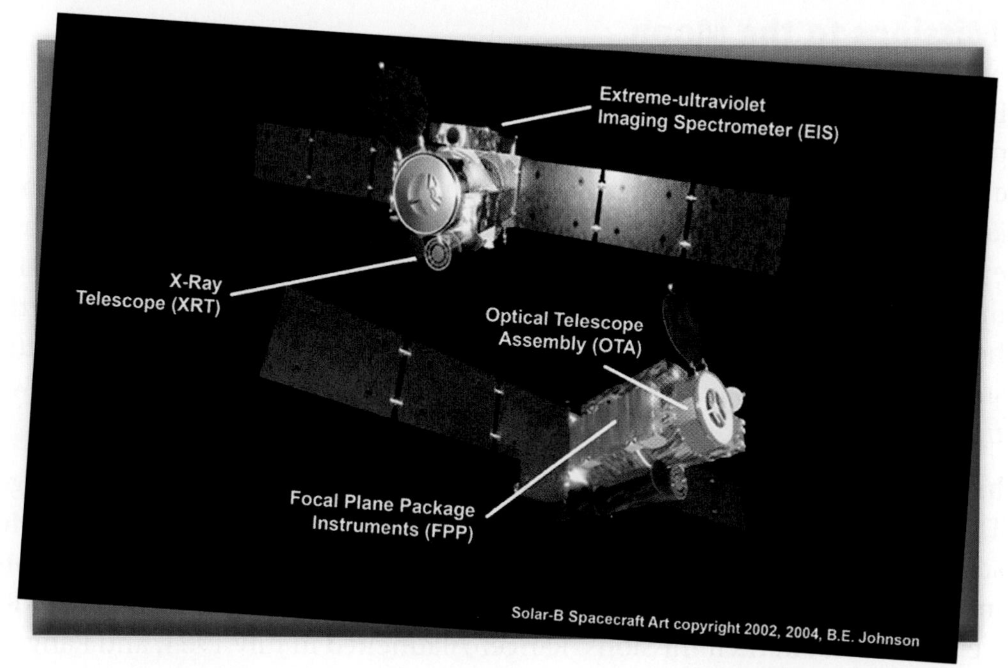

Hinode Spacecraft.

Courtesy of NASA

Parker Solar Probe

The Parker Solar Probe is NASA's latest space probe that has already made history. The Parker Solar Probe was launched on August 12, 2018, from Cape Canaveral, FL. The probe is set to revolutionize our understanding of the Sun. The Parker Solar Probe will use Venus's gravity to slowly get closer to the Sun. Over seven years, each orbit of the probe will be closer to the Sun's surface ,with the final flyby at 3.8 million miles from the surface. which is through the Sun's atmosphere.

The spacecraft will be close enough to the Sun to watch solar wind and investigate hazardous regions of intense heat and solar radiation. The Parker Solar Probe is part of NASA's "Living With a Star" program to explore aspects of the Sun-Earth system that directly affect life and society. In November 2018, the Parker Solar Probe completed its first flyby of the Sun at 15 million miles from the surface. This is the closest any spacecraft has ever gotten to the Sun. A flyby is *the close approach of a spacecraft to a planet, Sun, or Moon for observation.* The probe successfully maneuvered the orbit and protected the instruments aboard without control from Earth. The Parker Solar Probe also set a speed record. On November 5, 2018, the probe reached top speeds of 213,200 miles per hour. As the probe's orbit draws it closer to the Sun, it will continue to break its own speed records.

Illustration of Parker Solar Probe approaching the Sun.

Courtesy of NASA/Johns Hopkins APL/Steve Gribben

Missions to the Moon

Before sending a man to the Moon, the US and the Soviet Union attempted to send probes to the Moon. During 1958, several unsuccessful attempts were made by both countries to send a probe out of Earth's atmosphere. In January 1959, the Soviets launched the Luna 1 probe. Although it successfully made it out of Earth's atmosphere, errors in the control systems caused the probe to miss the Moon, and Luna 1 ended up orbiting the Sun. On September 13, 1959, Luna 2 successfully made it to the Moon; however it crash-landed on the Moon's surface. Then on October 7, 1959, Luna 3 successfully orbited the Moon and transmitted pictures back to Earth. This was the first time pictures were taken of the Moon.

In December 1959, NASA's Jet Propulsion Laboratory (JPL) initiated the Ranger project. The goals of the Ranger project were to send a probe to take pictures of the Moon's surface. The probe would continuously take pictures of the Moon's surface until it impacted with the surface of the Moon. Ranger 1-6 failed, but Ranger 7 and Ranger 8 were successful in their missions. Ranger 7 launched in July 1964, and Ranger 8 launched in February 1965. Both probes took images of the Moon's surface and sent them back to Earth. On March 21, 1965, Ranger 9 broadcast live TV footage to Earth before it crashed on the Moon's surface.

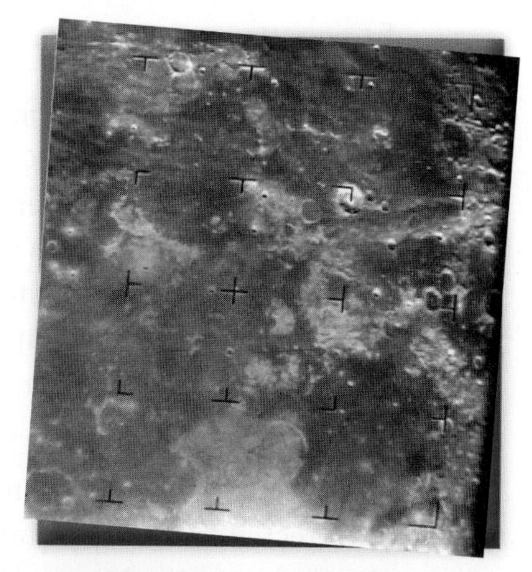

The first image of the Moon taken by Ranger 7.
Courtesy of NASA

The first successful US spacecraft to land on the Moon was the Surveyor. The Surveyor probes were used to capture close-up images of the surface of the Moon and determine if the surface was safe for manned landings. Each probe had cameras and a television camera for taking images. In addition, later Surveyor probes also had the ability to test soil samples and do chemical analysis of the lunar surface.

NASA began the Lunar Orbiter mission in 1966 and launched five orbiters over the next 11 years. The Lunar Orbiters were sent to the Moon to map the Moon's surface prior to the Apollo landings. All of the Lunar Orbiter missions were successful and NASA was able to photograph 99% of the Moon's surface.

Lunar Reconnaissance Orbiter (LRO)

The Lunar Reconnaissance Orbiter (LRO) was launched on June 18, 2009 with the goal of mapping the surface of the Moon to determine potential landing sites. The probe looked at the terrain roughness, usable resources, and radiation levels. The mission was broken into several phases:

- Phase 1: Exploration Phase – During the first year of the mission, the LRO spent one year in a polar orbit of the Moon.
- Phase 2: Science Phase – In September 2010, the LRO entered Phase 2. During Phase 2, the probe lowered its altitude to 30 kilometers and began acquiring images of the Moon's surface.

The LRO was determined to be a full mission success in 2011. The LRO provided stunning images of over 5.7 million square kilometers of the Moon's surface. The resolution of the images allowed scientists to see details of the Moon that had never been seen before.

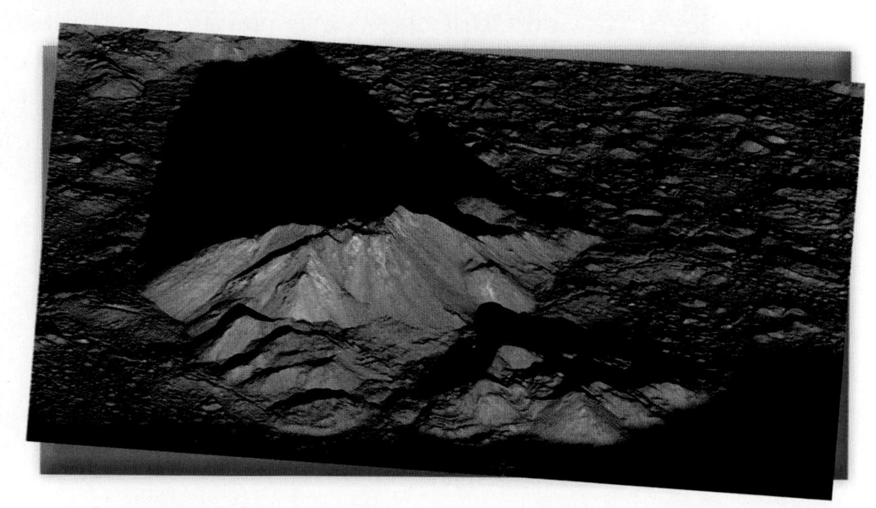

A dramatic sunrise view of the Moon's Tycho crater taken by the LRO.
Courtesy of NASA/Goddard Space Flight Center/Arizona State University

Lunar Crater Observation and Sensing Satellite (LCROSS)

The Lunar Crater Observation and Sensing Satellite (LCROSS) was launched in conjunction with the LRO. LCROSS was designed to search for water ice on the surface of the Moon. Shortly after arrival, NASA announced that LCROSS did indeed find water molecules on the Moon's surface. In addition to finding water, the LCROSS discovered useful minerals in the craters of the Moon. It also discovered water in the form of ice crystals. Finding these natural minerals and water molecules on the Moon's surface provides key information for future hopes of a lunar outpost.

At the end of its life, LCROSS strategically impacted the surface of the Moon; the impact intentionally created a dust plume so that soil underneath the surface could be resurrected and tested. The plume created vapor clouds and dust almost 10 miles above the surface of the Moon. The majority of the dust cloud was composed of pure water ice grains--which means water was either delivered to the Moon at some time by another object, or there is a chemical process on the Moon that creates water.

ARTEMIS

In 2017, NASA placed the ARTEMIS probes into orbit behind the Moon. The probe is in an Earth-Moon libration orbit, which is *an orbit that relies on balancing the Sun, Earth, and Moon's gravity so that a spacecraft can orbit a virtual location.*

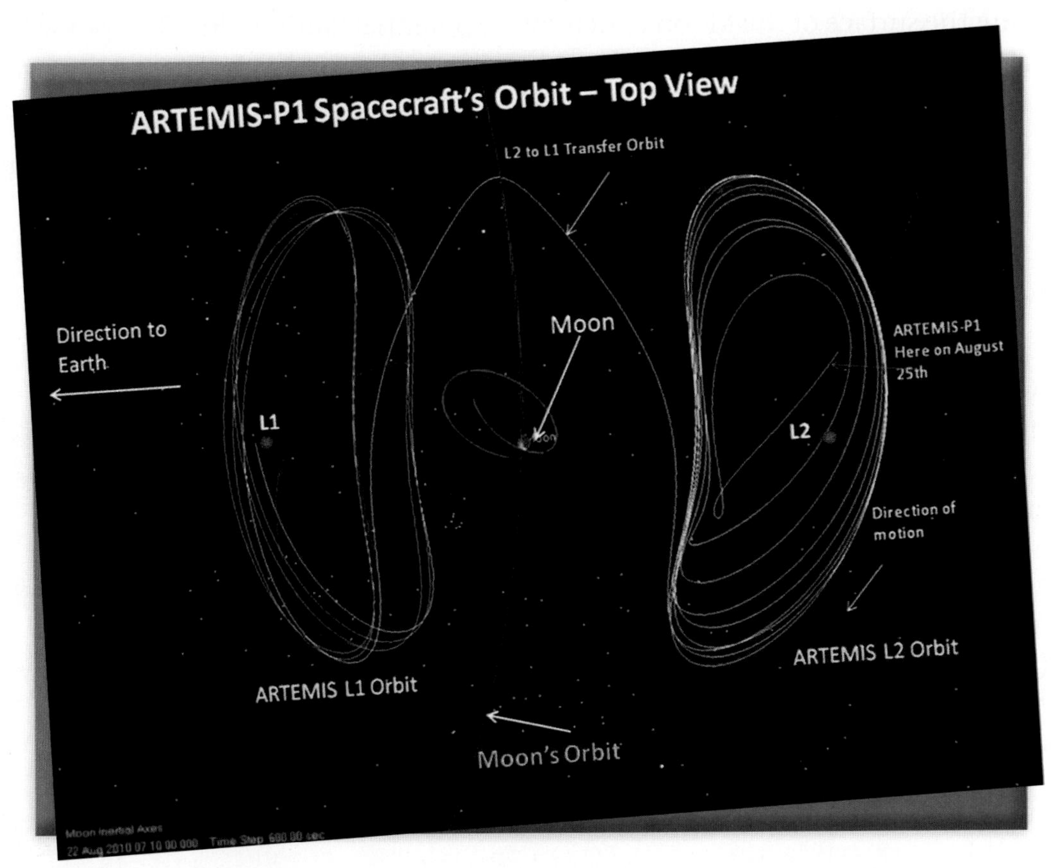

Illustration of Artemis-P1 liberations orbits..

Courtesy of NASA/Goddard

The ARTEMIS-P1 and ARTEMIS-P2 will take observations of the Moon's magnetosphere from opposite sides of the Moon for three months, then join up to fly orbits on the same side of the Moon. Magnetosphere is *the region surrounding a planet, star, or other body, in which the body's magnetic field traps charged particles and dominates their behavior*. This will allow scientists to observe space behind the far side of the Moon.

The ARTEMIS project, which stands for "Acceleration, Reconnection, Turbulence and Electrodynamics of the Moon's Interaction with the Sun" uses two of the THEMIS satellites already in orbit that completed their mission earlier in 2017. The re-use of spacecraft saved NASA billions of dollars and will produce an unprecedented amount of information.

Missions to Venus

Venus is a cruel planet. The clouds on Venus are made of sulfuric acid. And the surface temperatures are so hot that they could melt lead. And let's not forget the constant hurricane force winds. This unique situation also makes Venus an interesting planet to learn about. Landing a probe on Venus is a difficult task with the environmental conditions experienced by the planet. It takes approximately four months to reach Venus and scientists have created several unmanned missions to study the planet.

NASA's Mariner 2 flew past the surface of Venus in 1962 and transmitted information on its temperature, atmosphere, and rotation back to Earth. Mariner 2 became the first ever successful mission to another planet on December 14, 1962.

Venera

United States wasn't the only country interested in Venus. The Soviet Union initiated the Venera program in 1961 to send a series of probes to study Venus. Ten of the probes successfully landed and transmitted data from the surface of Venus. And 13 of the 14 probes successfully transmitted data from the atmosphere of Venus.

The right half of the panoramic view of the surface of Venus from the Venera 13 lander.

Courtesy of NASA

Venera 11 and 12 had gamma-ray burst detectors and identified many gamma-ray bursts during their lifetimes. Venera 13 and 14 studied Venus and the interplanetary medium. These probes discovered that gamma-ray bursts were happening every three days. Venera 13 was also the first spacecraft to transmit color photos of Venus back to Earth. The surface of Venus is very harsh and experiences extreme temperatures. The Venera 13 probe was designed to last about 30 minutes on the surface of the planet, but it transmitted images for more than two hours on March 1, 1982. No other probe has visited Venus since the 1980s.

Pioneer Venus

As part of NASA's Pioneer program, a probe was sent to orbit Venus on May 20, 1978. The main purpose of the Pioneer Venus project was to study solar wind on the planet, map the surface of the planet, and study the upper atmosphere.

Pioneer Venus was the combination of two spacecraft: the Orbiter and the Multiprobe. The Orbiter carried scientific instruments for measuring plasma in the upper atmosphere, observing sunlight and the wavelengths of light bouncing off the clouds, and a surface radar mapper. The Orbiter successfully began orbiting Venus on December 4, 1978. The instruments were intended to be operational for a year; however when the probe descended to Venus's surface in October of 1992, most instruments were still functioning. One of the key findings of the Orbiter was that its cameras detected constant lightning in the clouds of Venus. The Orbiter also supplied the first radar topographic or physical features map of the surface of Venus. Scientists discovered that Venus has a relatively smooth surface with the largest peak being seven miles high.

The Multiprobe was launched separately on August 8, 1970 and reached Venus on December 9, 1978. The Multiprobe consisted of five individual probes: the probe transporter, a larger atmospheric entry probe, and three smaller probes. The probes descended to the surface of Venus and sent information back to Earth as they descended. The probes confirmed that the clouds around Venus are composed of sulfuric acid.

The Magellan spacecraft is deployed from the cargo bay of the Space Shuttle Atlantis (STS 30) in 1989. Magellan was the first planetary spacecraft launched from the Space Shuttle.

Courtesy of NASA

Magellan

Magellan was the first US mission dedicated to studying the surface of Venus. The Magellan project was a low-budget project that used components left over from other missions. Magellan's purpose was to obtain near-global radar images of Venus's surface, obtain a topographical map of nearly the entire planet, research the planet's gravity field, and understand Venusian geological structures.

The Magellan spacecraft itself looked similar to an antenna dish that was about 3.7 meters high and 6.4 meters long. Large solar panels were attached to power the probe. Magellan successfully reached Venus on August 10, 1990. The initial phase began immediately and lasted for eight months. During this time, Magellan captured radar images of 84% of the planet's surface. The second phase of the project lasted for approximately seven months, and consisted of taking images of the planet's southern polar region. A third, eight-month phase collected data to produce stereo images of the surface. In Phase 4, Magellan acquired data on the gravity field of the planet. In the final phase of the project, Magellan used aerobraking, *a maneuver used to slow down a spacecraft by flying through a planet's atmosphere to produce aerodynamic drag,* to enter the atmosphere. For almost a month, Magellan collected high-resolution gravity data over 95% of Venus' surface. On October 11, 1994, Magellan began its final decent to Venus's surface. It is presumed that the spacecraft burned up in the atmosphere of Venus. As a result of the project, NASA successfully mapped 99% of the surface of Venus.

Venus Express

Venus Express was the European Space Agency (ESA) probe to study the Venusian atmosphere. It was launched on November 9, 2005, by the Russians. Venus Express was the ESA's first spacecraft to Venus. Remarkably, the ESA was able to define, prepare, and launch the mission in just three years. The spacecraft itself was designed to last about 33 months but operated for more than eight years.

Venus Express used a near-polar orbit, which meant it was almost locked into the space just below the planet. This allowed the spacecraft to observe almost the entire planet over a full Venusian day (243 Earth days). It also meant that the Venus Express flew very low over the northern hemisphere of Venus, which allowed high-resolution images of the clouds and surface to be captured. Venus Express focused on the structure, composition, and chemistry of the atmosphere and clouds covering Venus. Venus Express also was able to obtain a view of the complex vortex that covered the planet's south pole. Although the vortex previously appeared to remain constant, Venus Express revealed that the vortex changed shape every 24 hours and drifted around the south pole, making a full circle every 10 Earth days. Venus Express also discovered that the planet was very much like Earth in its early years. The atmosphere is still consuming water from the surface of the planet. In 2014, the Venus Express ran out of fuel and descended into Venus's atmosphere.

Akatsuki

Akatsuki is a Japanese probe launched in 2010. Akatsuki was intended to begin orbiting Venus in December 2010, but due to a failure of its orbital insertion rocket Akatsuki missed the opportunity. The probe spent five years wandering around the Sun. On December 7, 2015 Akatsuki successfully began its orbit around Venus. The goal of the project is to understand the motions of the Venusian atmosphere and identify the meteorology of Venus. Akatsuki is currently still orbiting Venus and collecting data.

Future Missions to Venus

Although no current missions are planned for Venus, that doesn't mean they aren't being discussed. NASA and Roscosmos are currently in discussions for a joint mission to Venus. The goal would be to land a probe on the surface of Venus that could survive for months. The committee is currently looking at possible mission objectives and equipment that might be needed for the probe to survive on the planet for any length of time.

Missions to Mars

Although a manned mission to Mars has not yet occurred, many unmanned probes have been sent to Mars. In the early 1960s both the United States and Soviet Union unsuccessfully attempted multiple flybys of Mars. The first successful flyby of Mars was the US spacecraft Mariner 4.

Early Exploration of Mars

Mariner 4 launched on November 28, 1964. It completed its flyby of Mars on July 14, 1965. During the flyby, Mariner 4 took 21 pictures of Mars and transmitted them back to Earth. Shortly after the successful launch of Mariner 4, the Soviets launched Zond 2. Zond 2 successfully passed Mars, but radio communications on the probe failed and no transmissions were sent back from the flyby. NASA continued sending probes to conduct flybys of Mars with the Mariner 6 and 7 probes. Each of the probes returned approximately a dozen photos.

In 1971, the Soviet Union's Mars 2 finally orbited Mars but crash-landed on its surface. The Mars 3 was more successful, as it landed on the surface and worked for a few seconds. The Soviet Union's Mars spacecraft project continued to send probes to Mars through 1973, but Mars probes 4 through 7 were plagued with failures. Only brief data was returned in 1974 before the probes failed.

Mariner 9 was the most successful of the Mariner missions. Mariner 9 was launched on May 30, 1971, and reached Mars in November 1971. Upon arriving at Mars, the Mariner 9 captured pictures of a giant dust storm covering the surface of the entire planet. But the images also showed something poking out of the top of the dust clouds. It was the tops of volcanos! Mariner 9 is also credited with discovering the Valles Marineris, which is a giant rift across the surface of Mars. Mariner 9 orbited the Red Planet for almost a year and sent back an astonishing 7,329 images. The Mariner 9 mission finally allowed scientists to view the surface of Mars. They discovered the planet was quite different from Earth and the Moon. Mariner 9 sparked even more interest in the planet and additional missions to Mars were planned.

Did You Know?

Early flybys of Mars flew over cratered areas of the surface, which provided an early impression that Mars was similar to the Moon. We now know that Mars is a multifaceted planet and very different from the Moon.

Mariner 9 spacecraft.
Courtesy of NASA/JPL

NASA continued its exploration of Mars with the Viking program. Viking 1 and Viking 2 were a pair of orbiters and landers designed to study the Red Planet. Viking 1 was launched on August 20, 1975. Viking 2 was launched on September 9, 1975. In 1976, both spacecraft arrived at Mars. The orbiters remained in Mars orbit while the landers were sent to the surface of Mars. The Viking landers survived for years on the surface of Mars and sent back an enormous amount of information and over 50,000 pictures.

The main goal of the probes was to search for signs of life on Mars. Unfortunately, the probes were not able to find any traces of microbes on the surface of Mars. Microbes would provide evidence of possible life on Mars. The Viking probes also studied the composition of the Martian atmosphere. It was determined that the composition of Mars's atmosphere was identical to some meteorites found on Earth. Thus, it was determined that the meteorites found on Earth were originally rocks formed on Mars and ejected by the impact of an asteroid or comet.

In the 1980s, the Soviet Union set its sights on Phobos, one of Mars's moons. Phobos 1 was launched on July 7, 1988, and Phobos 2 was launched on July 12, 1988. Unfortunately, both missions were unsuccessful. The landers were lost en route to Mars.

Missions to Study Mars

Seventeen years after the initial missions to study Mars, NASA launched a new mission to Mars. The Mars Observer launched on September 25, 1992. The Mars Observer was equipped with sophisticated science instruments to study the geology, geophysics, and climate of Mars. Just before the Mars Observer reached Mars's orbit, communication was lost. It is suspected that a fuel tank ruptured, causing the spacecraft to spin and lose contact with Earth. The failure of the Mars Observer was an especially hard loss due to the cost of the project. The project cost $813 million. NASA had a new mission to create "better, faster, cheaper" (BFC) missions.

MARS EXPLORATION PROGRAM

2017

EXOMARS (orbiter/lander) — 2016

ESA's trace Gas Orbiter had a successful orbit insertion.
The demonstration modular lander was lost on touchdown.

2015

2014

MAVEN (orbiter) — 2013

(Mars Atmospheric and Volatile Evolution) is obtaining critical measurements of the Martian atmosphere to help understand drastic climate change on Mars over its history and how fast gases are being lost to space today. Also able to provide communications relay support for landers and rovers on the Martian surface.

2012

CURIOSITY (rover) — 2011

Curiosity's scientific tools found chemical and mineral evidence of past habitual environments on Mars. Explores the rock records, acquires rock, soil and air samples for inboard analysis. Has 17 cameras, a laser to vaporize and study small pinpoint spots on rocks at a distance and drill to collect powdered rock samples. It hunts for and measures the chemical fingerprints present in different rocks and soils to determine their composition and history, especially their past interactions with water.

2010

2009

2008

PHOENIX SCOUT (lander) — 2007

High-resolution perspective of the landing site's geology. Provide range maps and identified local minerals. Checked samples of soil and ice for evidence whether the site was hospitable to life and scanned the atmosphere for data about the formation, duration, & movement of clouds, fog, dust, temperatures and pressure.

2006

MARS RECONNAISSANCE ORBITER (orbiter) — 2005

Detailed view of the geology and structure of Mars, identifying obstacles that could jeopardize the safety of future landers and rovers. Identifies surface minerals and studies the atmosphere. Carries a sounder to find surface water.

2004

2003

SPIRIT & OPPORTUNITY (rovers)

Field geology and atmospheric observations have found evidence of ancient Martian environments where intermittently wed and habitable conditions existed and could have supported microbial life. Provided high-res, full-colored images of terrain, rocks and soil. Analyzed chemical and mineralogical makeup of rocks and soil and examine the interior of rocks.

MARS EXPRESS (orbiter)

Participating with ESA and ASI exploring the atmosphere and surface from polar orbit. Conducted investigations to help answer fundamental questions about the geology, atmosphere, surface environment, history of water and potential for life on Mars. Discovered evidence of recent glacial activity, explosive volcanism, and methane gas. Provided information about features beneath the surface as well as coordination of radio relay systems.

2002

MARS ODYSSEY (orbiter) — 2001

Measurements to create maps of minerals and chemical elements and identified regions with buried water ice. Measured surface temperature and views of topography. Data regarding radiation in low-Mars orbit for eventual human exploration and potential health-effects. A communication relay for rovers and landers on Mars.

2000

1999

MARS POLAR LANDER (lander)

Lost on arrival.

DEEP SPACE 2 (probe)

Lost on arrival.

MARS CLIMATE ORBITER (orbiter) — 1998

Lost on arrival.

1997

1996

PATHFINDER (rover)

Returned images from the lander and rover, chemical analyses of rocks and soil, data on winds and other weather factors. Findings suggest Mars was at one time in its past warm and wet, with water existing in liquid state and thicker atmosphere.

MARS GLOBAL SURVEYOR (orbiter)

Studied the entire Martian surface, atmosphere, and interior. Observed that Mars has repeatable weather pasterns. Documented gully formation and debris flows. Showed the planet does not have a global magnetic field but localized magnetic fields in areas of the crust. Determined Phobos is covered by a large layer of powdery material from meteoroid impacts. Observed new boulder tracks, recently formed impact craters., and diminishing amounts of carbon dioxide ice within the south polar cap. Provided the first 3-D views of the north polar ice cap. Scientists created vertical profiles of atmospheric temperature and pressure from changes in radio transmissions. Shown that Mars has seasonal and long-term change recorded on the surface.

1995

1994

1993

MARS OBSERVER (orbiter) — 1992

Communication lost prior to orbit insertion.

1991

Timeline of the Mars Exploration program.

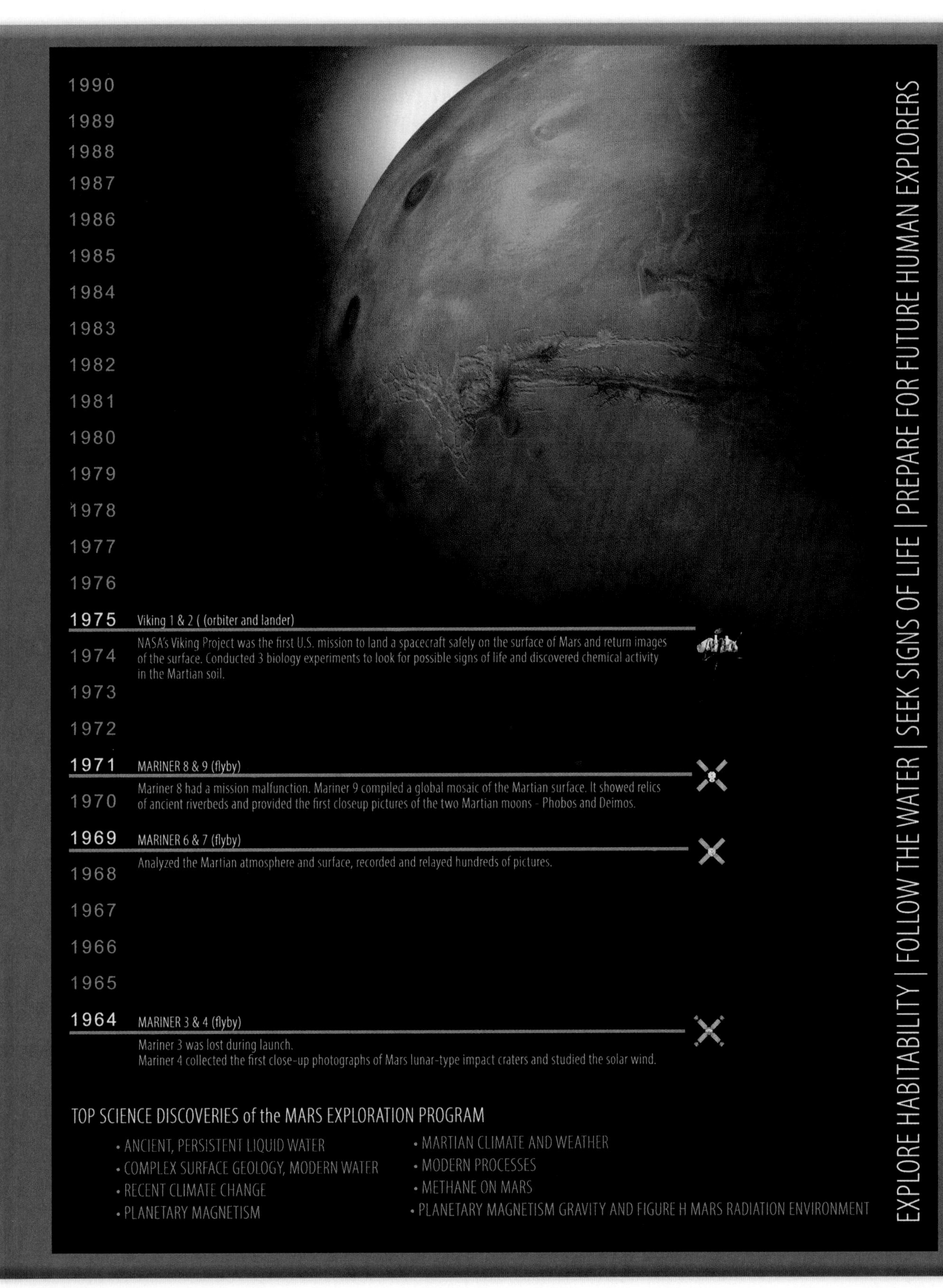

Courtesy of NASA

NASA conducted many missions to Mars using space probes that orbited the planet or with landers that landed on the surface to take images and measurements. But the next step would be to send a rover. A rover is *a space exploration vehicle that moves across the surface of the planet*. The first NASA rover was the Pathfinder mission. On July 4, 1997, Pathfinder arrived on Mars. Pathfinder consisted of a lander and a small rover named Sojourner. Pathfinder was NASA's first mission to reach the surface of Mars since Viking 1 and Viking 2. Sojourner was about the size of a microwave oven and traveled the surface of Mars for three months gathering data about the atmosphere, climate, geology, and interior of Mars.

On September 12, 1997, NASA's Mars Global Surveyor arrived at Mars. The Mars Global Surveyor was a huge success. The probe successfully mapped the surface of Mars from one pole to another. The most astonishing discovery was the sign of water. Minerals that form in water were discovered on the surface of Mars. The data garnered from the Mars Global Surveyor was also used to determine where to land rovers on the planet. In addition, the project produced many pictures of Mars for public distribution.

With the discovery of ancient water, the spark to explore Mars was reignited. Mars Odyssey became NASA's next mission. It launched on March 7, 2001, and reached Mars in October 2001. The orbiter is the longest-serving spacecraft on Mars and is still conducting science experiments on the planet. Mars Odyssey mapped elements and their locations across the planet and returned over 350,000 images of the planet. In addition, the Mars Odyssey orbiter is used to send back 95% of the data collected by the Spirit and Opportunity rovers. The Spirit and Opportunity were the first two rovers in the Mars Rover project. They were sent to the surface of Mars in 2004 to search for evidence that water was once present on the surface of the planet. They both successfully found evidence of water previously being on the surface of Mars. The Spirit rover died in March 2010, but the Opportunity rover is still functioning on the planet.

On August 12, 2005, NASA launched the Mars Reconnaissance Orbiter (MRO). MRO orbits Mars and sends back high-resolution images of the planet's surface. The MRO conducts searches for water and ice and also searches for landing locations for future rovers. MRO has produced more data on Mars than all the previous NASA missions to Mars. MRO is currently still orbiting Mars, and NASA plans to operate the spacecraft through the mid-2020s to support the Mars 2020 Rover.

On August 4, 2007, NASA launched the Mars Phoenix lander. The Phoenix would be a stationary lander that would study the history of water on the planet and search for a possible habitable zone. Phoenix landed on Mars on May 25, 2008. The lander successfully located water ice under the surface of Mars. Due to the harsh environment on Mars, the solar panels on the Phoenix suffered damage and become inoperable in November 2008. After several attempts to re-establish contact, Phoenix was considered a loss in May 2010. The $475 million project had a short life-span, but produced a great deal of information on Mars. In addition to digging and testing the soil, Phoenix took over 25,000 images of the planet. It also conducted scientific experiments on the planet on 149 of the 152 days it was operable. Additional findings include documenting a mildly alkaline soil environment unlike any found by earlier Mars missions; finding smallconcentrations of salts that could be nutrients for life; discovering perchlorate salt, which has implications for ice and soil properties; and finding calcium carbonate, a marker of effects of liquid water.

The next NASA project would be another rover. The Curiosity rover arrived on Mars in 2012 with the mission to search for signs of previous habitable areas of the planet. Curiosity discovered that there were areas of Mars that were previously covered by water. Curiosity also discovered methane on the surface of the planet and organic molecules that are often considered the building blocks of life. The Curiosity rover is still traveling the surface of Mars and sending data back to Earth. We'll learn more about the Mars rovers in Lesson 3.

Curiosity self-portrait at Okorusko drilling hole.
Courtesy of NASA

Interstellar Missions

As you have already learned, NASA and other space agencies have spent a great deal of time exploring the Sun, Moon, Venus, and Mars. But what about the other planets? It takes years to reach the outer planets of our solar system, but many missions have been successful. In total, six missions have been launched by NASA to study the outer planets.

- Pioneer
- Voyager
- Galileo
- Cassini
- New Horizons
- Juno

Pioneer

Pioneer 10 was launched March 2, 1972, and was the first spacecraft to go beyond the asteroid belt. It was also the first spacecraft to gain an up-close observation of Jupiter. During launch, rockets provided enough boost to allow Pioneer 10 to travel at 32,400 mph on its journey to Jupiter. With its launch Pioneer 10 became the fastest object to ever leave Earth. Pioneer 10 passed Jupiter on December 3, 1973, and obtained the first ever close-up images of the planet. In addition, Pioneer 10 mapped the radiation belts of Jupiter, identified the magnetic field of the planet, and discovered that Jupiter is mostly covered in liquid.

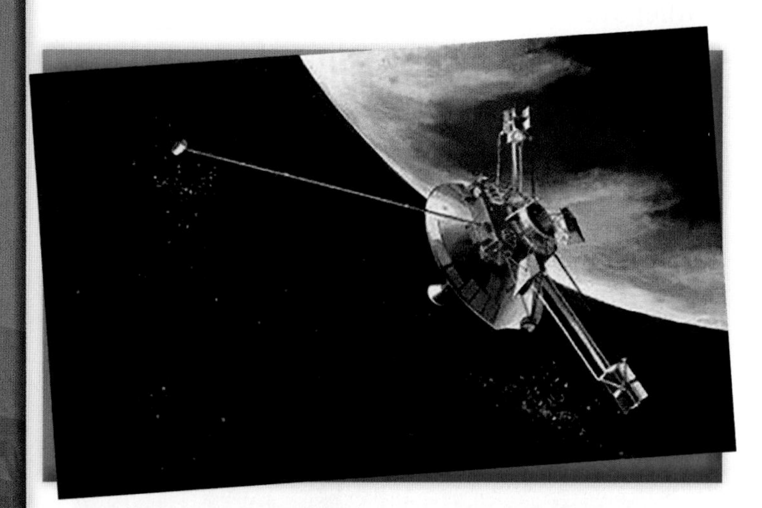

Artists illustration of Pioneer 10 orbiting Jupiter.
Courtesy of NASA

After Pioneer 10 passed Jupiter, it continued its journey into the solar system, studying energy particles from the Sun and cosmic rays entering the galaxy. The science portion of the Pioneer 10 mission came to an end on March 31, 1997. The signal from Pioneer 10 became too weak to continue operations. And after 30 years of transmissions, the last transmission from Pioneer 10 came on April 27, 2002. The power source for Pioneer 10 decayed over its lifetime and it no longer became possible to detect a signal from Pioneer 10. It is estimated that Pioneer 10 is now eight billion miles away from Earth.

Pioneer 11 launched on April 5, 1973 and was the sister spacecraft to Pioneer 10. Pioneer 11 also conducted a flyby of Jupiter and became the first spacecraft to directly observe Saturn. The Pioneer 11 mission also brought spectacular images of Jupiter. It became the first spacecraft to photograph the Great Red Spot and the poles of Jupiter. Then on September 1, 1979, Pioneer 11 passed over Saturn at 13,000 miles above the surface. Instruments on the spacecraft identified two small moons that were previously unknown and an additional ring around Saturn. Pioneer 11 also studied the magnetic field of Saturn.

Dramatic images of Saturn's rings were returned by Pioneer 11. The rings appear bright from Earth, but were actually very dark in the pictures provided by Pioneer 11. And the dark gaps between the rings that we see from Earth appear as bright areas up close. The scientific mission of Pioneer 11 concluded in September 1995. At that time, Pioneer 11 was four billion miles from Earth and communications took approximately six hours to be returned. Intermittent contact was received from the spacecraft until November 1995.

Did You Know?

It is believed that Pioneer 10 and Pioneer 11 are still hurtling through the solar system. Pioneer 10 is headed for the red star, Aldebaran, which forms the eye of Taurus (the Bull). Aldebaran is about 68 light years away and it will take Pioneer over two million years to reach it. Pioneer 11 is headed toward the constellation of Aquila (the Eagle). Pioneer 11 will pass near one of the stars in the constellation in about four million years.

Voyager

The twin Voyager spacecrafts that composed the Voyager mission were launched in the summer of 1977. The Voyager mission was designed for a specific launch period that would take advantage of the alignment of Jupiter, Saturn, Uranus, and Neptune. This specific planet arrangement only occurs every 175 years. By using that alignment of the planets, a spacecraft does not require large onboard propulsion systems. The spacecraft's flyby of the planet delivers enough force to propel it to the next planet.

A four-planet mission was deemed too expensive and thus the Voyager mission was approved to conduct flybys of Jupiter and Saturn. The flight path chosen for the twin spacecraft would allow a flyby of Jupiter and its moon Io as well as a flyby of Saturn and its moon Titan. Scientists were thinking ahead with the Voyager 2 probe and preserved the option to continue on to Uranus and Neptune with the chosen flight paths.

Images of Jupiter provided by Voyager.
Courtesy of NASA/JPL

Voyager 1 reached Jupiter on March 5, 1979, and then travelled on to reach Saturn by November 12, 1980. Voyager 2 arrived at Jupiter on July 9, 1979, and then continued on to reach Saturn on August 25, 1981.

The spacecraft were designed to last five years to complete their journeys to Jupiter and Saturn, but after the successful completion of their mission, the Voyager 2 spacecraft showed that it could successfully continue on to Uranus. NASA then authorized the additional funding to extend the project to conduct flybys of Uranus and Neptune. During the journey to Uranus and Neptune, NASA scientists used remote-control reprogramming to give the probes more capabilities than they had when they left Earth.

Voyager 2 conducted a flyby of Uranus on January 24, 1986. The probe sent back photos of the surface and data on the planet's moons, magnetic field, and dark rings. Twelve years after leaving Earth, Voyager 2 reached Neptune. In total, the Voyager mission explored all the outer planets and 48 of their moons.

Voyager 1 and Voyager 2 are still traveling through space. It was believed that Voyager 1 would be the first spacecraft to reach the heliopause, *the boundary between the end of the Sun's magnetic influence and the beginning of interstellar space*. Voyager 1, in fact, reached interstellar space on August 25, 2012. The project has since been renamed the Voyager Interstellar Mission. The probes will continue to study solar wind, magnetic fields, and the plasma wave at the edge of our solar system.

Galileo

The Galileo mission was designed as a two-year mission studying Jupiter, its moons, and magnetosphere. The Galileo probe was carried into orbit by the Atlantis space shuttle on October 18, 1989. Galileo spent two years studying Jupiter and produced a massive amount of data on the gas planet. But Galileo also set records for additional exploration. Galileo was the first spacecraft to visit an asteroid. Galileo visited Gaspra and Ida. And Galileo became the only spacecraft to observe a comet colliding with a planet. In addition, Galileo conducted a flyby of Venus in 1990 and provided images of the fascinating clouds surrounding the planet.

Galileo found evidence of a saltwater ocean beneath the surface of Jupiter's largest moon, Europa. On Io, another of Jupiter's moons, Galileo found volcanic processes. On September 21, 2003, Galileo was plunged into the atmosphere of Jupiter to prevent it from colliding with Europa.

Cassini

Ilmage of Saturn's rings taken by Cassini.
Courtesy of NASA/JPL-Caltech/Space Science Institute

Cassini was an ambitious mission to Saturn. Cassini was a joint mission between NASA, ESA, and the Italian Space Agency. Cassini itself was a sophisticated robotic probe that was the first spacecraft to orbit Saturn. In addition, Cassini carried a smaller probe, Huygens, which parachuted to the surface of Saturn's largest Moon, Titan, in January 2015.

Cassini was launched on October 15, 1997 and began its orbit around Saturn on June 30, 2004. The probe travelled seven years to begin its four-year mission. After the mission to study Saturn was completed, the project was extended twice to explore the deepest mysteries of Saturn. Cassini spent 12 years orbiting and studying Saturn. On September 15, 2017, the mission came to an end and Cassini was sent into Saturn's atmosphere for destruction.

Some of the key discoveries from Cassini include a global ocean of liquid water under the surface of Titan. It also identified rain, rivers, lakes, and seas on the surface of Titan. In addition, a nitrogen-rich atmosphere exists on the moon. It was also discovered that Saturn's rings were active and dynamic. The rings provided a great deal of data on how planets actually form. Cassini also observed Saturn's great northern storm that consumed the entire planet and lasted for months. And it discovered giant hurricane-like vortexes at the poles of the planet.

During the mission's "grand finale ," Cassini climbed in altitude over the poles and then plunged to the surface of the planet. During the plunge, Cassini collected data beyond the original plan of the mission. It measured the gravitational and magnetic fields, determined the mass of the rings, collected samples of the atmosphere, and captured amazing close-up pictures of the planet. The data provided from Cassini is still being analyzed and will continue to lead to scientific discoveries for years to come.

New Horizons

New Horizons launched on January 19, 2006 with its sights set on Pluto and the Kuiper Belt. This was the first mission to explore Pluto and the Kuiper Belt. Pluto is the largest object in the Kuiper Belt, which is the source of most comets with orbits shorter than 200 years. Scientists want to study the composition of Pluto and its moons in comparison to the nuclei of comets. The study of Pluto could also yield information on the origins of the solar system as well as what lies beyond our solar system.

Almost 10 years after launch, New Horizons conducted a flyby of Pluto on July 14, 2015. The data returned to Earth resulted in profound insights into Pluto and its moons. The data collected will continue to be studied for years to come.

In the opening vignette, you learned about the continued journey of New Horizons and the possible firsts this probe may have in store for us.

Four images from New Horizons' Long Range Reconnaissance Imager (LORRI) were combined with color data from the Ralph instrument to create this enhanced color global view of Pluto.
Courtesy of NASA

Juno

NASA's latest probe, Juno, launched on August 5, 2011. Juno's goal is to help us improve our understanding of the solar system's beginnings by studying the evolution of Jupiter. Specifically, the probe will study the atmosphere of Jupiter to identify water, cloud motions, temperature, and other properties. In addition, Juno will explore the magnetic and gravity fields of the planet.

Juno reached Jupiter's orbit in July 2016 and will continue to orbit and collect data through the end of its mission in July 2021.

Artist concept of Juno in front of Jupiter.
Courtesy of NASA/JPL-Caltech

Lesson 1 Review

Using complete sentences, answer the following questions on a sheet of paper.

1. Why is it so difficult to send a space probe to the Sun?
2. What was the first spacecraft to study the north and south poles of the Sun?
3. Why will the Parker Solar Probe revolutionize our understanding of the Sun?
4. What was the first probe to take pictures of the Moon?
5. What were the goals for the Lunar Reconnaissance Orbiter?
6. What does ARTEMIS stand for?
7. What was the first space probe to fly by Venus?
8. What did the Pioneer Venus orbiter detect on the surface of Venus?
9. What was so unique about the orbit of the Venus Express?
10. What did the images from the Mariner 9 mission show on the surface of Mars?
11. What was the main goal of the Viking project?
12. What were some of the findings of the Mars Phoenix mission?
13. What did the Pioneer 11 probe learn about the rings around Saturn?
14. What two firsts did Galileo achieve?
15. What was Cassini's "grand finale"?

APPLYING YOUR LEARNING

16. If you were in charge of a NASA space probe program, what space object would you want to explore? Why?

LESSON 2 Robotics in Space

Quick Write

Based on what you have read in the vignette, how do you think small, affordable robots like Andy or the CubeRover could revolutionize space exploration?

Learn About

- space robots
- robotic assistance on the ISS
- the future of robots in space

In 2018, Pittsburgh-based Astrobotic Technology in partnership with Carnegie Mellon University (CMU) won a NASA Phase II Small Business Innovation Research (SBIR) award to develop the CubeRover. NASA's SBIR provides funding to small businesses or universities to develop innovative products for NASA missions. The products must also have the potential to be used commercially. Astrobotic Technology is a private space organization that strives to provide affordable space technology and robotics. CMU is located in the heart of Pittsburgh, PA, and is known for its advanced engineering, robotics, and artificial intelligence initiatives.

As part of the Phase I SBIR, CMU developed "Andy." Andy is a four-wheeled prototype robot designed to explore the Moon. Andy can scale steep cliffs and survive in the extreme temperatures of the Moon. Andy is named after the namesakes of the university, Andrew Carnegie and Andrew Mellon. The robot was designed and developed over a nine-month period by a student workforce. Collaboration between the School of Computer Science, College of Engineering, College of Fine Arts, and the Mellon College of Science was required for the project. Andy was initially developed as CMU's contribution to Astrobotic's mission to the Moon as part of the Google Lunar XPRIZE competition. As you may recall, Google withdrew funding for the competition, and XPRIZE is currently searching for new sponsors.

SBIR Phase II provides $750,000 in funding to develop and deliver the CubeRover. CubeRover is a small rover weighing approximately 4.4 pounds that will perform small-scale lunar experiments. Over the next two years,

Astrobotic will continue to develop its CubeRover technology in preparation for a 2020 launch. The first mission will be short—probably less than 14 Earth days, which is the duration of one cold, dark lunar night. During those cold nights on the Moon, the robot will go into a hibernation mode. It is uncertain if the robot will be able to be powered back up. In future missions, the company will take advantage of landers to allow the robot to shelter during the night.

Vocabulary

- space robots
- humanoid robots
- artificial intelligence (AI)

CubeRover will revolutionize low mobile robotic access on the Moon.

Courtesy of Astrobotic

Space Robots

What are space robots? Space robots are *general purpose machines that are capable of surviving the rigors of the space environment and performing exploration, assembly, construction, maintenance, servicing, or other tasks in space*. Rovers, probes, and satellites can all be considered space robots.

Benefits of Using Robots in Space

A vast majority of space exploration has been done by robots. Sending a human into space comes with many risks, and robots can be used to alleviate some of those risks. Robots can provide many benefits to space travel. For example, space travel and microgravity carry great risks to the human body. By using robots NASA is ensuring human safety. Robots can explore areas prior to sending humans and also can go farther into space than humans. Space agencies do not need to worry about bathroom breaks or food supplies when using robots.

Robots are also less expensive to send into space and are designed to withstand the harsh conditions of space for many, many years. And a return trip to Earth is not required for robots. The use of robotics in space can save billions of dollars. It costs about $10,000 per kilogram (2.2 lbs.) to send something into space. When you compare a small robot to humans and all the supplies they need, the savings can be great.

Robots tend to complete tasks faster than humans. After all, robots do not need mental breaks or time to sleep. They can work around the clock to complete their tasks. Also, astronauts train for years and don't really want to go to space to perform menial tasks. Robots can help astronauts by performing menial tasks, such as switching levers or tightening bolts.

Issues with Space Robotics

When designing robots for space, robotics engineers must address four key issues:

- Mobility
- Manipulation
- Time delay
- Environments

Mobility refers to the ability of the space robot to travel from one point to the next without a risk of collision. The robot must be able to navigate itself around obstacles in order to minimize the chance of collision. The terrain on the Moon and other planets is rough. Robots must be able to navigate around or over obstacles. If a robot gets stuck in a ditch on another planet, the mission is over. There is no AAA for space robots.

Another key issue to address is manipulation. To do the tasks required of them in space, robots must handle tools or move objects around, which requires hands or arms for manipulation. For example, robots on the ISS need to be able to move supplies around the space station. This requires the robots to manipulate the supplies through the narrow rooms of the space station.

Space robots are typically controlled by humans on Earth or remotely by astronauts on the ISS. For a robot to be commanded remotely, it must operate under a time delay. This allows controllers to accurately send information to the robot and have it respond accordingly.

Finally, the robot must be able to handle the harsh environment of space. Robots on the ISS need to be able to operate effectively in a microgravity environment. Robots that are being sent to other planets or farther out into the solar system must be able to handle extreme temperatures, both hot and cold, extensive radiation, fine dust, corrosive atmospheres, and vacuum-like conditions.

Robotic Assistance on the ISS

As you have previously read, astronauts' time aboard the ISS is limited. The astronauts have many scientific research projects to complete while on board and adding in regular maintenance and risky spacewalks can take away from the time they have to complete their research. Robots in the form of personal assistants, robotic arms, and human-like robotics can help free up the astronauts' time.

The Astrobee is one example of an extremely helpful robot. As we all know, real bees are both busy and hard-working. The Astrobee, NASA's new free-flying space robots, will soon have the same reputation. Unlike bees that live on Earth, the robots will do their work flying alongside astronauts inside the International Space Station and will play a critical role in supporting innovative and sustainable exploration of the Moon, Mars, and beyond.

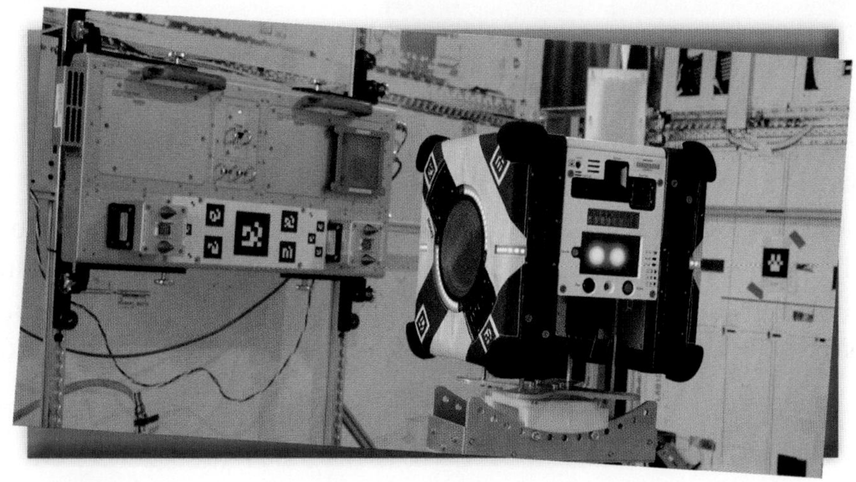

Kirobo

Courtesy of Sanrunya L/Shutterstock

Robotic Arms

One of the first uses of robotics in space was the robotic space arm. Canadarm1 was the first iteration of the robotic arm on the ISS and was installed during initial construction. Canadarm1 was permanently anchored to the space station. In 2001, Canadarm2 arrived on the ISS. Canadarm2 is approximately 55-foot long arm and not permanently mounted to the space station. Instead, it can be moved around the space station to complete its tasks. The robotic arm is responsible for most of the space assembly of the ISS. It now is used for space station maintenance, moving supplies and astronauts, and for "cosmic catches." A cosmic catch is when astronauts aboard the ISS use the robotic arm to grapple unpiloted cargo ships and dock them at the space station.

Canadarm2 has two hands called the Latching End Effectors (LEEs). The LEEs contain cables that pull together and allow the robot to firmly grip objects in space. Canadarm2 can be controlled by NASA, CSA (Canadian Space Agency), or by occupants of the ISS.

Canadarm2 is not permanently anchored to the space station and has the ability to "walk" around the station.
Courtesy of NASA

Did You Know?

Canadarm2's technology has led to the neruoArm on Earth. neuroArm is the first robot capable of performing brain surgery within a magnetic resonance imaging (MRI) machine. MRIs use strong magnets and radio waves to create detailed images of the organs and tissues within the body.

The Japanese space agency (JAXA) also developed robotic elements for the ISS. The Remote Manipulator System (RMS) has two robotic arms that are used to support the outside of the Japanese Experiment Module (JEM) of the space station. One arm handles large payloads, while the other is designed for more intricate work.

Robotic Assistants

Personal assistant technology is becoming increasingly popular. Even on Earth, many of us use Apple's Siri on our phone or Amazon's Alexa at home. So it's no surprise that astronauts also need personal assistants. However, the robotic assistants in space are designed to do more than just schedule appointments, answers questions, or play music. They have become essential helpers to space experimentation and missions.

The first robotic assistant project in space was SPHERES (Synchronized Position Hold, Engage, Reorient, Experimental Satellites). The ISS currently has three SPHERES on board: red, blue, and orange.

Did You Know?

SPHERES was created as the result of a challenge issued by MIT Professor David W. Miller to his students. The students were challenged to recreate the lightsaber training droid from Star Wars: Episode IV – A New Hope.

SPHERES arrived on the ISS on April 24, 2006, and quickly became popular. The SPHERES help astronauts test and record data. The robots perform flight formations inside the ISS to simulate the conditions for satellite servicing, vehicle assembly, and rendezvous and docking maneuvers that occur on the ISS.

NASA astronaut Shane Kimbrough is seen executing the SPHERES-RINGS experiment aboard the International Space Station.

Courtesy of NASA

NASA's next-generation robotic assistant will be the Astrobee. This free-flying robot is designed to help astronauts complete routine tasks on the ISS, test technology in zero-gravity, and provide visualization for flight controllers in Houston. Astrobee is a 12-inch cube covered in a soft material that will fly around the ISS un-supervised to complete its tasks. Astrobee is set to arrive on the ISS in 2019.

The Repair Robot

Dextre is a Canadian repair robot working on the ISS. Dextre works closely with Canadarm2 to offload supplies brought to the ISS. Dextre is also the handyman of the ISS. It completes dangerous spacewalks to repair the station or perform routine maintenance, allowing the astronauts to stay safely inside the space station.

Dextre arrived on the ISS in 2008. It is about 12 feet high and has two arms that each have seven joints, allowing it to move in many directions. The rotational capabilities of Dextre's arms allow it to move in ways that a human arm cannot. Each hand operates like a Swiss Army knife with a human-like sense of touch. Each hand is equipped with a retractable, motorized wrench; a camera and lights for close-up work; and a retractable connector to provide power, data, and video connectivity. Dextre's sensitivity allows it to grasp delicate equipment without causing damage, yet it is strong enough to lift a refrigerator. Dextre is controlled by on-ground controllers at NASA or the CSA. Most of the maintenance completed by Dextre is done when the astronauts are sleeping. Dextre is considered to be the most sophisticated space robot ever built.

Dextre at work on the International Space Station.
Courtesy of NASA

Did You Know?

What happens when a satellite is broken? Typically, it is either de-orbited or it just floats around as additional space junk. Many organizations, including NASA, are now working on robots that can repair other robots and mechanical assets in space. Restore-L is the $127 million NASA project to create a robot to refuel and relocate satellites in low orbit. Think of it as a robotic tow truck for broken satellites.

The Future of Robots in Space

The advancements made within space robotics are quite impressive and continue to stretch the imagination. Let's look at some of the space robotics projects in development.

One of NASA's latest robots is Puffer , Pop-Up Flat Folding Explorer Rover. Puffer is a lightweight robot inspired from origami designs. The robot has the capability to flatten itself completely so that it can explore tight areas.

BRUIE, Buoyant Rover for Under-Ice Exploration, is another newer robot for NASA. BRUIE floats under water and uses wheels to move itself along the underside of an icy surface. The robot will take pictures and collect data underwater. BRUIE will someday be used to search for signs of life on the icy bodies of water on other planets or their moons.

The BRUIE robot rolls its wheels on the underside of an icy surface.
Courtesy of NASA/JPL-Caltech

The Right Stuff

NASA, Stanford University, and MIT are currently working on the design for a robot named Hedgehog. Hedgehog is a spiky, cube-like robot that will explore small celestial bodies like asteroids or comets. Since there is very little gravity of these types of surfaces and the terrain is rough, Hedgehog is designed to hop or tumble across the surface.

The Hedgehog has the ability to move around on all sides of the cube. It also has a tornado maneuver it can use to get out of a deep crater.

The Hedgehog robot.
Courtesy of NASA/JPL-Caltech/Stanford

Humanoid Robots

NASA also has humanoid robots, which are *robots that look like humans*. Some tasks are quite difficult for the standard robot to complete. Having a robot that moves and operates like a human can assist in some missions. The robots are built and tested on Earth before a space mission is even considered for them. Many of the robots operating today are aimed to assist with building colonies on the Moon or Mars.

Robonaut

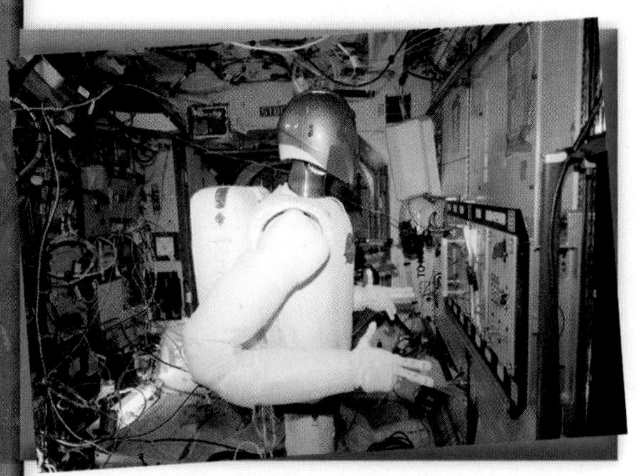

Robonaut is a testbed for exploring new robotic capabilities in space, and its form and dexterity allow it to use the same tools and control panels as its human counterparts aboard the ISS.

Courtesy of NASA

The first NASA humanoid robot was the Robonaut. Robonaut 1 was developed in 2000 and tested on Earth. In 2011, the second iteration of the humanoid robot, Robonaut 2 (also known as R2), was sent to the ISS. The goal for Robonaut 2 was to handle some monotonous tasks on the space station, such as flipping switches or levers. The initial version of R2 that was sent to the ISS looked like a human body without legs. In 2014, NASA sent a pair of legs to be connected to R2. The legs are a bit thin compared to the robot so it almost appears as though it has stick legs. The legs are about nine feet long and very flexible so that the robot can cling to areas of the space station; they are not designed for walking because R2 resides in a zero-gravity environment. After the upgrade to provide legs, R2 experienced some issues and was taken offline. In 2018, NASA announced it had discovered the source of the issue, and Robonaut 2 was brought home for repairs. NASA is hopeful that R2 will soon return to its home on the ISS once fully repaired.

Valkyrie

Valkyrie (also known as R5) is another humanoid robot created by NASA. Valkyrie is six feet tall and weighs in at 290 pounds. It is the first humanoid robot to stand on two legs. Valkyrie uses cameras, sensors, motors, and two computers to navigate its surroundings and mimic human movements.

The humanoid robot, Valkyrie.

Courtesy of NASA

Valkyrie was initially designed to provide disaster relief and search-and-rescue operations on Earth. But the possibilities are endless with the humanoid robot. NASA has provided one Valkyrie robot to Northeastern University and another to MIT to conduct research and development work. Although Valkyrie is not currently scheduled for any space missions, discussions are underway to determine how the robot can be used for space exploration. The technology used in the humanoid robot could be very useful on missions to explore Mars.

RoboSimian

RoboSimian was built for the 2015 DARPA Challenge. DARPA is the Defense Advanced Research Projects Agency, which strives to discover innovative technologies for national security initiatives. The goal of RoboSimian was to assist with natural and man-made disasters. RoboSimian looks a bit more like a spider than a human. It is a powerful robot with four legs that can travel over rough terrain. The robot uses 3-D technology to map its surroundings.

RoboSimian was developed at NASA's Jet Propulsion Laboratory in Pasadena, CA.
Courtesy of NASA/JPL-Caltech

During the 2015 Challenge, RoboSimian successfully drove a car through a slalom course, opened a door, turned a wheel to open a valve, cut a hole in drywall using a power drill, walked across uneven terrain, and walked up a set of stairs all within an hour. RoboSimian took fifth place in the competition.

Kirobo

Kirobo is a miniature humanoid robot designed by the Japanese space agency, JAXA, and Toyota. Kirobo is only 13 inches tall but is a powerful little robot. Kirobo's mission is to test human-robot interaction in space. Kirobo can recognize voices and faces. It can also read emotions. The goal of Kirobo is to offer companionship to those living alone or in isolated conditions.

Kirobo
Courtesy of Sanrunya L/Shutterstock

Kirobo traveled to the ISS in 2013 and had its first conversation in space with Japanese astronaut Koichi Wakata. Kirobo remained in space for 18 months before travelling back to Earth. He was so popular among fans that an app was created to track Kirobo's exact location in space. Kirobo only spoke Japanese so he was only available to communicate with astronauts who spoke Japanese.

His popularity grew so much that miniature versions of Kirobo are now available for purchase. The robots can be purchased at Toyota dealers across Japan for about $400.

Artificial Intelligence

Artificial intelligence (AI) is *the development of computer systems that are able to perform tasks that normally require human intelligence*. Those tasks include such things as visual perception, speech recognition, decision-making, and language translation. Artificial intelligence is a growing industry and has limitless possibilities when it comes to space exploration.

The first robot with artificial intelligence arrived on the ISS on July 2, 2018. CIMON (Crew Interactive MObile CompanioN®) is a small German robot equipped with AI. Upon arriving on the ISS, CIMON assisted German astronaut Alexander Gerst with a 90-minute experiment. CIMON was created by Airbus and IBM and works similarly to Apple's Siri or Amazon's Alexa. It has the ability to recognize faces, take photos and video, and move about autonomously using sensors.

German astronaut Alexander Gerst and CIMON pose for a picture aboard the ISS.
Courtesy of ESA/NASA

CIMON is currently in beta-testing to work out the kinks. He was designed to be a colleague and friend to the astronauts aboard the ISS. CIMON has had some glitches with his social interactions, and developers are working to ensure the robot interacts well with the astronauts. Artificial intelligence and robotics are evolving at a rapid pace and could change the course of space exploration. As humans travel farther away in space, having AI and robots to assist in their efforts could prove invaluable.

Lesson 2 Review

Using complete sentences, answer the following questions on a sheet of paper.

1. Why do robots typically complete tasks faster than humans?
2. What four key issues must robotics engineers address when designing robots for space?
3. What are the main tasks that Canadarm2 completes on the ISS?
4. What movie inspired the development of SPHERES?
5. What tools are included in Dextre's hands?
6. What does NASA's Puffer robot do?
7. Why is Valkryie so unique?
8. What are some examples of tasks that can be performed by artificial intelligence that normally require human intelligence?

APPLYING YOUR LEARNING

9. In this lesson, we discussed many advantages of using robotics in space. Do you see any disadvantages to using robotics in space exploration? Why?

LESSON 3

The Mars Rover Expedition

Quick Write

As this book is being written, NASA is running a naming competition for the next Mars rover, currently known as Mars 2020. What name would you give it and why?

Learn About

- Spirit and Opportunity
- Mars Science Laboratory: Curiosity
- Mars 2020

How do Mars rovers get their unique names? Sojourner, Spirit, and Opportunity rovers were all named by school-aged students.

Sojourner was named by 12-year-old Vallery Ambroise. She wanted to honor the 19th-century abolitionist and activist Sojourner Truth. Spirit and Opportunity were named by third-grader Sofi Collis. She found inspiration in looking toward the sky while living in an orphanage as a toddler.

In sticking with tradition, NASA held another naming competition for the current Mars rover, Curiosity. NASA received over 9,000 student entries in the quest for the name.

Sixth-grade Clara Ma from Kansas submitted a 250-word essay in response. In her essay, she explains the important relationship she has with her grandmother, who lives across the globe, in China.

Ma shares how she spends time with her grandmother telling stories and looking at the stars. The night sky represents a way of keeping her connected with her grandmother when they are apart. Along with this sentimental connection to the night sky, Ma explains that her own interest and wonder about space led to the name, Curiosity.

After winning the contest, Ma was invited to autograph the rover as it was being built. She also got to watch it launch from Cape Canaveral, Florida, in 2011. Former astronaut Leland Melvin even suggested to Ma that she could follow Curiosity to Mars.

Today, Ma is pursuing her interest in science, studying climate physics and political science at Yale.

Clara Ma, winner of the Mars Science Laboratory naming contest.

Courtesy of NASA/JPL Cal Tech

Spirit and Opportunity

Lesson 1 introduced a brief history of the Mars Rover Expedition. We saw the success of the Pathfinder and Mars Global Surveyor missions in 1997. The US experienced a renewed excitement in the exploring the Red Planet. Would we find signs of past life? How could we prepare for human exploration of Mars? To find out, NASA launched the Mars Odyssey mission in 2001.

As part of the Mars Odyssey mission, NASA designed two Mars Exploration Rovers (MER), Spirit and Opportunity, to go farther and collect more data than ever before. The main goal of Spirit and Opportunity? To search for a wide range of rocks and soils for clues to past water activity on Mars.

NASA did not know at the time, but the MER would outlive the agency's greatest projections. NASA originally planned a 90-day mission for the twin rovers. Instead, Spirit lived on until 2010 and Opportunity until 2019!

Did You Know?

In 2015, Opportunity broke the extraterrestrial driving record by driving more than 28.06 miles.

Spirit and Opportunity were spacecraft wonders! They gathered over 350,000 images and massive amounts of atmospheric, climatologic, and geologic data. Let's take a look back at their development to learn more about their important journeys and impressive findings.

Vocabulary

- Miniature Thermal Emission Spectrometer (Mini-TES)
- Mössbauer Spectrometer (MB)
- Alpha Particle X-Ray Spectrometer (APXS)
- Microscopic Imager (MI)
- Rock Abrasion Tool (RAT)
- aeroshell
- silica
- hematite
- gypsum
- ChemCam
- Mars Hand Lens Imager (MAHLI)
- Powder Acquisition Drill System (PADS)
- Collection Handling for Interior Martian Rock Analysis (CHIMRA)
- radioisotope power system (RPS)
- plutonium-238
- Dynamic Albedo of Neutrons (DAN)
- Radiation Assessment Detector (RAD)
- Sample Analysis at Mars (SAM)
- Chemistry and Mineralogy Instrument (CHEMIN)
- MSL Entry, Descent, and Landing Instrument (MEDLI)
- alluvial fan
- methane

MER Designs

Spirit and Opportunity were designed to be much larger and more powerful than their predecessor, Sojourner. These rovers had much bigger solar panels. The panels were attached like wings, instead of mounted on top.

Three generations of Mars rovers developed at NASA JPL. Front and center is a flight spare of Sojourner. Left is a working sibling to Spirit and Opportunity. Right is test rover Curiosity.

Courtesy of NASA/JPL Cal Tech

Each rover included a panoramic camera for studying the minerals, texture, and structure of the surface. The panoramic camera, or Pancam for short, actually used two cameras to take detailed, 3-D pictures of the landscape, the Sun and the sky. Scientists used Pancam to observe the horizon of Mars for landforms that would suggest a history of water. The Pancam was also used to map the area where the rover landed.

The rovers were equipped with navigation and hazard-avoidance cameras, with small cameras mounted on each robotic arm.

Did You Know?

Cameras provided human-like views. Robotic arms moved like a human's. A microscopic camera acted as a magnifying lens. Each rover appeared as a robotic geologist walking around Mars!

Each rover also held a Miniature Thermal Emission Spectrometer (Mini-TES). Mini-TES *used infrared scans to help identify different types of compounds based on different temperatures*. How did scientists use data from Mini-TES? They used it to locate minerals

The Right Stuff

Steven Squyres is probably not a name you recognize. His claim to fame? He is one of the most prominent people in the history of our mission to Mars! For over a decade, he served as the principal investigator for Spirit and Opportunity. NASA tapped him for this role based on his very impressive background in astronomy, geology, and mission management.

Steven W. Squyres is a professor of astronomy at Cornell University, Ithaca, N.Y. He served as the principal investigator for the science payload on NASA's Mars Exploration Rovers.

Courtesy of NASA

Squyres's journey to Mars all began with an interest in geology here on Earth. While in college, though, he became curious about rocks of the unknown, and the geology of Mars. But how could he study rocks, using traditional geology tools and techniques, on a planet that could only be seen and not touched? Wanting to learn more, he would lead the way in the study of Mars rovers.

Squyres was given a once-in-a-lifetime opportunity to participate in the Voyager mission to Jupiter and Saturn. Along with Voyager, Squyres worked on the Magellan Venus mission, the Near-Earth Asteroid Rendezvous-Shoemaker (NEAR-Shoemaker) mission, the Cassini Saturn probe, and a European Mars orbiter called Mars Express.

Under Squyres's direction, we now have evidence of water on Mars and the possible existence of a liquid water ocean on Europa.

for closer examination and for identifying the formation processes of rocks and soil. A main goal of the instrument was to look for minerals formed in water, like carbonates and clays. Mini-TES also provided temperature profiles and data on water vapor and dust of the atmosphere.

MER Robotic Arm Instruments

The MER spacecraft robotic arms included several state-of-the-art instruments. These instruments gathered information about Martian geology, atmosphere, and environmental conditions.

The Mössbauer Spectrometer (MB) *attached to the robotic arm, looked for the makeup of iron-bearing rocks and soils.* A sensor on the MB actually touched rock and soil samples for close-up study. Scientists used data from the MB to determine the magnetic properties of surface materials and early environmental conditions.

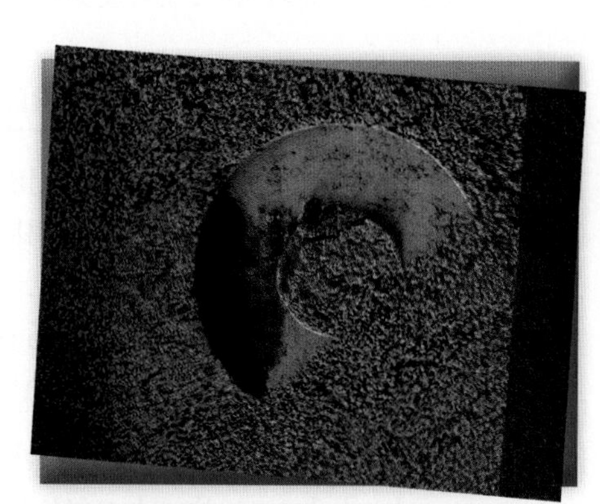

MB 3-D image of Martian soil from MER Spirit. Scientist must use 3-D glasses to view the images

Courtesy of NASA/JPL/Cornell/USGS

The robotic arm also held the Alpha Particle X-Ray Spectrometer (APXS). The APXS *calculated the amount of chemical elements in rocks and soils by measuring the way different materials respond to two kinds of radiation: x-rays and alpha particles.* What did scientists hope to discover by examining rocks and soil using APXS? They were trying to determine how materials formed and if they were ever changed by wind, water, or ice.

The Microscopic Imager (MI) *attached to the robotic arm of each rover provided high-resolution, close-up views rocks and soils*. It identified targets that could be studied and analyzed by other instruments. Scientists used data from the MI to identify more precise visual hints about rock and soil formation as well.

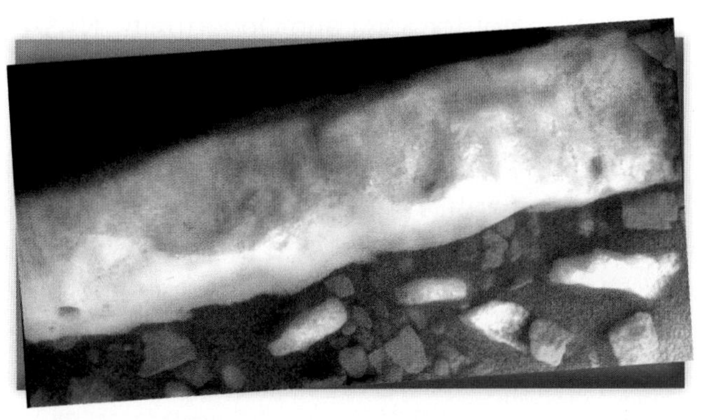

MER Opportunity inspected this mineral vein, called "Homestake," in November 2011. The area covered in this view spans about 2 inches (5 centimeters) across.

Courtesy of NASA/JPL-Caltech/USGS/Cornell Univ./Arizona State Univ

Opportunity's rock abrasion tool brushed dust out of the circular area the tool had ground into the rock named "Marquette Island" in January 2010.

Courtesy of NASA/JPL-Caltech/Cornell

The Rock Abrasion Tool (RAT) *attached to the robotic arm of each rover was simply a wire brush for dusting off rocks prior to examination*. Its rotating teeth would chew into the surface rock to uncover fresh mineral surfaces.

MER had solar panels, cameras, Mini-TES, and robotic arm instruments. They had wheels for moving around and magnets for collecting magnetic dust particles. And they had antenna for sending and receiving information.

Spirit and Opportunity Mission

On June 10, 2003, Spirit launched from Cape Canaveral Air Force Station in Florida. Opportunity's launch followed weeks later on July 7, 2003.The timing of these launches was significant. August 2003 would be the closest Mars came to Earth in over thousands of years. At 45 days into mission, the rovers were about to enter the Martian atmosphere. Mission control engineers took several critical steps.

First, they converted Earth time to "Mars time." This allowed them to coordinate their schedules on Earth with those of the rovers. Remember that a day on Mars is about 40 minutes longer than a day on Earth! As the rovers cruised towards Mars, engineers carefully tracked and corrected the flight path of the rovers. This helped to maintain communications and ensure the precise landing of each spacecraft.

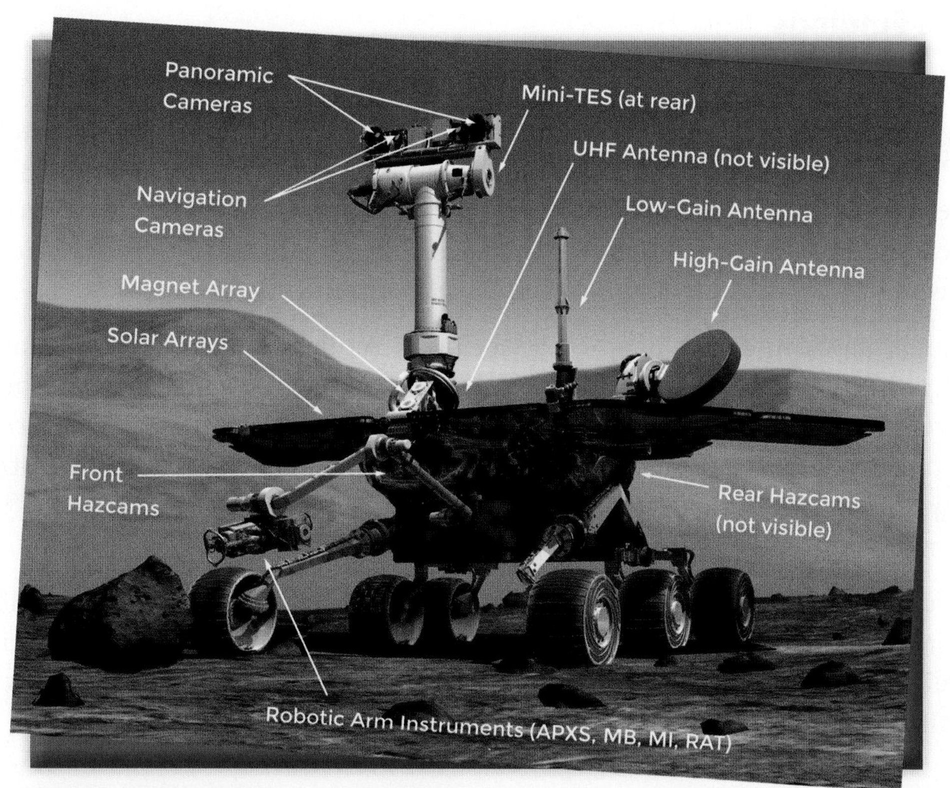

Spirit and Opportunity: Rover Parts.
Courtesy of NASA

Each spacecraft descended into Mars at a very high speed. They used an aeroshell and a parachute for deceleration. An aeroshell's *main purpose was to protect the lander, with the rover inside, from the intense heating of entry into the thin Martian atmosphere.* Retro-rockets further slowed the lander's speed before impact. And just before landing, an impressive bunch of over-sized beach-like balls inflated. The result? A bouncy, cushioned landing for our rovers!

Both rovers landed very near their targets. After the rovers stopped bouncing and rolling, their antennae transmitted tones to Earth. Mission control received these tones 15 minutes later. Engineers confirmed the landings as being close to perfect. The rovers landed on opposite sides of Mars. Spirit landed at Gusev Crater. Mineral deposits there suggest that liquid water in the past affected Mars. Opportunity landed at Meridiani Planum. This may be a former lake within a large impact crater.

MER landings involved gigantic air balls providing cushion as the spacecraft bounced up to nearly 100 feet high.
Courtesy of NASA

At landing, the landers retracted their airbags, and deployed their lander petals and solar arrays. And finally, they raised their panoramic cameras for an up-close view of the Martian world!

Spirit's Findings

A warmer climate, a watery past, and an ancient volcano are just some of the key findings gathered by Spirit. This rover used all its cameras, robotic arm tools, and the APRX to make a notable discovery. It uncovered rocks with large amounts of magnesium and iron carbonates. What do these chemicals tell scientists? Mars had a thick carbon-dioxide atmosphere and low acidic water. This suggests that Mars once had a warm and watery environment capable of supporting life.

Another fantastic finding happened by accident. While roving around the Martian planet, Spirit's right front wheel failed and ended up digging a trench. In doing so, the rover uncovered 90 percent pure silica, *a compound made of crystals that occur in quartz on Earth*. What did this mean to scientists? There may have been hydrothermal activity on Mars, possibly in hot springs. This finding again suggests the idea that Mars once supported ancient lifeforms.

Also, during its travels, Spirit discovered an ancient volcano at "Home Plate." This plateau was roughly 90 miles across with a shape like a baseball home plate. Scientists believe that explosive steam eruptions from hot underground water led to this volcanic formation.

Spirit's panorama of Home Plate reveals a rocky layered outcrop indicating previous volcanic activity.

Courtesy of NASA/JPL-Caltech/USGS/Cornell University

Opportunity's Findings

Spirit gave us some of the most impressive findings to support the theory of life on Mars. And Opportunity gave us even more evidence! In its landing spot, Meridian Planum, Opportunity revealed another important clue to life on Mars, the mineral hematite. Hematite *is made up of iron oxide and usually forms in water*. Fine-grained hematite also gives Mars its red hue.

Opportunity made a discovery near the rim of another crater, the Endeavor. The rover uncovered brilliant colored stripes of gypsum.

Gypsum *is a salt that precipitates from liquid water*. This tells scientists that water once gushed through underground fractures in the ground.

Along with gypsum, Opportunity also found clay mineral deposits in the Endeavor Crater. Clays form when there is neutral or low acid conditions. This gives us more support that Mars once had the necessary conditions for ancient lifeforms.

Martian "blueberries." Opportunity's investigation of the hematite-rich concretions in early 2004 provided evidence of a watery ancient environment.

Courtesy of NASA/JPL-Caltech/Cornell/USGS

Mission Completion

Spirit and Opportunity lived well beyond their scheduled 90-day missions. They provided us with a firm understanding that Mars could have sustained some forms of life, if any existed.

But in May 2009, Spirit became stuck in sand. It was already missing its right front wheel when it lost its right rear wheel. With two missing wheels, the rover could no longer drive onto the slopes that would tilt the solar rays towards the Sun. After months of planned corrections, NASA ended the mission on May 2011.

Opportunity continued to explore the Martian terrain. In June 2018, a severe planet-wide dust storm covered its location. After numerous attempts to restore communications, NASA finally ended the mission.

Opportunity's shadow was taken on July 26, 2004, by the rover's front hazard-avoidance camera as the rover moved farther into Endurance Crater in the Meridiani Planum region of Mars

Courtesy of NASA/JPL-Caltech

Mars Science Laboratory: Curiosity

The next NASA rover, Curiosity, was launched on November 26, 2011, and landed on Mars on August 6, 2012. It's mission? To see if small lifeforms, called microbes, could have ever existed on the planet. Scientists also aim to learn more about the climate and geology of Mars to prepare for possible human exploration.

As part of the Mars Science Laboratory mission, Curiosity was designed to be larger and even more capable than previous rovers. As this book is being written, Curiosity still roves around the Red Planet and sends data back to Earth. Let's look back at the development of Curiosity and its findings to date.

Curiosity's Design

Much of Curiosity's design is based on improving the designs created for Spirit and Opportunity. The solar panels powering those rovers were often coated in dust. Designers found a more reliable and continuous source of power for Curiosity, which will be discussed later in this lesson.

Spirit also experienced random computer rebooting throughout its mission. Designers of Curiosity found a better computer system backup to avoid such issues.

Lessons learned from the MER have led to other improvements. These include larger wheels, autonomous driving, and more specialized instruments. Let's take a closer look at some of these enhancements.

Curiosity's Profile.
Courtesy of NASA/JPL

Did You Know?

Curiosity is about the size of a small sport utility vehicle (SUV). It is 9feet ,10 inches long by 9 feet, 1 inch wide and about 7 feet high. It weighs 2,000 pounds! Spirit and Opportunity only weighed about 400 pounds each.

Curiosity is much larger than the MER spacecraft. But its body, navigation/hazard avoidance cameras, and Pancam (now called Mastcam) are similar in design.

Curiosity's cameras are much higher resolution, with Mastcam offering full 360-degree pans. Curiosity also includes laser-induced remote sensing for micro-imaging, using a camera called Chemistry and Camera (ChemCam).

From a distance, ChemCam *fires a laser and reviews the elemental composition of vaporized materials from areas smaller than 0.04 inch on the surface of Mars*. A spectrograph then gives details about the minerals and microstructures in the rocks. The laser also clears away dust from rocks to obtain very detailed images.

ChemCam uses the laser to clear away dust from Martian rocks and a remote camera to acquire extremely detailed images.

Courtesy of NASA/JPL/ LANL/ CESR/ CEA

A few new instruments also are attached to Curiosity's robotic arm:

- Mars Hand Lens Imager (MAHLI)
- Powder Acquisition Drill System (PADS)
- Collection Handling for Interior Martian Rock Analysis (CHIMRA)

Like a geologist's hand lens, the Mars Hand Lens Imager (MAHLI) *gives close-up views of the Martian rocks, debris and dust*. MAHLI allows scientists to identify the minerals in a rock, both day and night. It uses ultraviolet light to detect minerals like carbonates. This study helps to further understanding of the geologic history of the landing site.

The Powder Acquisition Drill System (PADS) *drills rock into powder*. The Collection Handling for Interior Martian Rock Analysis (CHIMRA) *then collects it for deposit into Curiosity's body for examination*.

Did You Know?

With MAHLI, scientists here on Earth can see Martian rock structures smaller than the diameter of a human hair!

Another new feature onboard Curiosity is a visual system called Mars Descent Imager (MARDI). It provided four-frame-per-second video when Curiosity landed. The camera also gathered useful information. It helped scientists examine geological processes and create relief maps of the landing site.

The most distinguished improvement of Curiosity over the MER may be its energy source. Remember, Spirit and Opportunity relied on multi-panel solar arrays for energy. Curiosity uses a radioisotope power system (RPS), *a type of nuclear energy technology that uses heat to produce electric power for operating spacecraft.*

The heat is produced by plutonium-238, *a material that emits steady heat due to its natural radioactive decay*. By powering Curiosity using RPS, the mission has an operating lifespan of a full Martian year (687 Earth days). It also provides the rover with enhanced operations and mobility.

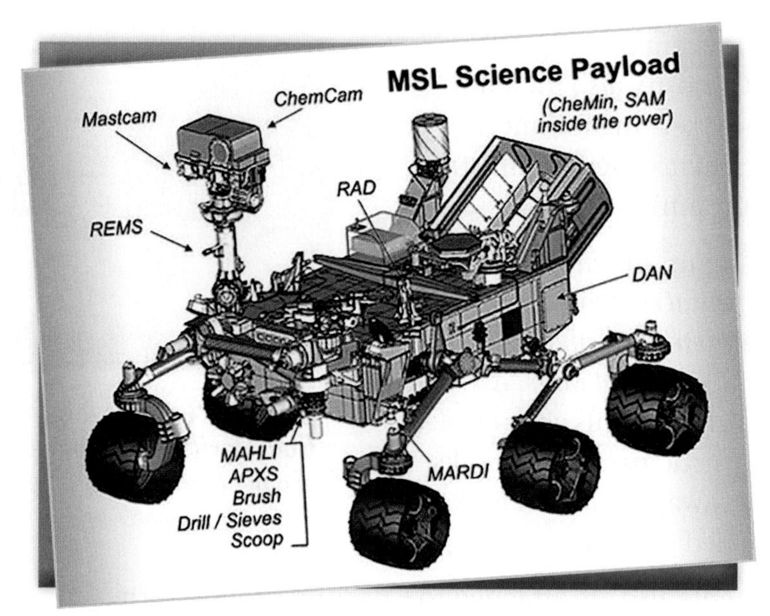

Curiosity: Rover Parts.
Courtesy of NASA

Several other instruments aboard Curiosity have been developed since the MER. Let's take a closer look at them now.

- The Dynamic Albedo of Neutrons (DAN) instrument *measures water content below the surface*. The Radiation Assessment Detector (RAD) instrument *measures radiation levels*.
- The Sample Analysis at Mars (SAM) instrument *is used to study carbon, hydrogen, oxygen and nitrogen*. The Chemistry and Mineralogy Instrument (CHEMIN) *uses x-ray beams to examine soil and drilled rock to identify minerals*.
- The MSL Entry, Descent, and Landing Instrument (MEDLI) *measures pressure and temperature during entry and landing*.

All of these instruments help scientists achieve the goals of the mission: to see if microbes could have ever existed on the planet and to prepare for possible human exploration.

The Right Stuff

How does NASA select landing sites on Mars? Of course, scientists and engineers consider a large number of factors. The landing site should be close enough to the equator, particularly for spacecraft powered by solar arrays.

For the actual landing, there should be enough atmosphere for deceleration above the site. To ensure a safe landing, there should be no hazards like steep slopes.

The landing site surface should be smooth enough for rovers to drive on once landed. And lastly, the mission goals are of huge importance when deciding where to land and what to discover.

So why was Gale Crater selected as the landing site for Curiosity? Scientists considered the usual criteria for selecting a safe and accessible landing site. But they also needed to focus on the mission goals to guide their decision. The landing site had to show clear evidence of both a strong geologic record and a past or present habitable environment.

Gale Crater met the criteria for the mission with its alluvial fan, a *fan-shaped deposit of sediment crossed and built up by streams*. And, at the center of the crater is Mount Sharp, an 18,000-foot-tall mountain made of layers of sedimentary rock.

Mount Sharp stands in the middle of Gale Crater, which is 96 miles (154 kilometers) in diameter. Spanning the center of the image is an area with clay-bearing rocks.

Courtesy of NASA

Curiosity's Mission

On November 26, 2011, Curiosity launched from the Cape Canaveral Air Force Station in Florida.

The cruise stage for Curiosity was like the MER missions. It involved several important steps to track and correct the trajectory of the spacecraft. This helped to maintain communications and to ensure a precise landing. An onboard flight computer communicated the health of the spacecraft to mission controllers.

> **Did You Know?**
>
> An onboard star scanner kept Curiosity's cruise stage on track by regularly monitoring its position relative to stars.

The entry, descent, and landing of Curiosity was known as "Seven Minutes of Terror." It involved many of the same techniques used by the MER mission, with a few new precision landing technologies.

Curiosity's Sky Crane Maneuver, Artist's Concept.
Courtesy of NASA/JPL-Caltech

This was a much heavier rover. To cushion the landing, a guided entry and a sky crane touchdown system was used instead of giant air balls. After the parachute slowed Curiosity and the heat shield separated, computerized steerable engines further decelerated the spacecraft. The sky crane system then lowered Curiosity to the Martian surface. Several small explosives fired that triggered cable cutters on the sky crane tethers.

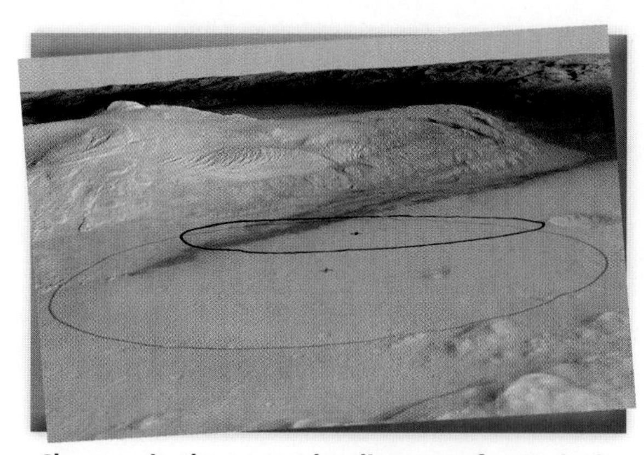

Changes in the target landing area for Curiosity. The larger ellipse was the target area before June 2012, when the project revised it to the smaller ellipse centered nearer to the foot of Mount Sharp.
Courtesy of NASA/JPL-Caltech/MSSS

Another key difference between Curiosity and its predecessors, was what happened upon landing. The MER first had to deploy their lander petals and solar arrays. Curiosity was ready to hit the ground running, and not just figuratively. Because Curiosity lost its cocoon on the way to landing, its wheels literally hit the ground and started roving.

On August 12, 2015, with the new guided entry and sky crane system, Curiosity landed within a 12-mile (20-kilometer) ellipse of its target in Gale Crater!

Curiosity's Findings

Recall that Curiosity's main mission is to determine if Mars is, or was, suitable for life. What are we finding? As early as 2013, Curiosity began making some impressive discoveries.

From its earliest drill samples, complex organic life-supporting molecules have been uncovered and analyzed. Findings include sulfur, nitrogen, hydrogen, oxygen, phosphorus, carbon, and clay. All of these elements are considered essential for life.

Curiosity's drilled 2.5 inches into an outcrop called "John Klein," marking the first time a rover drilled into a rock to collect samples on another planet.

Courtesy of NASA/JPL-Caltech/MSSS

Curiosity has also found evidence of atmospheric methane. This is another indicator that microbial life could have existed on Mars. Methane *a colorless, odorless flammable gas on Earth is produced mostly by microorganisms*. Scientists believe this finding on Mars may be the result of reactions between water and rocks. Perhaps even biological activity!

The RAD instrument also found significant amounts of radiation on the surface of the planet. This is an important find for NASA as they design safe missions for human exploration.

Finally, Curiosity reveals a thicker atmosphere and even more water in the planet's past. Atmospheric measurements indicate heavy isotopes of hydrogen, carbon, and argon.

Mars 2020

The next NASA rover, Mars 2020, is being designed to launch in ... you guessed it, 2020! Its mission? To learn more about the potential for human exploration. Mars 2020 will look for more signs of past microbial life itself.

The goals of Mars 2020 are:

1. Determine whether life ever existed on Mars.
2. Identify the climate of Mars.
3. Identify the geology of Mars.
4. Prepare for human exploration.

Its design is almost identical to Curiosity with a few new technologies. One innovation offers a technique for testing oxygen production.

Scientists want to address the goal of preparing for human exploration. They will use the carbon dioxide gas from Mars' atmosphere and turn it into oxygen. The oxygen could then be captured and made available for breathing and for refueling rockets.

Mars 2020 will also have the ability to set aside returnable storage tubes. They will hold rock core and soil samples for return to Earth by future missions. Having samples to work with on Earth will allow us to examine them using full-scale instrumentation. Just what we find in those samples could unlock even more great surprises from the Martian planet!

Mars 2020 includes other new technologies. They will serve to improve landing methods, help identify subsurface water, and classify weather and other environmental factors that could impact human exploration on Mars.

Instruments will be enhanced. These include several new cameras and a microphone to capture entry, descent, and landing. This will mark the first time in history that a Mars landing can be watched as it occurs! A second microphone will let us hear the sounds from Mars.

As this textbook is being written, a landing site for Mars 2020 has not yet been determined. There is a good deal of work left to prepare the spacecraft for final assembly.

The Mars Rover Expeditions have been hugely successful. We can hope that Mars 2020 (and other planned international and private missions) will share the same success. If that is the case, we will definitely continue expanding our understanding of the planet's evolution and geology. It could mean the potential for future human exploration. For certain, we will learn more about our own planet and the rest of our solar system. We will continue to inspire future generations of space explorations.

Lesson 3 Review

Using complete sentences, answer the following questions on a sheet of paper.

1. What was the main goal of Spirit and Opportunity?
2. What function did the Alpha Particle X-Ray Spectrometer (APXS) serve?
3. What was the role of the oversized air balls in landing Spirit and Opportunity?
4. What did Spirit and Opportunity tell us about the potential for life on Mars?
5. What are the goals of Curiosity?
6. What are some benefits of using the Radioisotope Power System (RPS) as the energy source for Curiosity?
7. What new system did Curiosity use to land the rover?
8. What are the goals of Mars 2020?
9. What will be seen for the first time with Mars 2020?

APPLYING YOUR LEARNING

10. NASA has established many new goals with Mars 2020. What do you think will be the most important goal if NASA wants to establish a human colony on Mars?

CHAPTER 7

An illustration of a satellite orbiting Earth.
Courtesy of Andrey Armyagov/Shutterstock

Orbiting, Space Travel, and Rockets

Chapter Outline

"Man must rise about the Earth—to the top of the atmosphere and beyond—for only thus will he fully understand the world in which he lives."

Socrates

LESSON 1 Orbits and How They Work

Quick Write

After reading the vignette, how do you think Newton's cannon ball theory helped develop the technology to put a satellite into orbit?

Learn About

- how orbits work
- types of orbits

In 1729, Isaac Newton told a story of orbiting. Imagine climbing to the top of Mt. Everest and, when you arrived, you placed a cannon on top of the mountain. You aim the cannon horizontally into the sky. When fired, the cannonball will travel horizontally until eventually gravity wins and pulls the cannonball to Earth's surface. Now add more gunpowder. The cannonball will travel horizontally along the plane of Earth a bit farther but will eventually succumb to gravity. What if you could add even more gunpowder? The cannonball will travel farther again but again eventually fall to Earth.

Newton asked this: is there a point where you could add enough explosives to the cannon so that the cannonball reaches the right velocity to stay in the air traveling completely around the planet. The gravitational field of Earth would constantly pull on the object, but the velocity of the object would be great enough to keep it from falling to Earth's surface.

In 1729, Isaac Newton wasn't taken very seriously with his proposition of orbits. But 228 years later, in 1957, Sputnik 1 was launched into orbit proving Newton's suspicions correct.

How Orbits Work

An orbit is *a regular, repeating path of one object around another object.* As you may recall from Chapter 5 Lesson 3, an object in orbit is called a satellite. There are two types of satellites: natural satellite and artificial satellite. The Earth and Moon are considered natural satellites. Artificial satellites are objects that are created by humans and put into orbit, such as the International Space Station (ISS).

To understand how an orbit works, we need to go back to Newton's First Law of Motion . Newton proposed that there is a force or gravity between two objects and that an object will remain in a uniform motion unless an external force changes the motion. So, what does this have to do with orbits? An orbit occurs when there is a perfect balance between the forward motion of an object and the pull of gravity from another object. In the case of the Moon, the Moon is a massive object moving forward, but the gravity from Earth pulls the Moon back toward the Earth, thus making the Moon orbit Earth. An orbit can be seen as a constant tug-of-war between the two objects.

Vocabulary

- orbit
- momentum
- orbital velocity
- escape velocity
- atmospheric drag
- launch window
- eccentricity
- perigee
- apogee
- dwell time
- highly elliptical orbit

In Figure 7.1, we can see how an orbit is created. The object has momentum, or *forward motion.* Gravity from another object pulls the object in. The object doesn't lose its momentum; it will continue to move forward but is pulled down by gravity. The result is a balance of momentum and gravity that creates a circular orbit.

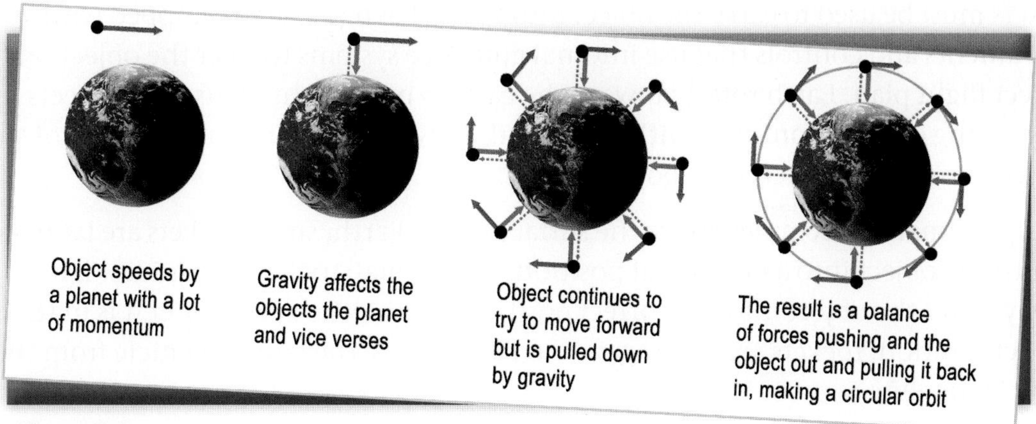

Figure 7.1

What happens if the forces are not perfectly balanced? If momentum is stronger than gravity, the object will speed past and continue its forward motion in space. If the momentum is too weak, the object will be pulled in by gravity and collide with the planet or other object. This concept is demonstrated in Figure 7.2.

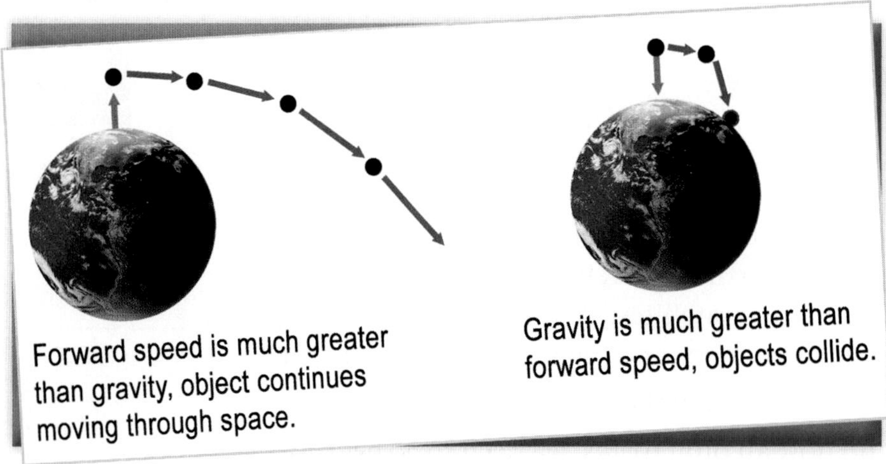

Figure 7.2

The perfect balance of gravity and momentum is referred to as the orbital velocity. Orbital velocity is *the velocity needed to achieve a balance between gravity and momentum.* If scientists want an object to orbit Earth at 150 miles above the surface, they require an orbital velocity of 17,000 mph. As the altitude above the Earth increases, the orbital velocity required decreases which makes sense because the farther you go away from Earth, the weaker the pull of Earth's gravity. Scientists must also take into account the escape velocity. Escape velocity is *the lowest velocity an object must have in order to escape the gravitational pull of another object.* When working with a probe set to explore outer planets, scientists use the escape velocity to calculate the speed needed to ensure the object does not get pulled into an orbit.

How Do we Get Objects into Orbit ?

Now that we understand what is required for an orbit to occur, let's discuss how an artificial satellite gets placed into orbit. To get an object into orbit from Earth's surface, rockets must be used to carry the object into space. Each rocket is equipped with instruments and controls that use internal guidance systems to steer the object on the correct flight plan. Earth rotates toward the east, so most flight plans send objects into space going east to coordinate with Earth's orbit. By using the natural rotation of the planet, the rocket can receive a boost.

At approximately 120 miles above the surface of the Earth, small rockets are launched to turn the object into a horizontal position. These small rockets are guides that slowly move the object into the correct position to orbit the Earth. After it is in the correct position, rockets are once again used to separate the launch vehicle from the artificial satellite.

At Vandenberg AFB in California, the gantry rolls back at Space launch Complex 2 in preparation for the liftoff of the Joint Polar Satellite System-1 (JPSS).

Courtesy of NASA/Kim Shiflett

At a minimum, a rocket requires a speed of 25,039 mph to successfully reach space and achieve orbit. The orbital velocity required depends on altitude. An object that is sent into a higher altitude orbit can stay in space much longer than an object that is at a lower orbit. At lower orbits, the object will experience atmospheric drag, or *drag from Earth's atmosphere, which will eventually decay the satellite.*

Another factor for launching an object into orbit is the launch window. The launch window is *the time period in which a particular mission must be launched.* When determining the launch window, experts evaluate the weather and the conditions required for the most efficient flight plan. For example, launching a probe to Venus when Venus is at its farthest point away from Earth would not be the ideal launch window. Scientists plan their launch windows based on the time of the day, rotation of Earth, and distance to destination at any given time. Of course, if the launch window is reached and the weather on Earth is not ideal, the launch may need to be rescheduled. For example, if a satellite was launched during a hurricane, the wind speed and hurricane conditions could damage or destroy the satellite, causing millions of dollars in damage or loss.

Shapes of Orbits

When discussing artificial satellites, there are two basic shapes to the orbit: circular or elliptical. Although some orbits are circular, most orbits are elliptical. Elliptical orbits are shaped like an ellipse or oval. All the planets, most satellites, and the majority of moons in the solar system have an elliptical orbit.

The shape of the ellipse can vary. It can be long and thin, or round to look more like a circle. The term eccentricity is *used to describe the shape of the ellipse*. The eccentricity of an ellipse falls on a scale from 0 to 1. The closer the eccentricity is to 0, the rounder the ellipse. If the eccentricity is close to 1, the ellipse is longer and skinnier.

Did You Know?

Earth's orbit around the Sun has an eccentricity of 0.0167 which means its orbit is almost a perfect circle. Comets on the other hand have high eccentricity values, and their ellipses are very thin.

The German astronomer, Johannes Kepler, made some important discoveries about orbits. Kepler's First Law of Planetary Motion states that planets move around the sun in elliptical orbits. Kepler also discovered that the planet's speed within the orbit varied based on the distanced to the Sun. Perigee is *the closest point in an elliptical orbit.* During perigee, the planets (and artificial satellites) move faster than when they are at apogee, *the farthest point away in an elliptical orbit.*

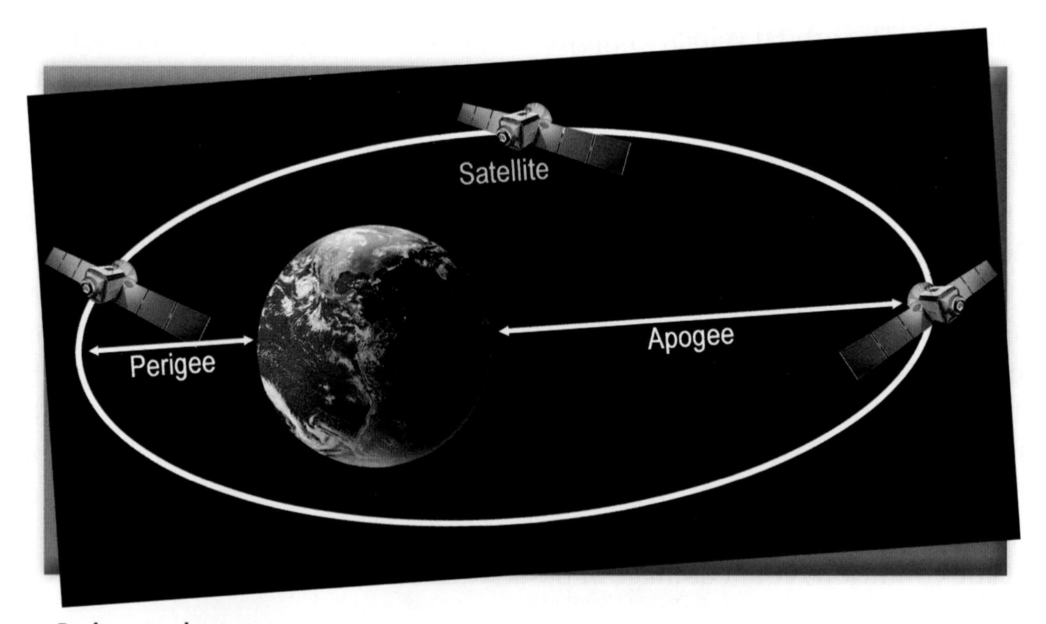

Perigee and apogee

Types of Orbits

Satellites are launched into several different orbital patterns based on the function of the satellite. The main types of orbit are low Earth orbit (LEO), highly elliptical orbit (HEO), geostationary orbit (GEO), or medium Earth orbit (MEO).

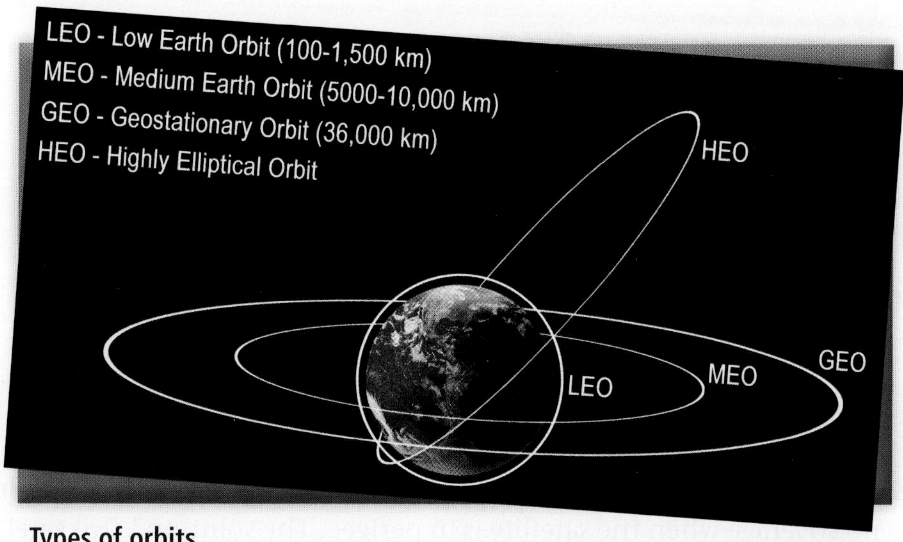

Types of orbits

Low Earth Orbits

A low Earth orbit (LEO) is an orbit between about 99 to 1,200 miles above Earth. The altitude of a LEO is designed so that the object is high enough to miss the tops of Earth's highest mountains and high enough to avoid the majority of atmospheric drag. Most satellites are in LEO. LEO is ideal for military satellites, human spaceflight, Earth observation, and weather satellites. LEO is convenient and has advantages for spacecraft, satellites, and the ISS . It is easy to send up crews for repairs or to conduct science experiments and return them to Earth rather quickly.

The disadvantages of LEO are atmospheric drag and speed. LEO objects do still experience some atmospheric drag, which means that over time their orbit will decay. The decay causes objects to slow down, meaning that gravity will eventually take over and pull the object back toward Earth. The decay caused by atmospheric drag on objects in LEO is slow, but it still exists.

The other disadvantage is the speed of the object orbiting the Earth. Objects in LEO need to travel at 18,000 mph or faster, which means they move around the Earth rather quickly. The rapid orbit provides very little time to study a specific area of Earth from space. *The time a satellite sits over one part of Earth* is called the dwell time. An LEO has a very short dwell time. To achieve a higher dwell time, a different orbital pattern must be considered.

Highly Elliptical Orbits

A highly elliptical orbit is *an orbit in which the satellite spends the majority of its time in apogee, moving very slowly*. This type of orbit allows a satellite to be above a specific region of the Earth for most of the time, before speeding around the Earth during perigee, as seen in Figure 7.3.

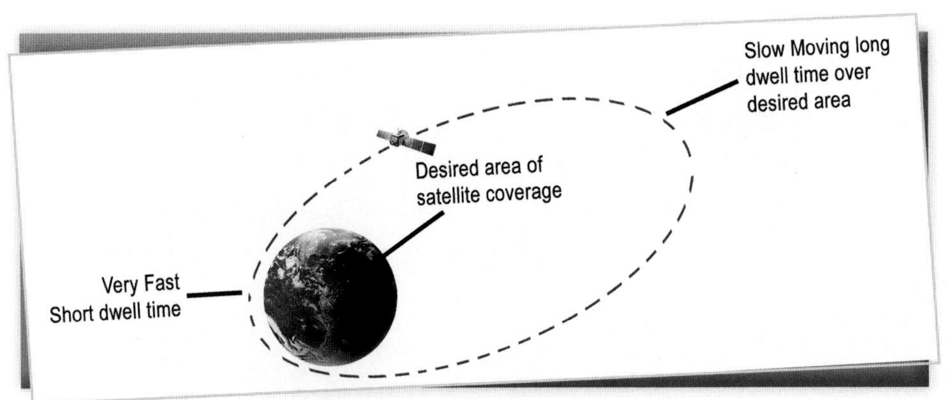

Figure 7.3

Using this type of orbit, the dwell time over a specific area is quite long. Of course, that leaves no coverage when the satellite is in perigee. The solution? Two satellites! A second satellite would be set up so that the two orbits would be on opposite sides, allowing for continuous coverage, as seen in Figure 7.4.

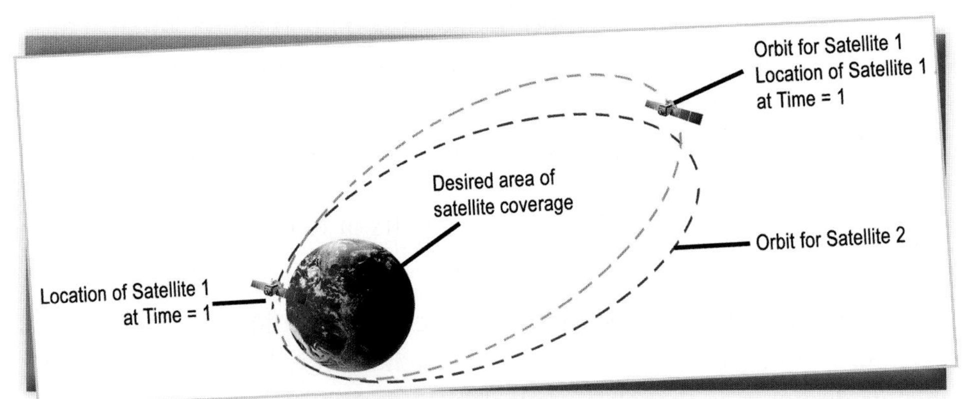

Figure 7.4

But what happens if we want coverage over the entire planet, rather than just one small area? A constellation of satellites would be used to set up multiple highly elliptical orbits around the Earth at different locations. As a result, the entire Earth would be (and is!) monitored by satellites.

Did You Know?

For GPS service, three or more satellites cover each location on the planet at any given time.

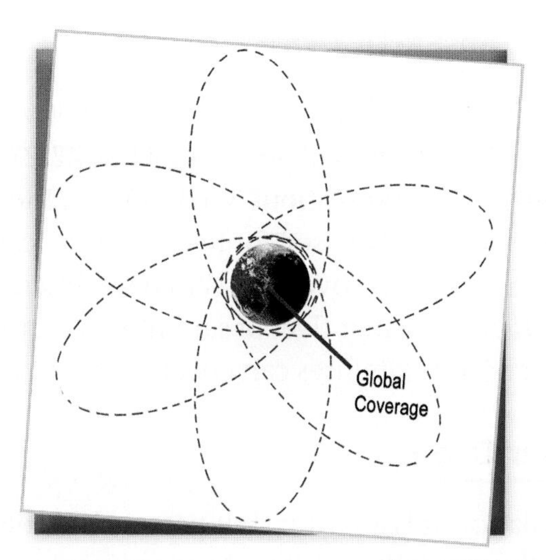

Figure 7.5

Geosynchronous Orbits

Geosynchronous orbits (GEO) travel at the same speed as Earth's rotation and occur at an altitude of 22,236 miles or higher above the Earth. Geosynchronous orbits have an orbital period of 24 hours. Because objects in geosynchronous orbit have such a high altitude, the atmospheric drag is avoided, and satellites can last for a very long time at this altitude. In addition, the dwell factor for geosynchronous orbits is infinite because the Earth's rotation and the satellite are travelling at the same speed.

Geostationary orbits are also included in the geosynchronous category of orbits. Geostationary orbits maintain a continuous presence over a large part of Earth. The main different between a geosynchronous and geostationary orbit is the angle at which they orbit. The geosynchronous orbit is tilted along the equator. A geosynchronous orbit appears at the same time and location on Earth every day, while a geostationary orbit appears as though it maintains a fixed spot above Earth.

Geostationary orbits are ideal for telecommunications satellites and satellites that monitor global weather patterns. Using a constellation of three geostationary satellites, coverage can be provided for the globe with the exception of the poles.

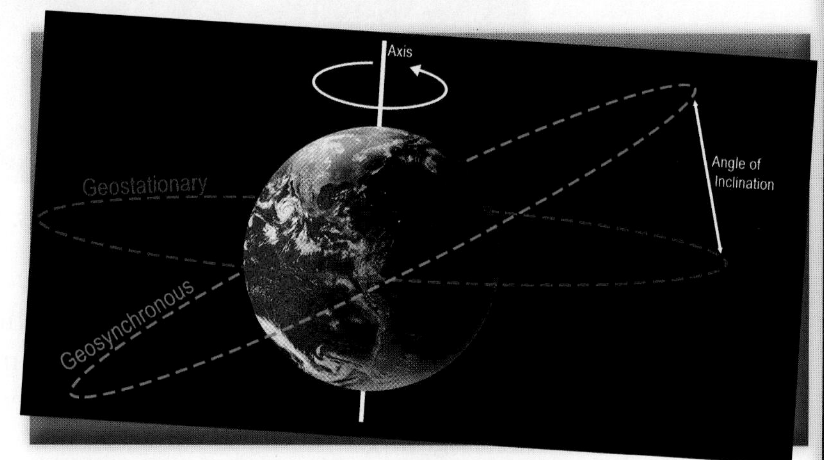

Geostationary Orbit vs. Geosynchronous Orbit

Medium-Earth Orbit

A medium Earth orbit (MEO) occurs between the LEO and GEO. Satellites in MEO operate at altitudes of about 1,243 to 22,232 miles. The orbital period for an MEO satellite is between about 2 to12 hours. Most navigation or GPS satellites are in MEO. MEO is also used for telecommunication satellites that provide global wireless communication coverage. When setting up a constellation of satellites for global coverage, a smaller fleet of satellites is required at MEO than at LEO because at the higher orbit, satellites can cover more area.

Polar Orbit

A polar orbit is an orbit that passes over the North and South poles at a low altitude. Satellites in polar orbit make many passes throughout the day and are used for reconnaissance and Earth observations.

Polar orbits are also referred to as Sun-synchronous orbits, but they are really a special polar orbit. When a satellite is in a Sun-synchronous orbit, its orbit is synchronized with the Sun, which means the orbit must shift by one degree per day to maintain the same angle as the Sun. Sun-synchronous orbits are used for solar studies, weather tracking, reconnaissance, and Earth observation.

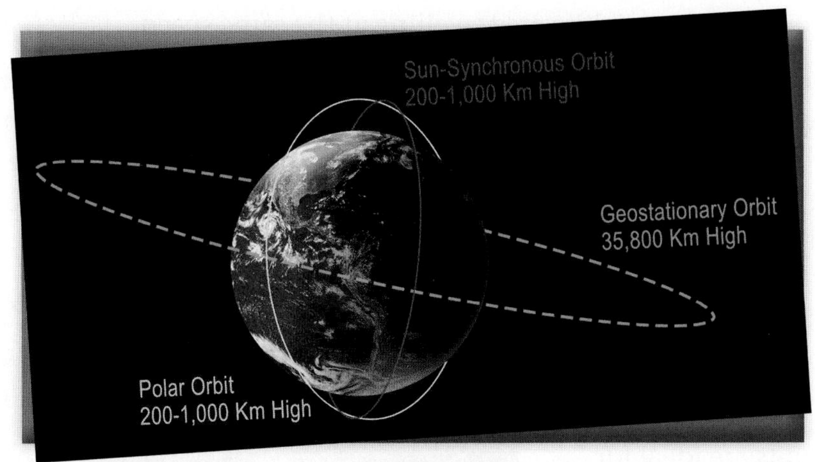

Polar orbit vs. Sun-synchronous orbit

Placing an object into orbit in space is just part of the process. Once an object is in space, the trajectory must be maintained, and it may need to be maneuvered to maintain the flight path. In our next lesson, we'll learn about trajectories, maneuvering in space, and navigating.

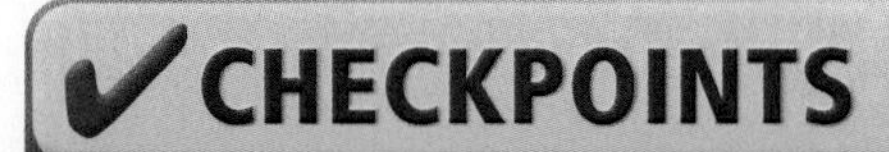

Lesson 1 Review

Using complete sentences, answer the following questions on a sheet of paper.

1. What does Newton's first law of motion have to do with orbits?
2. What orbital velocity is required to keep an object in orbit 150 miles above Earth?
3. What does eccentricity mean with regard to an orbit?
4. What are the disadvantages of a low Earth orbit (LEO)?
5. What is the orbital period of a geosynchronous orbit?
6. What is the difference between a polar orbit and a Sun-synchronous orbit?

APPLYING YOUR LEARNING

7. Using your own words, how would you explain how orbits work?

LESSON 2

How to Travel in Space

Quick Write

How do you think the Deep Space Atomic Clock can change the way we explore deep space?

Learn About

- trajectories in space
- maneuvering in space
- navigating in space

Accurate radio frequency is essential to any space mission to accurately determine the location of spacecraft. Ground-based atomic clocks have been utilized for deep space missions to provide the data necessary for precise positioning. A ground-based atomic clock works when subatomic particles called electrons are exposed to certain frequencies of radiation, such as radio waves, causing them to "jump" back and forth between energy levels. Clocks based on this jumping within atoms can therefore provide an extremely precise way to count seconds.

NASA's JPL is currently developing the Deep Space Atomic Clock (DSAC). This atomic clock is based on mercury-ion trap technology and will provide unprecedented time and frequency accuracy. **Mercury-ion trap technology** *uses electrically charged mercury atoms created in an electromagnetic trap for better control and accuracy.*

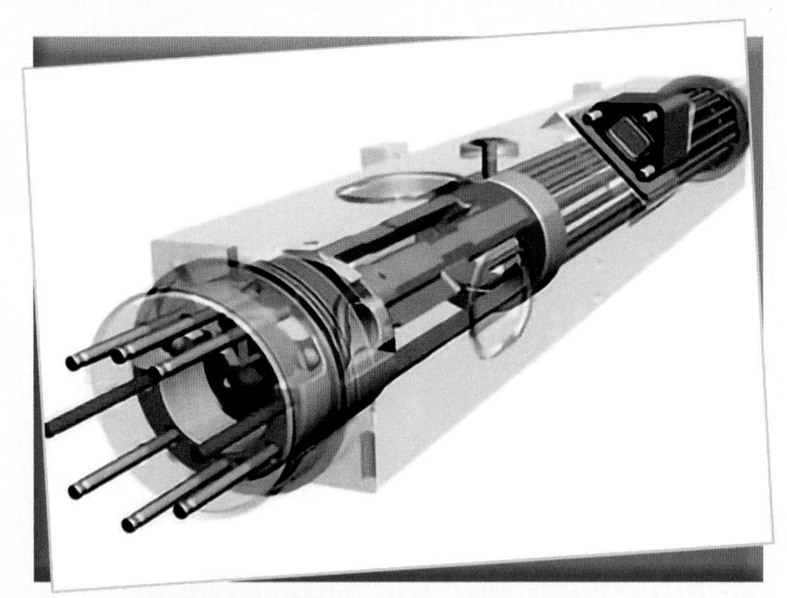

Deep Space Atomic Clock
Courtesy of NASA

The DSAC could ultimately change the way deep space missions are conducted since there is no need to "turn signals around" for tracking. DSAC will operate similar to GPS, using one-way signals to pinpoint a spacecraft's exact location. DSAC will account for the effects of gravity, space, and time that affect a spacecraft during its mission.

Engineers have been working on DSAC at NASA's JPL in Pasadena, California, for 20 years. They continue to improve and miniaturize the clock to prepare it for the harsh realities of space travel. Within the laboratory, the DSAC has been refined to permit a drift in time of no more than one nanosecond over a 10-day period. This provides more accuracy and stability than any other atomic clock sent into space. In addition to accuracy, the DCAS provides a cost savings for NASA by using GPS signals to track location.

Currently, a DSAC demonstration unit is being built by General Atomics Electromagnetic Systems in Englewood, Colorado. It is scheduled to be launched into orbit around Earth in 2019; it will spend a year in orbit to demonstrate its functionality.

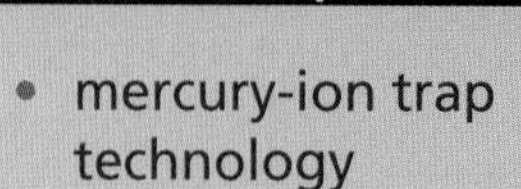

- mercury-ion trap technology
- trajectory
- gravity assist
- angular momentum
- thrusters
- ion propulsion
- protoplanet

Trajectories in Space

In Chapter 7 Lesson 1, we discussed orbits. In this lesson, we will focus on trajectories. A trajectory is *the path of a body in space*. Typically, the word trajectory is associated with paths that have a defined start and end point—for example, a probe that begins on Earth and travels to Jupiter. In comparison, an orbit is an infinite loop around another object.

In space, the path an object takes is determined by the gravity of objects near the object and Newton's Laws of Motion. Both of these factors must be considered when analyzing the path a spacecraft will take in space. A spacecraft can take three types of paths:

1. An elliptical path is a full eclipse or orbit.
2. A hyperbolic path is an open path that extends to infinity.
3. A parabolic path is a slightly open path that falls between an elliptical path and hyperbolic path.

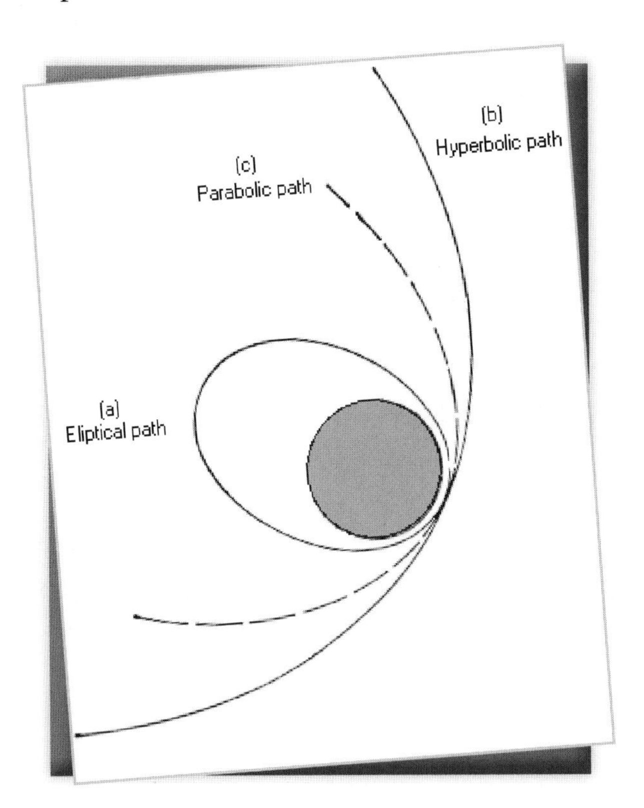

Types of paths
Courtesy of NASA

Escape velocity is the factor that determines if a flight path falls into an orbit or a trajectory. If the velocity of the spacecraft is greater than or equal to the escape velocity, then the trajectory of the spacecraft will take on an open-ended flight path like the hyperbolic path or parabolic path.

The Right Stuff

Escape velocity is calculated using the mass of the parent planet (Earth) and the distance between the center of the planet (Earth) and the spacecraft itself. A spacecraft that is at sea level on Earth has an escape velocity of 36,700 feet per second to escape Earth's orbit. Let's compare that to the escape body of other celestial objects:

Celestial Body	Escape Velocity
Moon	7,800 feet per second
Mars	16,700 feet per second
Jupiter	197,000 feet per second
Venus	33,600 feet per second
Asteroid Eros	50 feet per second

Types of Trajectories

There are two types of trajectories that can occur. Type I trajectory is when the trajectory carries the spacecraft less than 180 degrees around the Sun. Type II trajectory occurs when the trajectory carries the spacecraft more than 180 degrees around the Sun. Manned missions typically follow a type I trajectory, while cargo missions typically follow a type II trajectory.

Gravity-Assist Trajectories

Have you ever heard the news media refer to a spacecraft "slingshotting" around a planet? This refers to a spacecraft getting gravity assistance from another planet. Gravity assist *occurs when a planet's gravity is used to give a spacecraft additional momentum*. As a spacecraft moves around a planet, angular momentum, *the quantity of rotation of a body*, is transferred from the orbiting planet to the spacecraft as it approaches. This can approximately double the speed of the spacecraft. Most missions are carefully planned to take advantage of gravity assistance.

Most of the time a spacecraft moves because of gravitational influence from the Sun—except when the craft gets a gravity assist from other planets.

Did You Know?

The speed of a spacecraft may double with gravity assistance, but the craft itself feels no acceleration. This is due to the balanced tradeoff of angular momentum. If you were aboard a spacecraft that experienced gravity assistance, you would only feel the sensation of free-falling.

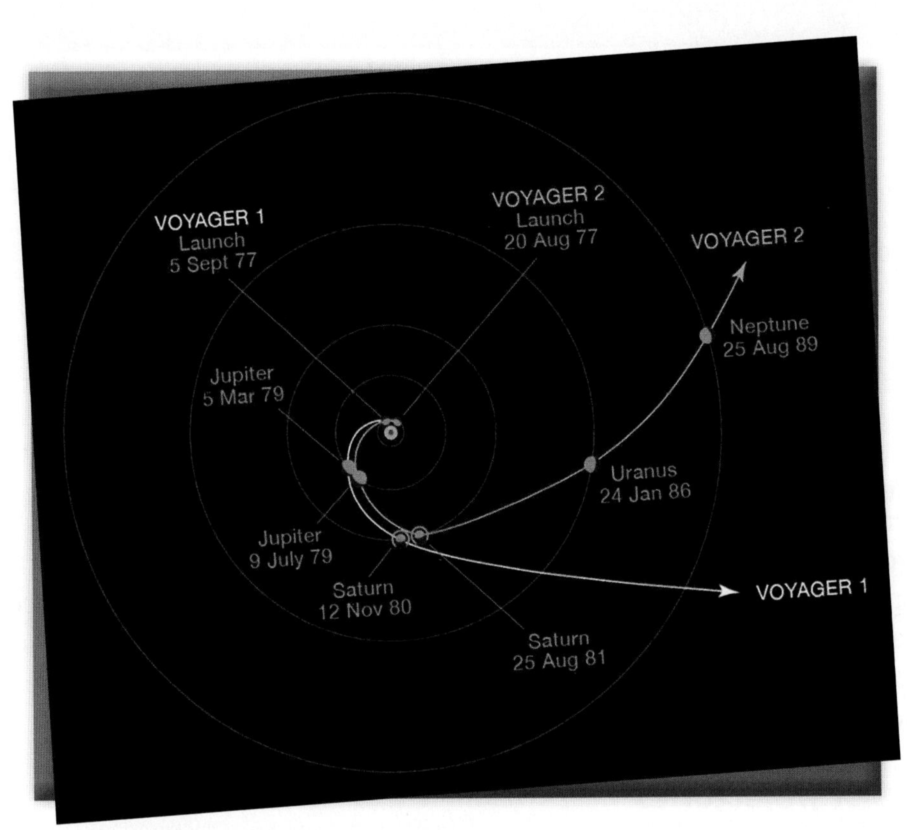

The Voyager missions used gravity assistance around Jupiter and Saturn to extend the missions to Uranus and Neptune.

Courtesy of NASA

Maneuvering in Space

We know that spacecraft use chemical rockets on Earth's surface to propel them into space. After the initial propulsion of the launch rocket, the spacecraft may use thrusters to adjust its trajectory. Thrusters are *small rockets on a spacecraft that are used to make alterations to the spacecrafts flight path*. Other than these thrusters, the spacecraft is free-falling through space and obtaining gravity assists wherever possible.

The Right Stuff

Early theories of space travel centered around Newton's Third Law of Motion—for every action there is an equal and opposite reaction. It was assumed that in order to explore space a combustible fuel must be burned to thrust a spacecraft into space. The theory, reaction propulsion, was the basic principle for space travel. Thus, it was assumed that only a small portion of the solar system could ever be reached for exploration.

In the 1960s, a graduate student at UCLA proposed a new radical theory for space travel—Gravity Propelled Interplanetary Space Travel. Michael Minovitch's theory did not require rocket propulsion in space and generated spacecraft velocities much higher than ever dreamed. In 1961, Minovitch proposed his theory to the head of NASA's Jet Propulsion Laboratory (JPL), and his theory was deemed impossible. As an employee of JPL, Minovitch continued his work and began solving other problems. He took on the unsolved Restricted Three-Body Problem, which, once solved, also made his Gravity Propelled Interplanetary Space Travel a reality. The Earth, Moon, and Sun are examples of a Restricted Three-Body Problem. This problem is created when a body of negligible mass (Moon or planetoid) moves under the influence of two massive bodies (Earth and Sun). The Moon exerts no force on the two massive bodies. Because of Michael Minovitch, the exploration of the entire solar system was now possible.

The Gravity Propelled Interplanetary Space Travel theory used escape velocity to send a spacecraft outside of Earth's orbit. Once outside of Earth's orbit, the spacecraft would basically be on a free fall in the solar system. However, when a spacecraft completed a fly-by of a planet, the gravitational force could be used as propulsion for the spacecraft and catapult the craft to a target farther into space.

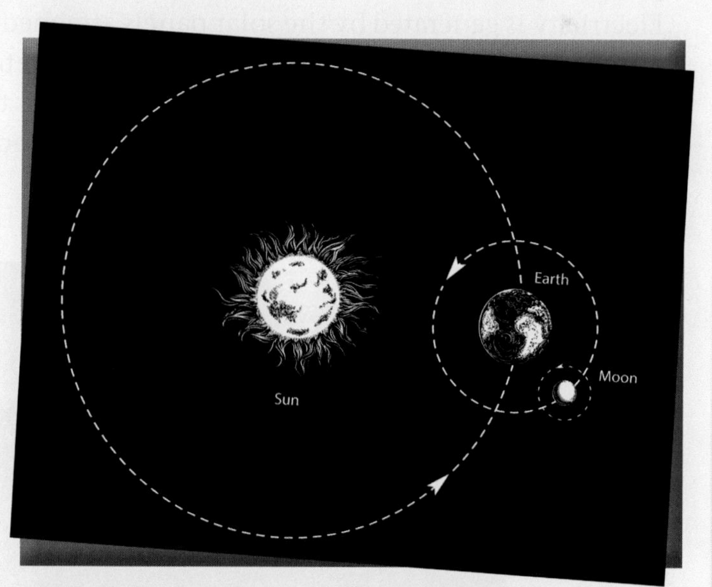

Orbit of the Sun, Earth and Moon - Vector

Courtesy of tutsi/Shutterstock

So how does a spacecraft change directions or its flight path in space? To understand this concept, we must first understand that a spacecraft is always orbiting something. It may be orbiting Earth, another planet, or the Sun. A spacecraft that is on the way to another planet is following its own orbit around the sun. Some spacecraft have even reached a velocity that puts them in a trajectory outside the solar system. These spacecrafts are orbiting the galaxy.

The speed of a spacecraft in space is directly related to the orbit it follows. The larger the orbit, the slower the spacecraft. To speed up the spacecraft, the orbital energy must be decreased to drop the spacecraft into a lower orbit, hence speeding it up.

In open space, thrusters are used to maneuver the spacecraft. If a spacecraft needs to speed up, a rear-facing thruster is fired. If a spacecraft needs to slow down, a forward-facing thruster is used. To alter the direction of the spacecraft, side thrusters are used. Scientists can also rotate the spacecraft by firing a pair of side thrusters.

Ion Propulsion

This concept of maneuvering in space is now changing with the success of Deep Space I and ion propulsion. Ion propulsion *uses magnetism and electricity to propel a spacecraft through space*. A gentle, continuous thrust over the course of months or year provide momentum to the spacecraft.

Have you ever tried to hold two magnets together? The magnets repel each other and push away from each other. This is similar to the concept of ion electric propulsion. Electricity is generated by the solar panels attached to the spacecraft. The positive electric charge created by the solar panels is given to the atoms inside the chamber. Magnetism is then used to pull the atoms towards the back of the ship. Magnetic repulsion is then used to push them out of the spacecraft to give it thrust.

Artist's rendering of ion propulsion.
Courtesy of NASA

Ion propulsion is a more efficient method for powering a spacecraft. A spacecraft using ion electric propulsion will travel faster than a spacecraft using chemical rockets. For example, the Deep Space I spacecraft reached speeds of 200,000 mph, whereas the Space Shuttle's speeds averaged 18,000 mph. Going faster means that the spacecraft can reach its destination faster.

One of the trade-offs of ion electric propulsion is low thrust. This type of propulsion can only be used in the vacuum of space. Ion propulsion cannot be used to launch a spacecraft from Earth. The thrust, however, adds up and builds the acceleration of a spacecraft leading to less fuel used and a shorter travel time.

In addition, the lower amounts of fuel used decrease the weight of the spacecraft and thus lower the cost of the launch. A smaller launch vehicle and less money goes into the launch of ion propulsion spacecraft.

Did You Know?

NASA's Dawn was a mission to explore the area between Mars and Jupiter and gain knowledge on how the solar system formed. Dawn successfully orbited the protoplanet Vesta and the dwarf planet Ceres as part of its mission. A protoplanet is *a large body of matter in orbit around the sun or a star and thought to be developing into a planet.* Using ion propulsion, Dawn became the first spacecraft ever to orbit two destinations.

Navigating in Space

Navigation is a key component of any space mission. After all, the mission would be a failure if the spacecraft did not go to the location desired. There are three parts to navigation of any mission.

Mission design is the first step. Mission design is responsible for planning the trajectory of the spacecraft. NASA's JPL utilizes the Mission-Analysis Operations and Navigation Toolkit Environment (MONTE) software to analyze and design the trajectory. MONTE is a sophisticated software suite that was designed using over 50 years of space travel data, abstract mathematics, and the formulations of many astronomers, such as Galileo, Newton, Kepler, Tsiolkovsky, Einstein, and others.

The next component of navigation is orbit determination. Orbit determination tracks the spacecraft position during the mission. The orbit determination team is constantly tracking where the spacecraft has been, where it is presently, and where it will be in the future. A spacecraft in deep space is constantly drifting from its trajectory. The orbit determination team must take this drifting into account to determine the accurate path of a spacecraft.

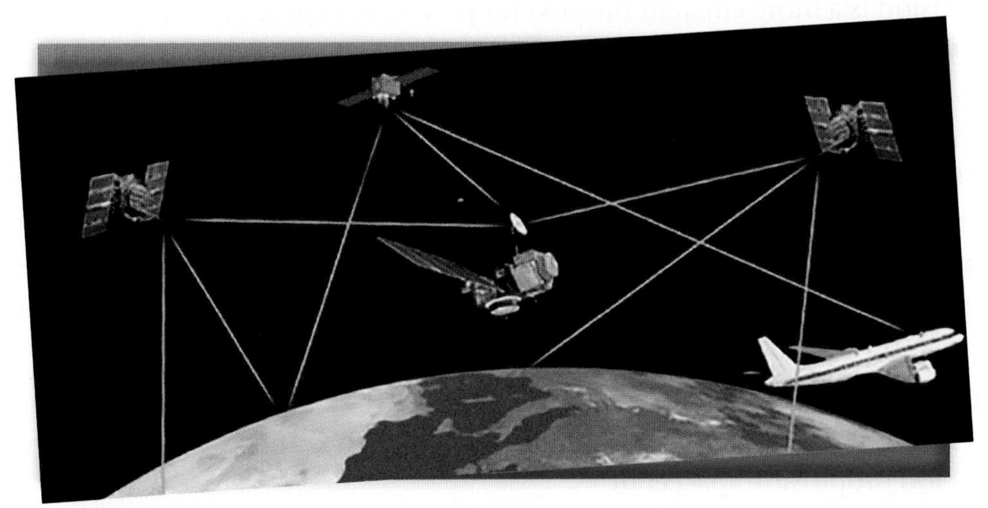

GIPSY, is the GNSS-Inferred Positioning System and Orbit Analysis Simulation Software package by JPL used to help determine a spacecrafts location.
Courtesy of NASA/JPL

Finally, we have flight path control. Flight path control creates maneuvers to keep spacecraft on the planned trajectory. This team determines how far a spacecraft has drifted from its planned trajectory and designs maneuvers to get the spacecraft back on track. The flight path control team can initiate one of the following maneuvers to get the spacecraft back on track:

- Trajectory Correction Maneuver (TCM)—This is a minor flight-path control maneuver.
- Orbit Trim Maneuver (OTM)—This is used for spacecraft that are repeatedly orbiting another planet or object to help maintain the ideal orbit of the spacecraft.
- Deep Space Maneuver (DSM)—This is a large TCM that is used when gravity assist is not available.

The navigation tracking process is an endless cycle during a spacecraft's mission. The teams are constantly tracking the location and trajectories of a spacecraft and performing maneuvers to keep the spacecraft on track.

Lesson 2 Review

Using complete sentences, answer the following questions on a sheet of paper.

1. What are the two factors that must be considered when determining the path of an object in space?
2. What factor determines if a flight path falls into an orbit or trajectory?
3. What is the difference between a type I and type II trajectory?
4. How does the orbit of a spacecraft affect its speed?
5. How does ion electric propulsion work?
6. What are the responsibilities of the orbit determination team?
7. What are the three maneuvers used by a flight path control team?

APPLYING YOUR LEARNING

8. Why do you think it's important to understand how a spacecraft maneuvers in space?

LESSON 3 How Rockets Work

Quick Write

Read the vignette about how Robert Goddard's rocket research continued despite the resistance he encountered from fellow scientists. How far do you think American rocketry would have advanced if he had not continued his research?

Learn About

- how a rocket works
- the evolution of rocket development
- different types of rockets
- launch logistics

The most influential American on rocket technology was Robert Goddard. Goddard founded an entire field of science and engineering with his developments. In 1902, Goddard wrote "Navigation of Space," which outlined a tactic of using several cannons arranged like a nest to navigate in space. Then, in 1912, he developed the mathematical theories of rocket propulsion for which he was awarded two patents in 1914.

Dr. Robert Goddard and a liquid oxygen-gasoline rocket fired on March 16, 1926.
Courtesy of NASA Goddard

Did You Know?

Goddard initially wrote "Navigation of Space" as an article that he submitted to "Popular Science News," they chose not to publish the article. Goddard was in high school when he wrote "Navigation of Space."

In 1915, Goddard proved that rockets could produce thrust in a vacuum, making spaceflight a possibility. The Smithsonian Institution offered Goddard a grant in 1916 to devise a theoretical base to use rockets to explore the upper atmosphere. Goddard discovered that if velocity is 6.96 miles per second with no air resistance, an object could escape Earth's gravity, thus establishing the escape velocity.

Goddard was ridiculed for his research because many thought space travel was impossible. He became very secretive about his research, but never abandoned his studies. Overall, he was granted 214 patents that mostly focused on liquid-fueled rockets.

On March 16, 1926, Goddard launched his first rocket. It rose 41 feet in 2.5 seconds. The launch caught the eye of Charles Lindbergh and Daniel Guggenheim. Guggenheim provided Goddard a $50,000 grant to set up an experiment station in Roswell, New Mexico. His developments continued from 1930 to 1941. He was able to develop systems for steering a rocket in flight, and navigation systems to keep the rocket going in the correct direction. His research resulted in the successful launch of a rocket to 9,000 feet in 1941.

When World War II began, Goddard entered the Navy and spent his time in the service developing jet-assisted takeoff (JATO) for military aircraft. Goddard died in 1945, but many of his theories became standards in rocket development. NASA named the Goddard Space Flight Center in honor of Robert Goddard. NASA's Goddard Space Flight Center in Greenbelt, Maryland, is home to the nation's largest organization of scientists, engineers, and technologists who build spacecraft, instruments, and new technology to study Earth, the Sun, our solar system, and the universe.

Vocabulary

- thrust
- booster
- expendable launch vehicles (ELVs)
- heavy lift launch vehicles
- multistage
- Thor missile
- hypergolic fuels
- propellant
- oxidizer
- cryogenic propellants
- space launch system (SLS)
- reusable launch system (RLS)
- solid-fuel rocket
- combustion chamber
- plasma rockets
- fission rockets
- nuclear-electric rockets
- nuclear-thermal rockets
- fusion rockets
- fusion
- antimatter rockets
- antimatter
- launch sites
- clean room
- launch vehicle stage adapter (LVSA)
- launch pad
- crawler-transporter

How a Rocket Works

Understanding how a rocket works isn't rocket science! It's actually a simple concept: exhaust gases come out of an engine and push the rocket forward. Let's look at an example. If you blow up a balloon and then let the balloon go, the air is expelled and the balloon moves in the opposite direction. This is the basic principle of how rockets work. They use Newton's principle "for every action, there is an equal and opposite reaction." For that reason, rocket engines are referred to as reaction engines.

Let's imagine you are an astronaut floating in space in your space suit. You throw a ball in space, which causes your body to move in the opposite direction. Now how fast will you move? Well, that depends on the mass (weight) of the ball and the acceleration of the ball. According to Newton's theories, the force that is applied to the ball will equal the force applied to move you backwards. But, how do we calculate force:

mass (weight) x acceleration = force

In this scenario, let's say the mass of the ball is one pound and you weigh 100 pounds with your spacesuit on (remember you weigh less in a zero-gravity environment). You throw the ball at 20 mph, but you weigh 100 times more than the ball so you must take your mass into consideration. So, if the ball is thrown at 20 mph, then you move backwards at .20 mph. (force (.20 mph = weight (1 pound/100 pounds – your weight) * acceleration (20 mph) Now what if you want to create enough force to go faster? Using our formula for force, we need to either increase the mass or acceleration to increase the force.

Rockets use high-pressure gas as their mass. The engines produce the mass and expel it at high speeds causing the rocket to move in the opposite direction. Thrust is *the strength of a rocket measured in pounds*. One pound of thrust is what is needed to keep a one-pound object motionless against the force of gravity on Earth. The acceleration of gravity on Earth is about 21 mph per second (32 feet per second). If we go back to the scenario of throwing a ball in space, you would produce one pound of thrust by throwing a ball 21 mph.

One of the difficulties of putting a rocket in space is the weight. A rocket is designed to carry a payload into space and that payload has a mass associated with it. This weight needs to be taken into account when determining how much thrust is needed. In a typical launch, the fuel ends up weighing much more than the payload. The rocket will need a booster to get into Earth's orbit; *a booster is the first stage of a multistage launch vehicle.*

The Right Stuff

How much fuel was required to launch the Space Shuttle?

When calculating the fuel required to launch a Space Shuttle into orbit, the mass (weight) of each component must be accounted for. The Space Shuttle had three components:

- Orbiter–165,000 pounds empty
- External tank–78,100 pounds empty
- Rocket boosters–185,000 pounds each empty

We have a total weight of 428,100 pounds for an empty space shuttle. Each orbiter was equipped to hold up to 65,000 pounds of payload, which gets us up to 493,100 pounds. Now let's add in the fuel:

- External tank–143,000 gallons of liquid oxygen (1,359,000 pounds) and 383,000 gallons of liquid hydrogen (226,000 pounds)
- Rocket boosters–1,100,000 pounds of fuel each

The Space Shuttle Columbia launch.
Courtesy of NASA

Our total weight of the space shuttle is over 3.1 million pounds! A mass of 3.1 million pounds is required to get 493,100 pounds of payload into orbit. We needed 20 times more fuel than the payload for a Space Shuttle to reach orbit.

What Are Launch Vehicles?

Launch vehicles are rockets that are used to place objects in orbit or space. When discussing space, launch vehicles and rockets are interchangeable. They come in a variety of shapes and sizes based on the goals of individual missions.

A launch vehicle is a complicated piece of equipment that must accelerate itself and its payload to a minimum velocity of 25,000 mph to leave the protective atmosphere of Earth. That's about 32 times the speed of sound!

Vice President Mike Pence tours the Blue Origin Manufacturing Facility near NASA's Kennedy Space Center in Florida on Feb. 20, 2018.
Courtesy of NASA

There are two types of launch vehicles:

- Expendable launch vehicles (ELVs)
- Reusable launch systems

Expendable launch vehicles (ELVs) *are used only once to carry payloads into space.* An ELV usually consists of several rocket stages, discarded one at a time, as the ELV gains altitude and velocity. Reusable launch systems are launch vehicles in which one or more components can be reused. Until recently, the partially reusable space shuttle was the only reusable launch system in existence. All other launch vehicles had been designed as ELVs.

Another designation used for launch vehicles is heavy lift launch vehicles. Heavy lift launch vehicles *launch larger payloads into space.* Heavy lift launch vehicles are used to send large satellites into higher geostationary orbits.

Launch vehicles evolve and continuously improve in capacity, reliability, and purpose. Early launch vehicles delivered payload to other locations on Earth. Others sent various satellites into orbit and, of course, have sent crews to the International Space Station (ISS) and to the Moon.

The Evolution of Rockets

While it's unknown who first invented rockets, the evolution of rockets can be traced back to the Chinese discovery of gunpowder. It wasn't until the nineteenth century that rockets made space travel a possibility.

In 1903, Russian theorist Konstantin Tsiolkovsky wrote the "Rocket Equation," that included calculations for space flight. His research provided the groundwork for space travel but would take years to develop.

Hermann Oberth and Wernher von Braun

You read about Dr. Robert Goddard in the opening vignette, now we'll discuss two other influential scientists that were key to the development of rockets as we know them today. Hermann Oberth (1894-1989) was born in Romania but relocated to Germany. His interest in rockets began early. At age 14, Oberth imagined a recoil rocket that moved through space using its own exhaust. As an adult, Oberth continued his studies by focusing on multistage rockets and how rockets could escape Earth's gravity. In 1923, Oberth published "The Rocket into Planetary Science," which discussed every phase of rocket spaceflight.

As a young man, NASA scientist Wernher von Braun (1912-1977) was inspired by Hermann Oberth. Braun was one of the greatest rocket scientists of all time. Fascinated with space exploration, he mastered calculus and trigonometry at a very young age. These led him to the German Society for Space Travel in 1928. Just a few years later, von Braun began work with the German army to develop liquid-fuel rockets. During World War II, Oberth and von Braun used a secret laboratory to develop the V-2 rocket for Germany.

Professor Hermann Oberth and Dr. Wernher von Braun are briefed on satellite orbits by Dr. Charles A. Lundquist at Army Ballistic Missile Agency, Redstone Arsenal, Huntsville, Alabama.
Courtesy of NASA

The V-2 rocket was 46 feet long and weighed in at 27,000 pounds. The rocket was designed to reach speeds over 3,500 mph and carry a 2,200-pound warhead. The V-2 was a sophisticated rocket and used gyroscopes to control its altitude. The first test flight of the V-2 occurred in 1942. By September 1944, Germany began using the rockets in the war. By the end of World War II, over 2,800 V-2 rockets had been launched at England and Antwerp, Belgium.

During World War II, the US was also working on rocket development. It captured V-2 rockets so that its components could be studied as part of a secret project known as Project Paperclip. After World War II, captured German scientists immigrated to the US or Soviet Union to continue their research. Von Braun and his colleagues chose to surrender to US forces and continue their work in the US. A V-2 test program was set up at Fort Bliss in El Paso, Texas for von Braun and his team.

Did You Know?

When Von Braun surrendered, the US government chose to overlook his Nazi past and war crimes. Von Braun used concentration camp workers to construct his rockets in Germany. According to the Smithsonian Air and Space Museum, at least 10,000 workers died in the construction of the V-2 rocket. Which is more than those killed by being hit by the V-2 rocket.

After World War II, the US initiated a new rocket project dubbed the Bumper Project. The project used a smaller missile on top of the V-2 rocket to test two-stage rockets. On February 24, 1949, the test launch successfully reached 244 miles in altitude at a velocity of 5,150 mph.

For the next 15 years, von Braun worked for the US Army to develop ballistic missiles and to assist in V-2 launches.

During those years, von Braun and his Army Redstone Arsenal team designed several ballistic missiles and launch vehicles. In 1958, one of the Redstone rockets, the Jupiter C, orbited the first US satellite, Explorer I.

Did You Know?

The final launch of the Bumper Project occurred on July 24, 1950, and was the first successful takeoff at Cape Canaveral, Florida.

A Bumper-WAC, a combination of the V-2 rocket with a WAC Corporal upper stage, awaits launch on July 24, 1950. It was the eighth in the Bumper Project; the vehicle reached an altitude of 393 kilometers (244 miles).

Courtesy of MSFC/NASA

Holding a model of Explorer I, from left, Director William Pickering and scientists James Van Allen and Wernher von Braun announce the satellite's successful launch on Jan. 31, 1958.

Courtesy of NASA

In 1960, von Braun became director of NASA's Marshall Space Flight Center. The Saturn V launch vehicle was developed under his direction, and it successfully launched the Apollo 11 crew for the first Moon landing. This was a personal mission and dream come true for von Braun.

In 1970, von Braun retired after briefly heading up strategic planning initiatives for NASA. He died in Alexandria, Virginia, on June 16, 1977.

Important Rockets Throughout History

Trying to figure out how to get rockets to escape Earth's atmosphere was a major challenge at the time. Rocket engineering was just developing, and computers were in their infancy.

The US and Soviet Union seized many of the V-2 rockets left over after World War II. Both countries studied the V-2 to help develop the design of future launch vehicles, such as the Russian R-7. The R-7 would become the next milestone in rocketry.

The R-7 Intercontinental Ballistic Missile (ICBM)

While von Braun and his team developed ballistic missiles for the US Army, the Russians were also at work. They were creating the first Intercontinental Ballistic Missile (ICBM). An ICBM is a guided ballistic missile intended for payload delivery (including nuclear weapons) to destinations over 3,000 miles away.

Rollout of Soyuz TMA-2 aboard an R-7 rocket.
Courtesy of NASA

Although the Soviet R-7 was intended for weapon delivery, it turned out to be impractical as a weapon. Instead, it became a most important ELV for the Russian space program. On October 4, 1957, the Soviet R-7 launched Sputnik, the world's first artificial satellite, and placed it in low Earth orbit.

The R-7 was much bigger than the V-2 and weighed just over 275 tons. It included a multistage design. Multistage *refers to an ELV having two or more engines, stacked one on top of another and firing in succession.*

The R-7 used four boosters on the first stage. The second stage used a single core booster. Each engine burned kerosene fuel, instead of the alcohol fuel used by the V-2. Both stages would ignite during launch. The four boosters of the first stage would separate from the core two minutes after liftoff.

Vostok/Voskhod, Soyuz, Tsiklon and Proton

Improving on the R-7, Russia would go on to create Vostok, Voskhod, Soyuz, Tsiklon, and Proton ELVs. Engineers used the original R-7 core and four booster rockets for Vostok and Voskhod. They also added a third stage (called Block 1) to the top, for the purpose of delivering payloads to the Moon.

An engine with one main thrust chamber and four extra chambers powered the Vostok Block. The Voskhod Block used four large combustion chambers and four additional chambers. These provided a more powerful engine and larger fuel tanks.

The launch and first two stages of Vostok and Voskhod followed the same sequence as the R-7. After reaching velocity for entering orbit, the third stage would also separate from the spacecraft.

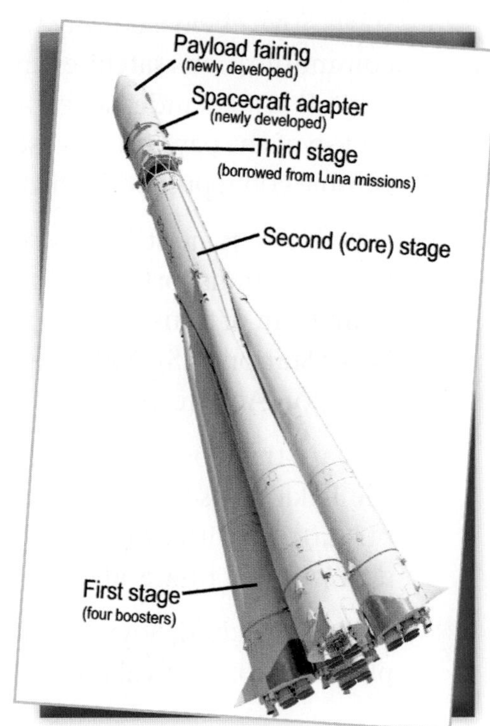

Vostok rocket stages.

The Right Stuff

We credit the Vostok missions with many "firsts in space" for Russia and the world. On April 12, 1961, Yuri Gagarin flew into space aboard Vostok 1. Gagarin became the first person ever to orbit the Earth.

Other Vostok mission "firsts" followed. These include the first full day in orbit, the first two-spacecraft mission, the first five days in orbit, and the first woman in space, Valentina Tereshkova.

Voskhod also established a few firsts. Voskhod 1 launched on October 12, 1964 with the first three-person crew. On March 18, 1965, Voskhod 2 carried Alexei Leonov, who would make the first spacewalk.

The rocket that powered the world's first manned spacecraft, Vostok 1, at an exhibition in Moscow, Russia.

Courtesy of vastram/Shutterstock

By 1966, the larger Soyuz vehicle surpassed Voskhod. This ELV incorporated yet another stage. Soyuz held a booster that would help place communications satellites and early probes into orbit. It also included an escape tower, mounted on top of the payload. In case of emergency, the capsule could pull away with the crew.

Did You Know?

Soyuz is the longest-lived line of rockets, with an estimated 1,700 flights. It has had minimal failures, making it a very dependable rocket.

Soyuz rockets are still in use today for cargo and astronaut transportation. In fact, they are the main form of transportation to the International Space Station (ISS). NASA began purchasing seats on Soyuz spacecraft in 2011 to transport American crews after the retirement of the US Space Shuttle Program.

Perhaps the most dependable of all Russian (now Ukrainian) rockets, though, is the Tsiklon (also referred to as Cyclone). Starting in 1967, Tsiklon made hundreds of launches with only a few failures. Tsiklon 3 last launched in 2009. Tsiklon 4 is in development now and scheduled to be in service by 2020.

Vostok, Voskhod, Soyuz, and Tsiklon led to many firsts in space. But no other ELV has played a more critical role in Russian space exploration than Proton.

Early in the space race, Proton was designed as a mega-sized ICBM specifically for space missions. It has three or four stages, depending on mission goals. With four stages, Proton uses a restartable engine and can launch payloads of almost five-tons.

The Proton line of launch vehicles deserves high marks! It has launched many cosmonauts, scientific satellites, and several components of the Salyut and Mir space stations.

The launch of a Proton M rocket in Baikonur, Kazakhstan, on December 11, 2011. Proton M includes a number of improvements over earlier models, including a digital flight control system.

Courtesy of Nostalgia for Infinity/Shutterstock

Delta

Just after NASA was created, it contracted with the Douglas Aircraft Company (now Boeing) to create a new ELV called Delta. It used a version of a Thor missile, *an intermediate range ballistic missile (IRBM)*. The missile used a single-stage, liquid-oxygen rocket motor.

A second stage used hypergolic fuels. Hypergolic fuels *can be stored as liquids at room temperature and are easily ignited*. By using storable propellants, Thor could remain constantly fueled. A third stage of the ELV used solid propellant. A propellant is *a substance, usually a mixture of fuel and oxidizer, used to accelerate an object*.

In 1960, NASA launched Delta 1. Its first successful launch involved a passive communications satellite called Echo. Delta 1 placed the first satellite to observe solar gamma rays. It also launched the famous Telstar satellite discussed in a previous lesson.

Delta 1 was only able to deliver about a 100-pound payload. For that reason, Douglas Aircraft needed to make several changes. Engineers improved the engine and added solid rocket boosters. They widened and lengthened propellant tanks. What did these changes achieve? They increased Delta's payload capacity and thrust several times over. Delta I evolved into the next generation ELV, Delta II.

Delta II could successfully launch two tons of payload into geostationary orbit. Some of Delta II's most memorable missions include NASA's rovers Spirit and Opportunity, and the Phoenix Mars Lander.

Since Delta I and II, we have seen continued growth in payload capacity. Today, Delta IV Heavy can launch a payload of over 25 tons into LEO. Delta IV Heavy is credited for launching a test flight of the Orion spacecraft in 2014 and for launching the Parker Solar Probe in 2018.

United Launch Alliance Delta IV Heavy common booster core is transported by truck to Cape Canaveral, Florida.

Courtesy of NASA/Kim Shiflet

Gemini Titan II rocket at New York Hall of Science Rocket Park in Flushing on September 2, 2013. Titan was a family of US expendable rockets used between 1959 and 2005.

Courtesy of Leonard Zhukovsky/ Shutterstock

Titan

The Titan ELV was also in development in the 1950s. This rocket was conceived by the US Air Force and built by Lockheed Martin as a two-stage ICBM. It proved to be a significant ELV for the military during the Cold War. Its first stage used two rocket engines, and the second stage used a single engine. Both stages used kerosene as fuel and liquid oxygen as an oxidizer. An oxidizer is *a type of chemical used to burn fuel in space where an atmosphere containing oxygen is absent.*

Titan 1 launched in 1962 with 54 launches to follow. The early missions provided a great foundation for later models, such as the Titan II. Titan II made great improvements to ELV components and replaced the Titan I in 1965. Enhancements to Titan II made 10 Gemini spacecraft missions possible.

Titan III came on the scene in 1964, equipped with a pair of solid rocket boosters. A restartable engine was added to increase payload capacity for large military payloads. One Titan III model used an upper stage fueled by cryogenic propellants. Cryogenic propellants *are gasses chilled to subfreezing temperatures to create very combustible liquids*. The payload of Titan III included Gemini spacecraft and spy satellites. It also held solar probes, both Viking Mars probes, and both Voyager probes.

The Titan III further evolved in size, payload capacity, and technology to Titan IV. With its enhancements, the ELV could deliver extra-large military payloads. It was used to launch the Cassini and Huygens probes. The last Titan rocket launch, a Titan IVB, lifted off on October 19, 2005.

Launch of Friendship 7, the first American manned orbital space flight. Astronaut John Glenn was aboard the modified Atlas Intercontinental Ballistic Missile. May 5, 1961.

Courtesy of Everett Historical/Shutterstock

Atlas

Development of the Atlas family of ELVs also began in the 1950s. Atlas was the first American ICBM, conceived by the US Air Force and built by Conair (now Lockheed Martin). Atlas delivered the first communications satellite and the first Americans to orbit Earth. In fact, John Glenn's first Earth orbit was the result of an Atlas ELV!

Like many other ELVs of the time, Atlas represented constant evolution. Improvements allowed for larger and heavier payloads. Unique to the Atlas design is its very lightweight body. Atlas has launched nearly 600 times. It completed launches for national security, space exploration, and commercial missions. All achieved 100 percent mission success!

Saturn V

Saturn V was designed under the leadership of Wernher von Braun at NASA's Marshall Space Flight Center. The primary goal was to carry humans to the Moon. Saturn V was also developed to increase reliability and payload capacity over previous ELVs.

First launched in 1967 for Apollo spacecraft, Saturn V used a three-stage design. The first stage included a rocket booster cluster of five engines using hydrogen-oxygen. The first stage would lift the rocket to an altitude of nearly 42 miles.

The second stage would then go from there into orbit. The third stage placed the payload into Earth's orbit with a push toward the Moon. All three stages would burn off their engines and then separate.

Kennedy Space Center, FL. Visitors looking at the Saturn 5 rocket, November 6, 2018.

Courtesy of John Silver/Shutterstock

Saturn V was used to carry a crew of three on Apollo spacecraft 8 through 17, including the first mission to land a man on the Moon. Saturn V also took astronauts to the Moon on Apollo 12, 14, 15, 16 and 17. The last Saturn V launched in 1973 when it carried the Skylab space station into orbit.

Saturn V is the largest launch vehicle ever built by the US. Its first stage used four and a half million pounds of propellant at 15 tons per second. Its capacity (to LEO) was 130 tons. In its entire history, Saturn V experienced no losses.

Space Launch System

NASA's very own Space Launch System (SLS) is in development! It may be the most powerful it has ever built. According to NASA, the Space Launch System (SLS) is *a heavy lift launch ELV rocket that is part of NASA's deep space exploration plans.*

The SLS is being designed as part of NASA's Orion mission. Orion is a spacecraft designed for deep-space missions with astronauts. Orion will serve as the exploration vehicle that will carry the crew to space, provide emergency abort capability, sustain astronauts during their missions, and provide safe reentry from deep space.

NASA plans to deliver the first SLS launch vehicle, called Block 1, to Kennedy Space Center in Florida for launch in 2020. It will use a core stage with four RS-25 liquid-propellant engines (the same type of engine used in the space shuttle) and two solid rocket boosters. After reaching space, an Interim Cryogenic Propulsion Stage (ICPS) will send Orion on to the Moon.

The next planned SLS vehicle, called Block 1B, will use a more powerful Exploration Upper Stage (EUS) for more complicated missions. It is being designed to carry the Orion crew vehicle and other exploration systems and science spacecraft. After that, Block 2 will be designed to carry even more payload (more than 45 tons) to the Moon, Mars, and other deep space destinations.

The primary goal of SLS and its evolving design is to lay the foundation for human exploration beyond Earth's orbit. The SLS will also offer new options for robotic missions to places like Mars, Saturn, and Jupiter.

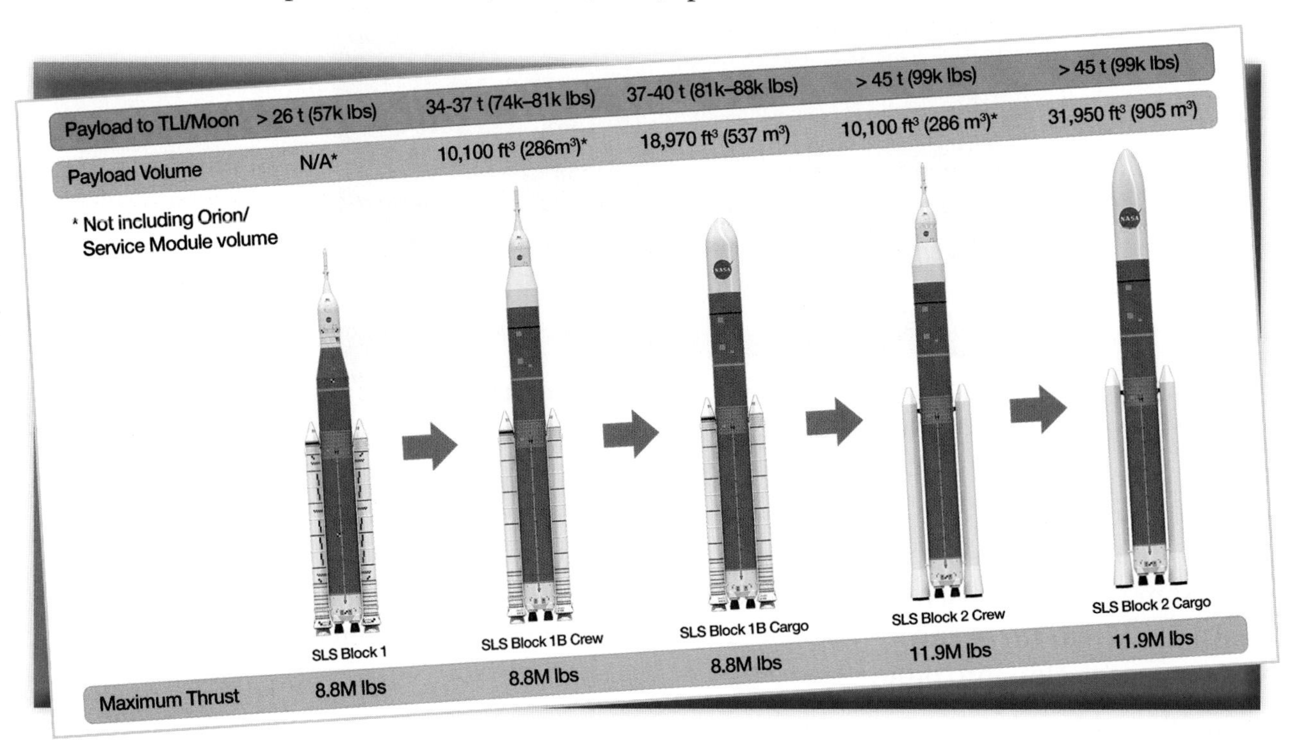

SLS Evolution

Courtesy of NASA

Reusable Launch Systems

In 1969, President Richard Nixon formed the Space Task Group. It proposed a more cost-effective means of going to and from space. Its proposal was the Space Shuttle Program (see Chapter 4). Unlike an ELV, the space shuttle consisted of three reusable components. These were the spacecraft (orbiter), the rocket boosters, and the engines. While the external fuel tanks of a space shuttle burn off after launch, scientists and engineers wanted to reuse the remaining components. They carefully inspected, disassembled, cleaned, reassembled, and sometimes replaced, each part.

Did You Know?

The process of reusing components was intended to save money. Instead, the Space Shuttle Program ended up being more expensive.

A reusable rocket sounds like a myth from a science fiction novel. After all, previous rockets typically propelled a launch vehicle into space and then separated, falling to the ocean to be deemed trash. A reusable launch system (RLS) is one that *allows for the recovery of all or part of the system for reuse of launch components, including rockets*. NASA originally planned for the shuttle boosters to be reusable after being recovered in the ocean. However, the cost of refurbishing and recertifying the boosters made it cheaper to build new booster rockets to launch the shuttle.

Today, several private companies are developing rapidly reusable launch systems intended to dramatically cut costs. So far, SpaceX is the only company that has successfully launched this technology. In 2017, SpaceX launched Flight 32 of its Falcon 9 rocket. This marked the first relaunch and landing of a used orbital rocket. SpaceX went on to launch its previously used Dragon spacecraft using a Falcon 9 the same year.

SpaceX is also credited with launching the most powerful rocket to date, the Falcon Heavy. The Falcon Heavy's 2018 launch represented the first time a pair of recycled rocket boosters were used to send a heavy payload (a Tesla roadster) into space. Both rocket boosters landed at the same time in Cape Canaveral, Florida.

What is next for SpaceX? Engineers are currently working toward the goal of a 24-hour turnaround between a rocket's use and reuse with their Block 5 version of the Falcon 9.

Along with Space X, other companies, like Blue Origin, are making progress toward reusable rockets.

Different Types of Rockets

Today, rockets are divided into different types based on the fuel source for the rocket. The four types of rockets are solid-fuel rockets, liquid-propellant rockets, ion rockets, and plasma rockets.

Solid-Fuel Rockets

Solid-fuel rockets have been around for quite a long time. Solid-fuel rockets use *a fuel and oxidant mixed together as fine powders and then pressed into a solid "cake."* The goal is to use something that burns quickly but will not explode. A typical rocket solid-fuel is composed of 72% nitrate, 24% carbon, and 4% sulfur. Of course, different fuels and oxidants can be used to obtain the ideal thrust.

The solid propellant is packed into a cylinder. Many cylinders use an 11-point star shape to increase the surface area in which the propellant burns, which therefore increases the thrust. The cylinder also contains a space that serves as a combustion chamber, or *an enclosed space where combustion takes place*. Upon ignition, the solid fuel begins to burn. High temperature and pressure exhaust gas is produced. The hot exhaust gas is passed through a nozzle, which accelerates the flow. Thrust is then produced. The nozzle is a key component to any rocket because it causes the acceleration of the gases and allows for more of the energy to be extracted from the cylinder. By increasing the acceleration, the thrust is also increased.

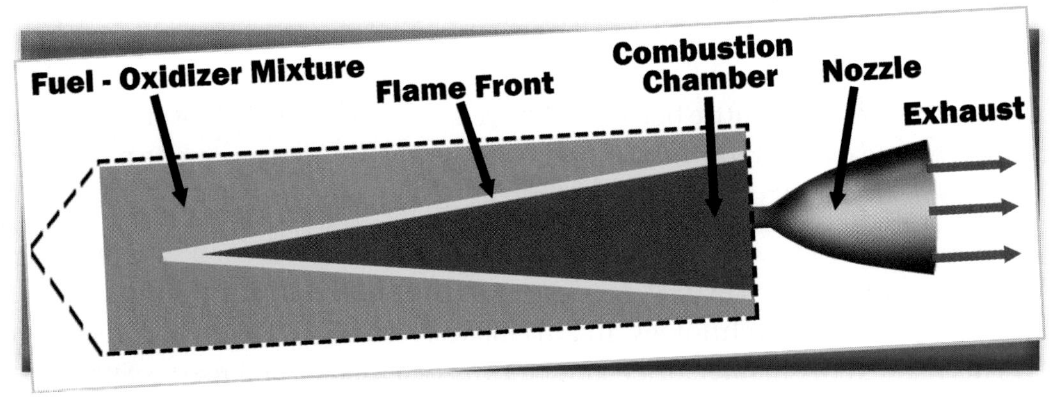

Solid rocket engine

The space shuttle's rocket boosters used solid-fuel. The boosters burned for approximately two minutes and created 3.3 million pounds of thrust each. Today, solid-fuel rockets are mostly used for air-to-air or air-to-ground missiles, and for booster engines in satellite launches.

The advantages of a solid-fuel rocket are that it is simple, lower in cost, and relatively safe to use. The disadvantages are that the thrust is uncontrolled, and once the rocket is ignited there is no way to stop it. This type of fuel is very useful on tasks that require a short thrust time.

Liquid-Propellant Rockets

Liquid propellant was first tested by Robert Goddard in 1926 using gasoline and liquid oxygen. In a liquid-propellant rocket, fuel and an oxidizer are pumped into the combustion chamber. When they burn, they create high-pressure, high-velocity gasses. The gases are then directed into the nozzle to provide additional acceleration. The typical exit velocity for a liquid-propellant rocket is 5,000 to 10,000 mph.

There are several types of liquid propellant used in rockets. Liquid hydrogen and liquid oxygen were used in the space shuttle's main engines. Gasoline and liquid oxygen were first used by Goddard as he conducted testing on liquid-propellant engines. Kerosene and liquid oxygen were used on the Apollo program. Alcohol and liquid oxygen were used on the German V-2 rocket. And nitrogen tetroxide/monomethyl hydrazine were used in the Cassini engines.

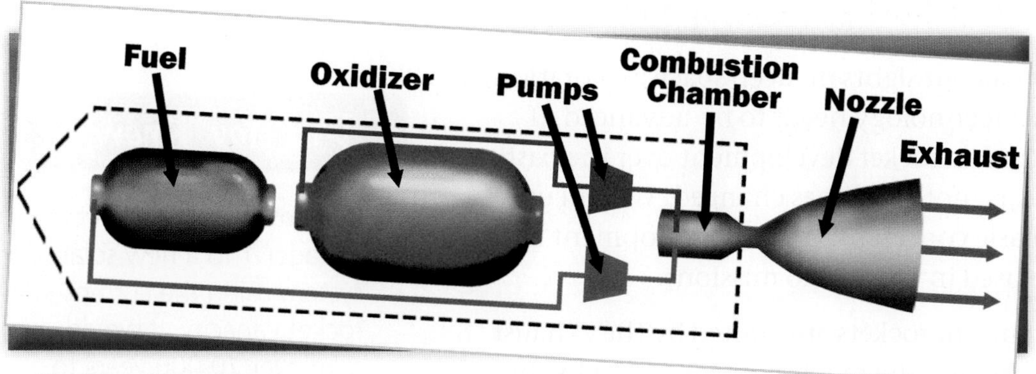

Liquid rocket engine

The specific liquid propellant selected for a rocket is determined by the specifics of the mission. Not all of the liquid fuel combinations burn the same. The rate at which they burn, the thrust produced, and the mass of the fuel must be accounted for when designing a rocket engine.

Ion Rockets

In Chapter 7 Lesson 2, we learned about the ion electric rocket that uses magnetism and electricity to propel a spacecraft through space. Ion electric propulsion is more efficient, but it can only be used in the vacuum of space. The thrust produced by an ion rocket is lower, but it builds up acceleration, leading to less fuel used and a shorter travel time.

Plasma Rockets

Plasma rockets are in the development stages. Plasma rockets *accelerate by using plasma that is produced by stripping negative ions from hydrogen atoms in a magnetic field and pushing it out of the engine.* NASA's plasma Variable Specific Impulse Magnetoplasma Rocket (VASIMR®) is being built by the Ad Astra Rocket Company near Houston, Texas. It is the first plasma rocket in development. It is expected that the plasma rocket could shorten the trip to Mars into a few weeks versus the current eight-month trip.

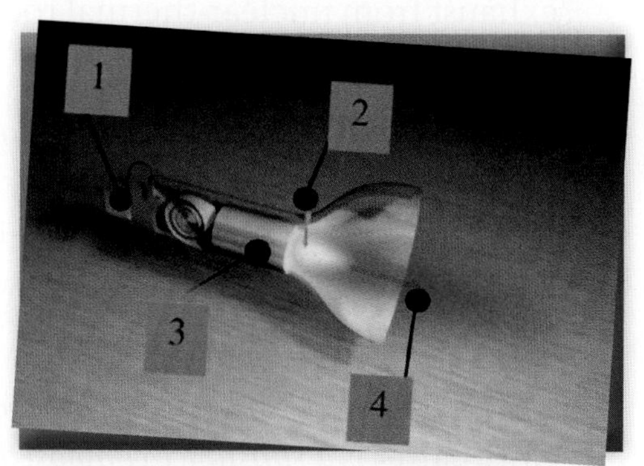

(1) Energy storage unit. (2) Ignitor. (3) Fuel rod. (4) Plasma acceleration region.
Courtesy of NASA

As you have read from the lessons in this chapter, the concept of how a rocket works is quite simple. However, the process of designing and building a successful rocket is quite complex. NASA employs rocket scientists who analyze each factor of a mission to determine the appropriate rocket and fuel that should be used to exit the force of Earth's gravity.

Future of Rocket Development

As we set our sights on deep space exploration, rocket technology needs to be advanced. If you look at rocket development over the past 30 years, not much has changed. We still use the basic concepts of rocket development that were used in the Apollo missions.

Our current rockets are limited by the exhaust velocity, and the maximum exhaust velocity cannot take us much farther than the Moon. Remember, deep space exploration probes are free floating and use other planets' gravity to slingshot them deeper into space.

> **Did You Know?**
>
> If scientists would redirect Voyager 1 to a new solar system using the current rocket capacity, it would take over 70,000 years to reach the closest star.

Alternative propulsion systems are currently being investigated as sights are set on interstellar exploration. Some of the alternatives being explored are fission rockets, fusion rockets, and antimatter rockets.

Fission rockets, or nuclear thermal rockets, are *an efficient method of using nuclear power to propel a rocket*. There are two types of nuclear fission rockets: nuclear-electric and nuclear-thermal. Nuclear-electric rockets *use a nuclear reactor to generate electricity, then use the electricity to make high-velocity ions*. The end result would be ion propulsion to help propel the rocket. This is an extremely efficient method of creating propulsion. Nuclear-thermal rockets *use a reactor to create heat to generate turbine gas and thrust*. The exhaust from nuclear-thermal rockets is pure hydrogen, and the thrust provided is quite high.

Fusion rockets are *small rockets that use fusion propulsion*. Fusion *happens when the nuclei of two or more atoms combine*. When this happens, energy is released. A fusion rocket would use plasma injected into a chamber where a magnetic field would produce fusion. The energy released would be funneled out the back of a spacecraft, creating thrust.

Antimatter rockets are *a class of rockets that uses antimatter as the energy source*. Antimatter is *a substance composed of subatomic particles that have the mass, electric charge, and magnetic moment of the electrons, protons, and neutrons of ordinary*. Antimatter is the most potent fuel known. Just 10 milligrams of antimatter is needed to propel a mission to Mars, compared to the tons of chemicals needed today. Antimatter rockets are a concept of rocket engineering that is still being researched. In reality, antimatter is very unstable. Antimatter reactions can produce blasts of high energy gamma rays. Remember gamma rays are like x-rays on steroids and are not healthy to be around. The NASA Innovative Advanced Concepts (NIAC) Program is funding a team of researchers who are working on a new design for an antimatter-powered spaceship to avoid the gamma-ray blasts.

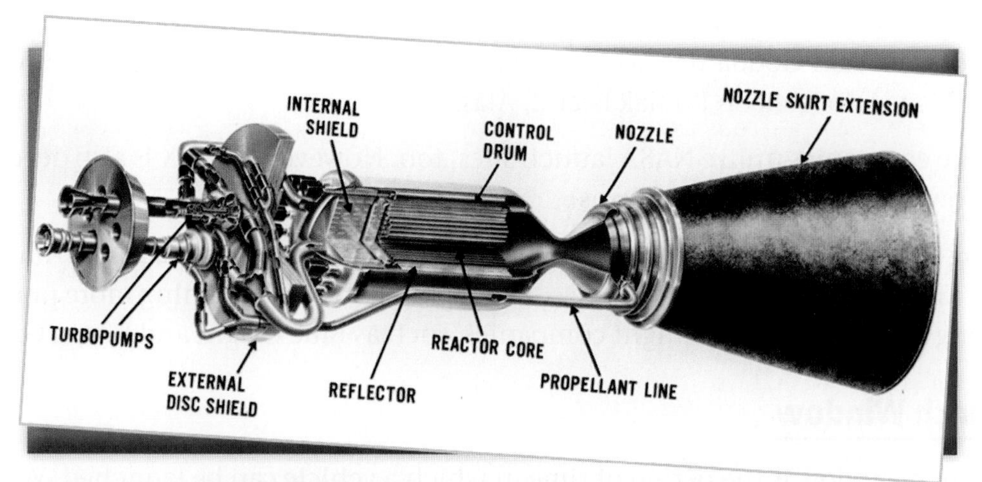

New work on Nuclear Thermal Propulsion builds upon NASA's Nuclear Engine for Rocket Vehicle Application (NERVA) program of the 1960s.
Courtesy of NASA

Launch Logistics

Many factors go into a successful launch. These factors include launch sites, launch dates and times, and other pre-launch preparations.

Launch Sites

Launch sites are *the physical locations from which rockets take off*. Launch site locations depend on a number of factors. Sites must allow for launches that will not fly over populated areas. They must also be located in areas with acceptable wind, weather, and lighting. NASA currently has the following launch sites:

- Kennedy Space Center in Cape Canaveral, Florida
- Vandenberg Air Force Base in California
- Wallops Flight Facility in Virginia
- Pacific Spaceport Complex on Kodiak Island in Alaska

The orbital requirements of a mission are used to help determine which launch site is used. The main rocket launch sites in the US are Kennedy Space Center in Florida and Vandenberg Air Force Base (VAFB) in California. Most equatorial orbits launch from Cape Canaveral. Remember, equatorial orbits use Earth's rotational speed. This means rockets need less propellant for launch. Missions requiring polar orbits most often launch from Vandenberg. In fact, Vandenberg Air Force Base is the only military installation in the US for launching satellites into polar orbit.

Missile row at Cape Canaveral Air Force Station, 1964.
Courtesy of Everett Historical/Shutterstock

Wallops Flight Facility is typically used to launch suborbital missions. This facility is part of the NASA Goddard Space Flight Center on Virginia's eastern shore. A secondary launch site is located on Kodiak Island, Alaska.

Private companies utilize NASA launch sites, too. However, SpaceX is also developing its very own site in Brownsville, Texas.

Along with major US launch sites, smaller civilian sites, such as Spaceport America in New Mexico and California's Mojave Air & Space Port, are becoming more prevalent among other private spaceflight companies, such as Blue Origin and Virgin Galactic.

Launch Window

A launch window is the period of time in which a vehicle can be launched. What factors determine a launch window? First, there's the mission itself. The needs of the spacecraft, types of rockets, required trajectories, and other orbital needs must also be considered. Launch windows can be daily or monthly. Daily launch windows provide a window of a few hours during a 24-hour period. Monthly launch windows provide a few days during a month or lunar cycle.

Did You Know?

Apollo 8 had a launch window of approximately five hours. If the launch was delayed past the launch window, the launch would have been cancelled.

Let's examine this closer with an example. Consider the launch of the Cassini mission to Saturn. As you may recall, the Cassini probe completed flybys of several planets prior to reaching Saturn. Engineers had to carefully study how and when to most effectively launch the probe and utilize the boost from Earth's gravity and a boost from Venus and Jupiter.

Given the positions of the planets and the trajectory Cassini would fly, the launch window was determined to be October 6, 1997, to November 15, 1997. Remember, this calculation was based on placing Cassini at Saturn seven years later.

The absolute best conditions for launch were calculated as being between October 6, 1997, and November 4, 1997, so the launch was scheduled for October 13, 1997. Although Cassini could still have successfully launched between November 5 to 15, 1997, this was a less desirable launch period because it would delay the arrival to Saturn by months or even years, and additional propellant would be needed.

In case Cassini couldn't be launched during that period, a secondary launch window and backup trajectory was calculated, which eliminated the gravitational boost from Jupiter. This secondary launch window was calculated to occur between November 28, 1997, and January 11, 1998, over a month later than the primary launch window; and the journey would take 8.8 years instead of the originally planned seven-year journey.

Just to make sure all bases were covered, NASA calculated an additional backup launch window for the secondary launch window. The backup launch window was from March 19, 1999, to April 5, 1999. This launch would place Cassini at Saturn on December 22, 2008 after a 9.8-year journey, adding almost three years.

As you can see, by missing the initial launch window NASA could have waited over two years longer for Cassini to reach its destination.

In addition, NASA needed to calculate a daily window to determine the time that was best to launch Cassini. For the scheduled launch on October 13, 1997, the daily window was 140 minutes in length starting at 4:55 a.m. Eastern Daylight Time. Each day that the launch was pushed back, the daily window moved forward by about six minutes per day.

Launch Preparations

After a launch vehicle, launch site, and launch window are identified, the launch vehicle must undergo testing prior to launch. Components of the launch may be built and pretested at various locations across the country, but then need to be tested as a whole.

All instruments end up at the same location in a clean room. A clean room is like a hospital operating room. It is *a sterile environment, completely free of contaminants, with a strong ventilation system and highly advanced air filtration*. We use clean rooms to test spacecraft, using environmental, electrical, software, and systems tests.

What looks like a teleporter from science fiction being draped over NASA's James Webb Space Telescope is actually a "clean tent." The clean tent protects Webb from dust and dirt when engineers at NASA's Goddard Space Flight Center in Greenbelt, Maryland, transport the next generation space telescope out of the relatively dust-free cleanroom.

Courtesy of NASA/Chris Gunn

Today, NASA builds and assembles components of SLS and Orion spacecraft all over the US. When components are ready to go, they make the trip to the launch site. Some SLS components are very large, such as the launch vehicle stage adapter (LVSA). It travels by way of truck or barge from Marshall Space Flight Center in Alabama to Cape Canaveral. The launch vehicle stage adapter (LVSA) *connects two major sections of the launch vehicle's stages*. Engines and the core stages also travel by barge.

(A steel mock-up of the SLS core stage at Michoud Assembly Facility in New Orleans preparing for travel to Stennis Space Center in Mississippi for testing.

Courtesy of NASA MSFC. Michoud image: Jude Guidry

A NASA Railroad train transports solid rocket booster segments during the space shuttle era.

Courtesy of NASA

Other components of SLS have a slightly more complicated trip. The Interim Cryogenic Propulsion Stage (ICPS) travels by barge from Alabama to Florida, but only after going through a clean room first. Meanwhile, trains carry components, such as the booster segments, from Utah to Florida.

In preparation for the STS-87 mission, space shuttle Columbia rolls out to Launch Pad 39B atop the crawler-transporter.

Courtesy of NASA/KSC

Once assembled and tested, the launch vehicle must be transported to the launch pad. The launch pad is *the circular structure from which the launch takes place*. A mammoth machine called a "crawler-transporter" is *used to transport the launch vehicle, payload, and spacecraft to the launch pad*. NASA has two crawlers at Kennedy Space Center. Each crawler is the size of a baseball field and is powered by locomotive and electric engines. They weigh in at 6.6 million pounds each and can haul approximately 18 million pounds. That's about the equivalent of 20 fully loaded Boeing 777 airplanes!

Final preparations begin to ensure that all systems are a "go" for launch. Throughout the countdown to liftoff, there are "polls" where various launch team members announce if they are a "go" for launch. With 5 or 10 minutes left in the countdown, the launch process is almost completely computer-controlled. If there are no aborts at T-minus zero, we have lift-off!

Lesson 3 Review

Using complete sentences, answer the following questions on a sheet of paper.

1. How do rockets work?
2. How many pounds of thrust is needed to keep a one-pound object motionless against the force of gravity on Earth?
3. What are the two types of launch vehicles?
4. How did the R-7 ICBM compare to the V-2 ballistic missile?
5. What was the purpose of the third stage (called Block 1) of Vostok and Voskhod ELVs?
6. What kinds of propellants have been used in Delta ELVs?
7. What major milestones are credited to Atlas ELVs?
8. What are the four types of rockets?
9. Why do many cylinders use an 11-point star shape?
10. What are the types of liquid propellant used in rockets?
11. How do nuclear-electric and nuclear-thermal rockets work?
12. What factors determine a launch window?
13. How does the launch vehicle get to the launch pad?

APPLYING YOUR LEARNING

14. Based on the information covered in this lesson, what lessons could be learned from past rocket launch technology for developing future launch systems?

CHAPTER 8

Cybersecurity.
Courtesy of MaHa1/Shutterstock

Cybersecurity

Chapter Outline

LESSON 1
Foundations of Cybersecurity

LESSON 2
Principles of Cybersecurity and Computer Basics

LESSON 3
National Cyber Policy

"Passwords are like underwear: make them personal, make them exotic, and change them on a regular basis."

Overheard at SecureWorld Atlanta

LESSON 1 Foundations of Cybersecurity

Quick Write

Think about the scenario in Matanuska-Susitna, how would your daily life be affected without having access to the internet for 10 weeks?

Learn About

- what is cybersecurity?
- vulnerability of US assets
- requirements of cybersecurity in protecting US assets

Matanuska-Susitna (Mat-Su) is a remote region of Alaska the size of West Virginia with a population of approximately 100,000. The local government manages an animal shelter, ski chalet, ice arena, swimming pools, libraries, and various community outreach centers. They employ more than 300 people and have over 70 facilities. On the morning of July 23, 2018, employees began their day normally; but then, one by one, their computers began showing signs of malware. **Malware** is *software that is designed to disrupt, damage, or gain unauthorized access to a computer system*. The IT department began the necessary steps to remedy the situation. They asked all employees to change their passwords, and then launched automated software to remove the malicious files.

When the automated software launched, it triggered a second attack. Government offices were now under a cyberattack. A **cyberattack** is *an attack that attempts to damage or destroy a computer network or system*. Employees were locked out of their files, and a ransom was demanded. Leadership called in the FBI to help investigate and prepared to live without the internet.

All government offices were affected, including the libraries, animal shelters, swimming pools, trail offices, etc. More than 700 computers and printers had to be unplugged and gathered up to be "scrubbed." Office forms had to be completed manually; old typewriters were dusted off; and people had to stand in line at offices just as they did before the internet. All public Wi-Fi was shut down in the region. Animal shelters no longer had access to medication records for the animals in residence. Construction was halted, as forms could not be completed. This nightmare continued for 10 weeks!

That's right. It took 10 weeks to scrub each computer and remove all of the malicious files. The municipality discovered that the malware files had been on the affected computers since May. Although they are not sure who initiated the attack, they know that it stemmed from a seemingly harmless email. Once opened, the "ransomware" used the government's networks to travel through all of its computers. The price to fix this was expensive, costing Mat-Su taxpayers over $2 million.

Vocabulary

- malware
- cyberattack
- cybersecurity
- cloud computing
- jamming
- spoofing
- hacking

The Pioneer Peak and the Mat-Su Valley.
Courtesy of Rob Conrad/Shutterstock

What is Cybersecurity?

Cybersecurity refers to *the precautions taken to guard against crime that involves the internet, especially unauthorized access to computer systems and data.* Data is collected and stored online by government agencies, the military, financial institutions, private corporations, educational institutions, and medical facilities. The data stored online includes sensitive data, such as social security numbers, bank account numbers, medical history information, etc. Cybersecurity is concerned with keeping this data safe and ensuring that your identity and sensitive information is private.

Challenges of Cybersecurity

Cybersecurity is a challenge for most organizations and individuals. Technology is constantly evolving and cyberattacks evolve right along with it. Criminals are developing new methods for attacking online systems and networks. It is a challenge to keep up with the constantly evolving threats. Think about your smartphone: does it seem like it is constantly reminding you to update the system? These updates often offer patches and protections against the latest cyber threats.

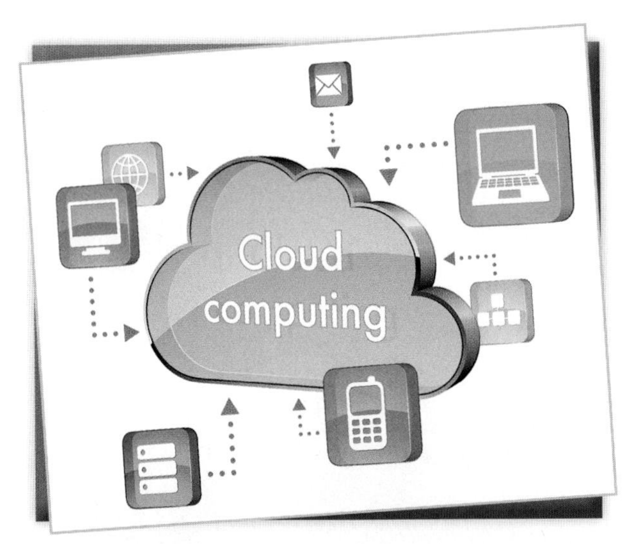

Cloud computing services are commonly used for storing data online so that it is available on any device.

Courtesy of beboy/Shutterstock

Another challenge for cybersecurity is that a system can be attacked from anywhere in the world. Cloud services are abundantly used for storing data online. In fact, over 60% of organizations store their data using cloud computing. Cloud computing refers to *internet-based computing where large groups of remote servers are networked to allow for virtual data processing, data storage, and online access.* The advantage of remote servers allows the user to store and access information from anywhere in the world that has internet access. The major disadvantage is that this information can also be accessed, altered, or retrieved by a successful cyberattack.

Did You Know?

In 2017, the global credit ratings agency, Equifax, suffered a data breach leaving 147 million people's data exposed. The cost of recovery from this attack was $439 million

Today's society is more technologically advanced than in the past, and this adds to the challenge of providing cybersecurity. More than 2.7 billion people own a smartphone, which is a jump of about 1 billion in the past five years. Organizations make it easy to access our data, pay our bills, and review sensitive information using our smartphones, thus creating many areas of vulnerability for cyberattacks. Combine that with the fact that cyberattacks are on the rise and it's no wonder cybersecurity is such an important part of our daily lives.

Did You Know?

Techcrunch.com expects smartphone ownership to shoot up to 6.1 billion people, by the end of 2020.

Cybersecurity must maintain a role in our everyday life so that we can properly prepare for cyberattacks. So how can you keep yourself safe? Keep informed of cybersecurity initiatives and understand threats that may exist. Be aware of tools available to protect your sensitive information. Unfortunately, antivirus software is no longer enough. Take steps to protect your data. When a website or organization asks for your data, make sure you know why they need it. Don't give out sensitive data on the web to unsecured or untrusted sites. And think about a disaster recovery plan for yourself. What would you do if you were attacked? Do you have the ability to back up and recover your data?

Cyber Statistics

$1.5 trillion profit/year in the cybercrime economy

$300 billion cybersecurity market by 2024

$15 billion provided for cybersecurity in the 2019 US government budget

30% of all top websites are unsecure

68% of all companies have no disaster recovery plan in place for a cyberattack

$6 trillion in damages from cyberattacks annually by 2021

2% of a company's budget is spent in cybersecurity

23% of all Americans are victims of cybercrime

60% of all Americans are exposed in cyberattacks

$1 is all that is needed to purchase cyber tools and kits on the dark web

Cyber security, data protection, Internet and technology concept.
Courtesy of Wright Studio/Shutterstock

Vulnerability of US Assets

Throughout this book, we have discussed US assets used in space. Many critical components of our defense systems, communication systems, navigation systems, and weather monitoring rely on satellites orbiting the Earth. These satellites are vulnerable to cyberattacks just as your PC is.

Why are Satellites Vulnerable?

As we have discussed in the past, there are many satellites orbiting the Earth and many of these satellites were not designed with security in mind. Satellites have components that use high-end technology, and those components are sourced from all over the world. Satellites are designed for long lives, but technology changes during a satellite's lifespan, so updates are needed. To update the software on a satellite, remote connections from Earth to the satellites are used. Although a remote connection is useful in updating a satellite, it also provides a point of vulnerability. Organizations uplink, or send information to a satellite from Earth, or downlink, receive information from the satellite. During the uplink or downlink, opportunities for cyberattacks arise.

Think of it this way, hundreds of satellites are orbiting the Earth, moving faster than the speed of sound, and they are using an outdated operating system, Windows 95. There are many old satellites in space, and while the cost of satellites is dropping, it is far more expensive to bring a satellite down from orbit and replace it than it is to keep it orbiting. These older satellites are a prime target for a cyberattack.

Types of Cyberattacks on Satellites

Let's begin by discussing the types of cyberattacks that can occur on a satellite. There are three main types of cyberattacks: jamming, spoofing, and hacking.

Portable wireless mobile signal jammer.

Courtesy of rumruay/Shutterstock

Jamming *occurs when the signal from a satellite is degraded or disrupted by interference.* If you have ever seen the movie "National Treasure," which features actor Nicolas Cage, you have seen an example of jamming. In the movie, a device is pointed towards Nicolas Cage's character to disrupt the signal from his wire to the FBI. Jammers have been used for decades in various applications to disrupt a signal. A simple jammer emits noise so that the signal is lost; a more sophisticated jammer can home in on the signal and block specific networks or frequencies.

Did You Know?

All wireless communication systems are susceptible to jamming. Simple jammers are relatively inexpensive. and can be used to disrupt the signal in your cell phone

A signal that is jammed on the uplink is referred to as "orbital jamming." A signal that is jammed on the downlink is referred to as "terrestrial jamming." Terrestrial jamming affects a specific geographic region. For example, Russia, Cuba, and China have all used terrestrial jamming to prevent their citizens from viewing unauthorized TV or radio broadcasts. Another example of terrestrial jamming occurs when mobile phone network access or internet access is restricted. For example, some organizations use jammers in their buildings to restrict mobile phone usage of anyone in the building.

Did You Know?

The equipment required to jam a satellite is easy to purchase, use, and conceal. It costs from $4,000 to $30,000. And jammers for cell phones can be purchased for around $200.

Spoofing *occurs when the signal from a satellite is replaced with a false signal*. These situations are particularly troubling because the satellite is functioning appropriately and a signal is being received, making it hard to realize a problem is occurring. When spoofing occurs, it reduces the integrity of the data provided by the satellite. Spoofing goes beyond jamming because it jams the original signal and replaces it with a different signal.

Spoofing is one way a satellite can be compromised.

Courtesy of Stuart Miles/Shutterstock

To demonstrate the seriousness of spoofing, Dr. Todd Humphreys of the University of Texas at Austin conducted an experiment in 2013. Using his lab, he broadcast GPS signals stronger than those of the space satellites, thus replacing the satellites' signal. Using his spoofing experiment, he took control of a luxury yacht's navigation system and navigated the ship based on an incorrect GPS signal. The ship and captain believed they were following an accurate GPS signal and had no idea they were actually being spoofed. (The yacht's captain had agreed to participate in the experiment in advance.)

Hacking *occurs when unauthorized access to data or a computer system is obtained*. Satellites are quite vulnerable to hacking because of the inexpensive high-power antennas they use.

Did You Know?

Many satellites don't even use data encryption! Early satellite engineers never envisioned that people on the ground would want to hack into satellites. Therefore, security wasn't originally a concern when developing satellites.

For example, let's look at the Hubble Telescope. The Hubble Space Telescope is a large telescope orbiting Earth. Launched in 1990, the telescope is the size of a school bus and takes pictures of planets, stars, and galaxies. The pictures provided by the telescope are not only beautiful, but they provide scientists with valuable information about space. Hubble has seen stars being born and stars dying. Hubble also has the ability to see galaxies trillions of miles away. The Hubble Telescope is an invaluable tool for scientists, yet it is also vulnerable to hackers. If hackers took control of the Hubble Telescope, they could open the camera hatch when the telescope lens was pointed toward the sun and destroy the camera optics.

Let's look at another example: SpaceX and OneWeb were recently authorized by the FCC to launch thousands of satellites to create a global internet. This initiative will triple the number of satellites in space. If an older, compromised satellite is attacked and thrown out of orbit it could potentially collide with thousands of newer satellites orbiting the Earth, simply because of the satellite overpopulation issue. Debris would scatter across space, putting the ISS at risk of collision and Earth at risk of falling debris.

Requirements of Cybersecurity in Protecting US Assets

Any effort to protect satellites in space must be made at an international level. After all, countries all over the world are responsible for the increased population of satellites in space. And satellites orbit over every part of the globe. Unfortunately, the international community is not addressing the threat as seriously as it should.

If an attack occurs in space, how is the culprit located and punished? No regulations currently exist for this scenario. If a cyberattack occurs on a satellite, who's responsible for investigating? And once a suspect is identified, how would international laws be applied? Currently, no international policies or laws exist with regard to space crimes.

The 1967 Outer Space Treaty dictates that nations cannot conduct a war in space or place nuclear weapons or weapons of mass destruction in orbit. It does not clearly outline a situation in which a cyberattack occurs on a satellite. In retrospect, the reason cyberattacks weren't included in the Treaty makes sense: the first PC had only been introduced in 1953, and computers didn't have enough memory or power to be considered a risk at the time.

Today, the US has several initiatives for cybersecurity of space assets. NASA's Independent Verification and Validation (IV&V) Program provides a cybersecurity program aimed at assuring the safety and success of NASA software. The IV&V program includes the following:

- The Vulnerability Assessment Program (VAP) compiles a team of skilled assessors to evaluate and minimize threats against space assets.
- NASA's Assessments and Authorization (A&A) performs independent assessments of NASA and non-NASA systems.
- The Risk Assessment program calculates risk levels for vulnerabilities based on the likelihood that the weakness will be exploited, along with its corresponding impact.
- NASA's Third Party Assessment Organization (3PAO) performs inspections under the Federal Risk and Authorization Management Program (FedRAMP). The inspections cover security assessment, authorization, and continuous monitoring for cloud products and services.
- The IV&V also offers security training with a hands-on lab that allows security novices and experts alike to act as "white-hat" hackers.
- NASA's security testing initiative offers a variety of approaches in order to obtain an in-depth and accurate representation of a computer system's assurance level.

Each branch of the military also has a cyber command or cyber forces division to manage its cybersecurity. The 24th Air Force (Air Forces Cyber) is the operational warfighting organization that establishes, operates, maintains, and defends Air Force networks to ensure warfighters can maintain the information advantage as US forces conduct military operations around the world.

US Air Force Staff Sgt. Joshua Foster, a radio frequency transmissions technician with the 379th Operations Support Squadron, uses a spectrum analyzer to perform diagnostics on a satellite at Al Udeid Air Base, Qatar, March 30, 2018.

Courtesy of Master Sgt. Phil Speck /US Air Force

Did You Know?

China's Micius satellite combines quantum physics with cybersecurity to establish a secure communication network. In 2017, the team sent intertwined—or more technically, "entangled"—quantum particles from the satellite in space to ground stations more than 1,200 km apart, Because of the nature of entangled particles, transmission of them as "encryption keys" makes it virtually impossible to hack into data without being detected. Additional tests are planned for the satellite, but it may prove to be the stepping stone for a secure communication network and even the beginning of a quantum internet.

On November 16, 2018, the Cybersecurity and Infrastructure Security Act of 2018 was signed into law. This act redesignated the Department of Homeland Security's (DHS) National Protection and Programs Directorate as the Cybersecurity and Infrastructure Security Agency (CISA). The goal of CISA is to provide cybersecurity tools, incident response services, and risk assessments to government agencies. This includes websites that end with the .gov domain. Not only does the new initiative aim to protect federal networks and critical infrastructure, it provides information sharing and alerts for other organizations so that many can benefit from the cybersecurity methods developed.

The homepage of the official website for the United States Department of Homeland Security.

Courtesy of chrisdorney/Shutterstock

Lesson 1 Review

Using complete sentences, answer the following questions on a sheet of paper.

1. What are the challenges of cybersecurity?
2. How can you prepare for a cyberattack?
3. Why are satellites vulnerable to cyberattacks?
4. What are the three types of cyberattacks on satellites?
5. What is the difference between orbital jamming and terrestrial jamming?
6. What does the 1967 Outer Space Treaty dictate?
7. What does the Cybersecurity and Infrastructure Security Act of 2018 provide?

APPLYING YOUR LEARNING

8. Provide a brief explanation about why you think the US and other countries should make cybersecurity a top priority.

LESSON 2 Principles of Cybersecurity and Computer Basics

Quick Write

Think about the story of Floyd's Coffee Shop. What online activities do you engage in while logged into free public Wi-Fi? What should you be doing to protect yourself?

Learn About

- computer basics and internet security
- strategies of cybersafety
- cyber strategies against cyberweapons

Floyd's Coffee Shop in Portland, OR, is known as a casual, relaxing environment for customers to come relax and enjoy a cup of coffee. A perk offered to its customers is free public Wi-Fi.

After a few months of slow Wi-Fi, the owner called in an information technology (IT) expert. While sitting in the coffee shop, a hacker had accessed its network. The hacker changed passwords, gained access to fellow customers, and accessed their surveillance cameras. Floyd's Coffee Shop was no longer a safe place for customers to enjoy a cup of coffee while perusing the news on their phone or checking their email.

Experts suggested that the coffee shop protect itself by setting up two Wi-Fi networks: one password-protected network for all coffee shop business; and another free public Wi-Fi network for customers. In addition to the two separate networks, the coffee shop set time limits for how long users can be logged into the free public Wi-Fi.

Although the coffee shop can protect itself with cybersecurity measures, customers accessing the free public Wi-Fi are still vulnerable. Customers were reminded to think about what they were doing on the public Wi-Fi. They shouldn't be visiting sites that provide access to their banking or health information. In addition, customers were encouraged to purchase a virtual private network (VPN) that protects them from hackers by launching an encrypted network that prevents others from seeing what they are doing online in the coffee shop.

This incident is just another reminder that we must be diligent and always be thinking about cybersecurity measures that we can take to protect ourselves.

Computer Basics and Internet Security

Before exploring how to keep our data secure, let's look at how computers and the internet work.

How Computers Work

A computer is basically an electronic machine that processes information. It takes in information, stores that information in memory, processes it, and then produces output. The basic components of a functioning computer include:

- Input–Data is put into the computer using a keyboard and mouse.
- Memory–Storage is available for saving data. A computer uses a hard drive for memory; your cell phone or camera use flash memory cards. Flash drives are more convenient for smaller devices because of their size and durability.
- Processor–A computer has a central processing unit (CPU) that does all of the processing work. This little chip gets hot fast, so it's typically accompanied by a fan to help cool it off.
- Output–This is the information that comes out of the computer based on your input.
- Video card–This ensures that the output from the computer can be displayed to the user via a monitor.
- Random-access memory (RAM) – This provides your computer with speed. It prepares the input for the processor and helps your processor move through input quickly.
- Power supply–A computer has an integrated power supply that is provided with electrical power. A laptop uses a built-in battery that can be recharged with electrical power.

Vocabulary

- computer programs
- macroinstructions (macros)
- virtual machine
- cloud services
- cloud computing
- internet
- router
- local area network (LAN)
- domain name system (DNS) server
- internet of things (IoT)
- smart device
- data management
- multi-factor authentication (MFA)
- security token
- Universal Serial Bus (USB)
- geotagging
- pass phrase
- external hard drive
- network attached storage (NAS)
- phishing
- cryptojacking
- cryptocurrencies

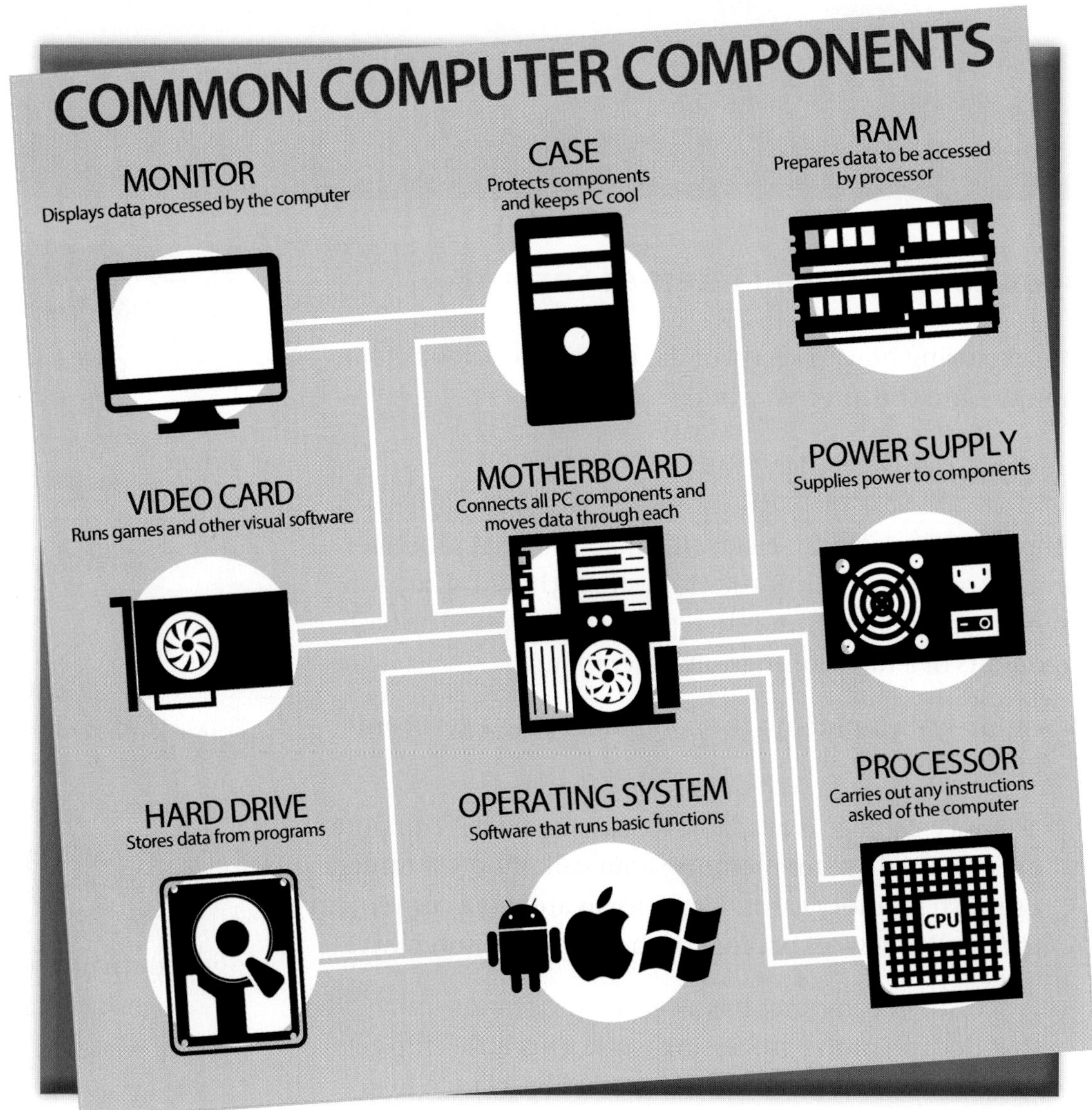

Diagram of the common components of a computer.

All computers use an operating system, which is the core of the computer. It controls the input, memory, processing, and output for a computer. The most common computer operating systems are Microsoft Windows and Apple's Mac operating system. The operating system enables all components of a computer to communicate with each other. It also allows the computer to communicate with other hardware, such as the keyboard, mouse, monitor, etc. The operating system is what you view each time you log into your computer; it allows you to access software applications and perform work on your computer.

Your computer isn't the only device with an operating system. Your phone has an operating system, too. If you use an iPhone, your phone utilizes a mobile iOS operating system from Apple. There are many types of mobile operating systems available for phones and tablets; another popular OS is Android.

Did You Know?

In the 1940s, Thomas Watson, the head of IBM Corporation, predicted that the world would need no more than five computers. Today, there are over one billion computers in the world!

Early on, computers performed complicated mathematical equations and operated as very large calculators. Now computers have endless possibilities. Computer programs are *used to give the computer instructions on how to process the data*. For example, Microsoft Excel provides instructions on performing calculations, creating charts, or setting up macroinstructions (macros). The term "macroinstructions" (macros) refers to *a programmable pattern that translates a sequence of input into a preset sequence of output*.

Computer programs are written for specific operating systems and work only on those systems. The difference between computer programs and operating systems can be a bit confusing. An "operating system" is really a series of computer programs that make a computer accessible, allowing you to use it. A computer without an operating system is unusable. However, a computer "program," such as Microsoft Excel, is what allows you to perform specific functions on the computer, such as word processing, database compilation, and even game playing.

A computer programmer writing code on his computer.

Courtesy of wutzkohphoto/Shutterstock

Computer programs are written in a variety of "languages" that your computer can understand, such as C, C++, Java, or JavaScript. Developers use many different programming languages to create computer programs. There are even tools that allow you to create small, easy-to-use computer programs.

Virtual Machines and Cloud Computers

One type of computing used frequently involves a virtual machine. A virtual machine is *an operating system or application that is installed on software that imitates hardware*. It's basically the creation of a computer inside a computer. Virtual computers are popular because large operating systems or programs can be housed on computer servers in a remote location and then be accessed from various computers on the network. The setup allows the system or virtual computer program to be completely separate from the computer system that accesses it.

Virtualizing a machine is a cost-saving initiative because it reduces the need for physical hardware, lowers the cost of the program, and lowers the maintenance costs of the application. Virtual machines simplify backups, disaster recovery, and upgrades. In addition, virtual machines are more efficient for large applications and provide redundancy.

Let's look at an example. A large pharmacy chain is interested in purchasing a new system to help fill prescription medications. The pharmacy chain has more than 250 locations nationwide and therefore must examine the cost of each system carefully. A virtualized system may be the most economical decision for the pharmacy chain. Instead of purchasing servers and installing the application at each pharmacy location, the business can allow each location to use its existing computers and hardware to access the virtualized system. This selection would provide for an easy transition to the new pharmacy system and save the organization money.

Multiple devices connect to the cloud.
Courtesy of Ye Liew/Shutterstock

A virtual machine is considered to be a basic cloud computing model. Cloud services *allow you to easily access data and information stored on the internet*. There are no physically distant computer servers that you are logging into, merely your own, local computer, which then accesses data stored in "the cloud.". Cloud computing *uses a network of servers hosted on the internet rather than a local server*. Cloud computing has many advantages.

- It eliminates the cost of purchasing local physical hardware and software.
- It allows organizations to easily scale (increase or decrease in size and usage) an application or services.
- It provides self-service functionality and is able to access vast amounts of information in remarkable speeds.
- It allows IT teams to focus on important business goals rather than setting up hardware and software.
- It is a reliable service that offers data backup and built-in security to its users.

How the Internet Works

What exactly is the internet? The internet is *a global computer network using a standardized communication protocol*. The internet is basically a network of computers that are connected together. But do you know how the internet really works?

The internet requires an intricate network of routers. A router is *used to transfer or route information between computers*. In your home, you may have a Wi-Fi router that creates a mini-network of devices attached to your home internet. This may include a computer, iPad, smartphone, etc. This smaller network can be referred to as a local area network (LAN). A local area network is *a system of connecting devices within a building or cluster of buildings*. Your LAN cannot directly reach a computer hosting a website you want to visit. It requires many steps to reach the final destination.

Let's say you want to visit a website. Your computer would send a packet of information from your LAN to your service provider's routers. The router then sends the packet of information on to another set of routers. This process continues until the information reaches the destination and a packet of information is returned to you. The entire process takes place extremely fast. The time it takes for a packet of information to travel to its destination is measured in milliseconds.

Just as if you were on a road trip, there are many directions a packet of information can travel. So, if one route is blocked or down, the packet can take a different route to reach its final destination. But how do computers know where to go? Each computer or device attached to the internet has a unique IP address. The IP address is like your device's Social Security number. IP stands for "internet protocol." The IP address is used by computers to route packets of information through the network.

Now let's say you want to access Google.com. There are probably thousands of IP addresses registered to Google, so how does your computer know which one to use? Google.com is a domain name. A domain name is a name owned by a person or organization that is used as an internet address. The domain has a domain name system (DNS) server, which *contains a list of the domains' alphabetic or alphanumeric sequence followed by a suffix*. The DNS server routes information designated for Google to the appropriate location in its network.

One of the most important aspects of communicating over the internet is making sure the individual computer systems are speaking the same language. The internet uses standard protocols so that computers can talk to each other. The most common protocol is TCP/IP (Transmission Control Protocol/Internet Protocol). TCP/IP is a transportation protocol that allows computers to transport packets of information to each other. There are also application protocols that web browsers use to display information to you. HTTP or HTTPS are application protocols.

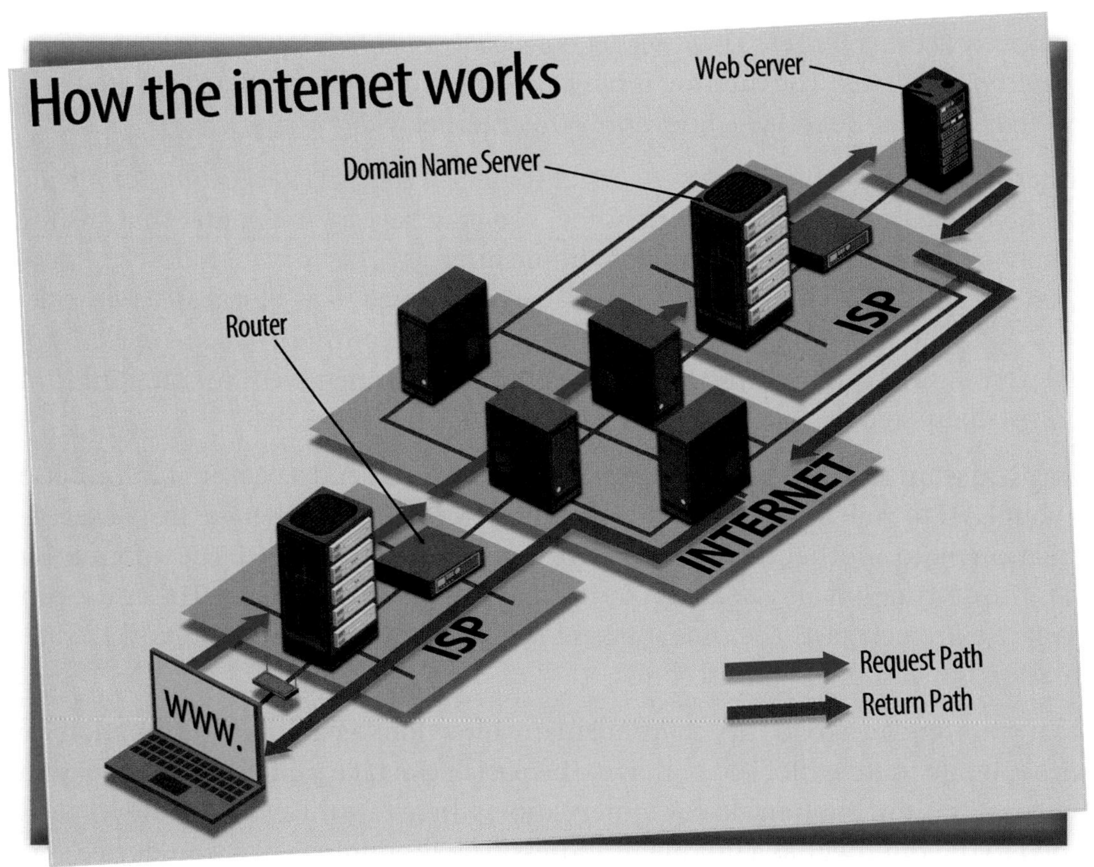

Diagram of how a packet of information travels over the internet.

Did You Know?

Wi-Fi is the most common method of connecting to the internet. Although the consumer uses a wireless method, most internet is ultimately provided through underground wires or above-ground wires on telephone poles. Satellite internet is provided through satellites in space, but currently it is mostly used for military applications and communications. Satellite phones that use the satellite internet are available for purchase, but they are quite expensive. Someday, the internet may use satellites. Aerospace company SpaceX plans to have 12,000 high-speed internet satellites in orbit around Earth by the year 2027.

Strategies of Cybersafety

As we read in Chapter 8 Lesson 1, cyberattacks are becoming more frequent, and protecting data should be a strategy employed by every individual. The personal data and information of anyone who uses the internet is available via multiple sources online; developing a data management strategy is essential to cybersafety.

Internet of things (IoT) refers to *any device that is connected to the internet*. We are living in a world of IoT where many common devices that we access each day are connected to the internet. Most people utilize a cell phone, computer, and/or tablet that are connected to the internet, but what about other devices? Smart home security devices, personal assistants such as Echo Dot, and Smart TVs are all connected to the internet. Devices are constantly evolving, and smart versions are appearing on the market. A smart device is considered *an electronic device that is connected to other devices or networks*. For example, you can purchase a mirror application that connects to the internet and shows the weather or your daily to-do list as you're getting ready for the day. Or what about a smart egg tray that can text you when you are running low on eggs?

In the IoT, anything is possible, but it also means you are becoming increasing vulnerable to a cyberattack. Each smart device becomes a vulnerability point.

Managing Data

Both business and individuals need to manage the data stored on computers and the internet as part of a successful cybersecurity strategy. Data management is *the process of acquiring, storing, and protecting information*. There are several data management strategies that you should consider when managing your data.

Multi-factor Authentication

Multi-factor authentication (MFA) is *an authentication process that uses two or more factors for authentication*. MFA adds another layer of protection to ensure that you are the only person to access an account. Think about your smartphone. Do you have MFA on your phone? What if someone guessed the password on your phone? What would they have access to? Your social media accounts, your email, and your personal information? As you advance your career, you may add apps to access bank account information, mortgage payments, or car lenders. The amount of data available on your personal devices is enormous.

Many devices already have MFA built-in. iPhones use both password and fingerprint technology to achieve multi-factor authentication. Many new computers and cell phones have password protection along with facial recognition.

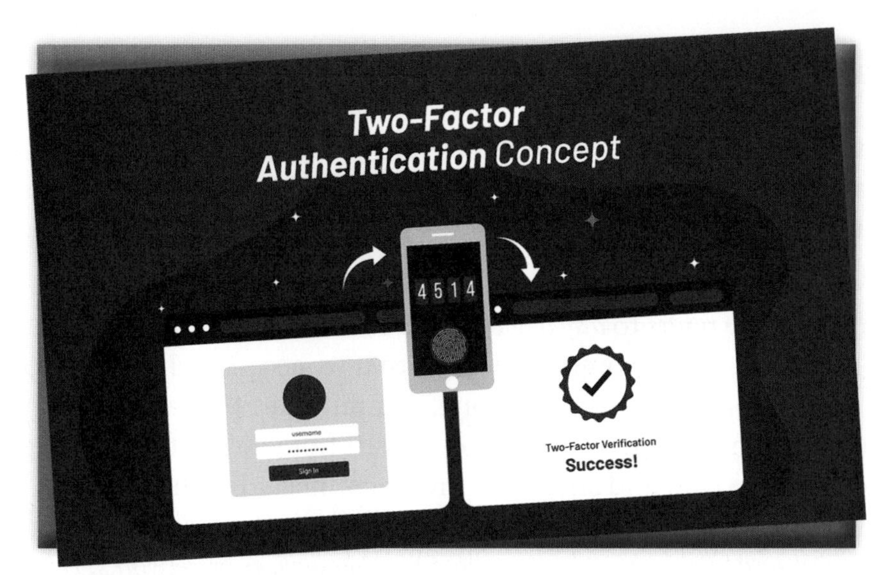

Two-factor authentication concept: account log in from web browser, confirmation code received in mobile, access successfully gained after entering code.
Courtesy of Allies Interactive/Shutterstock

In addition, there are many apps that can be used for MFA. 1Password is a free password management app that can generate six-digit, one-time passcodes for accessing your accounts online. LastPass Authenticator is another free app that provides MFA for devices.

Another solution used by many businesses is a security token. A security token is *a small device typically attached to a key chain that randomly generates a unique code at regular intervals*. The security token code is then required when logging into the business's secure internet. These devices are relatively inexpensive and can be purchased for personal use as well. Newer security tokens use Universal Serial Bus (USB) technology to attach to your computer and provide MFA. Universal Serial Bus (USB) is *a common interface that enables communication between devices and a host controller, such as a personal computer*.

By using MFA, if a hacker were to figure out your password, they still would not be able to access your system or data because a second level of authentication is required.

Protect your Wi-Fi

Anytime your devices are online, their vulnerability is increased. Do you ever walk into a restaurant or store and immediately connect to their free public Wi-Fi? This may not be the best idea when trying to keep your data secure. Public Wi-Fi is an unsecured method of accessing the internet. Any data being sent to or from your device over public Wi-Fi is unencrypted. That means anyone can read it! Hackers prey on public Wi-Fi locations where they can eavesdrop and collect personal data.

Avoid public Wi-Fi when possible. If you must log into public Wi-Fi, make sure to do the following:

- Do not log into apps or sites where your sensitive information is stored. You should never log into any sites that require a username and password on public Wi-Fi.
- Look for the green lock icon in your browser, it will tell you when you are connected to a secure site.
- Keep your software and operating system up-to-date with the latest versions. The updates for your operating system contain important security patches that can help keep your data secure.
- Think about installing a virtual private network (VPN) on your computer or phone to keep your connection private. Many are available for free.

When thinking about Wi-Fi, also review your Wi-Fi settings at home. Do you have a password set up on your home Wi-Fi? Is your home Wi-Fi router secure? If not, anyone within the vicinity of your house can access your network and your personal information. Here are some steps you can take to ensure your Wi-Fi router is secure:

Think carefully about what applications you use and what websites you visit when connected to free public Wi-Fi.
Courtesy of StNewton/Shutterstock

- Make your password difficult to remember. You may have visitors that come to your house and ask to connect to your Wi-Fi network. If you use a password that is difficult to remember, they won't easily be able to provide the password to others. Ideally, you should create a password that is 20 characters long.
- Limit access to your Wi-Fi. Don't feel obligated to provide your password to everyone that comes to your house. If a friend shows up and is waiting 10 minutes for you to get ready to leave, they don't need the password. If a contractor is in your house doing work, they don't need the password. It's okay to be picky about who you give your password.
- Change your password often. Remember, once you've provided someone the password their device is typically set up to remember the password and automatically log back into the network when they are in the vicinity. By changing the password often, you can prevent unwanted access to the network. Ideally, you should change the password monthly.

- Change the router's administration (admin) credentials. Any device that is connected to the network can access the admin console for the router. Most manufacturers use the same admin username and password for every device they sell. This makes it easy for hackers to access your admin console and lock you out of your own network. You should access the admin console and change the username and password so that it's something only you know.
- Change your network name. Most consumers use their house number or last name to identify their network. This is an advertisement for hackers on whose information they are about to steal. Using the admin console, you can easily change the name of the Wi-Fi network so that it does not contain any identifying information.
- Enhance your Wi-Fi encryption. In the admin console of your Wi-Fi router, you can add wireless security to your router. There are typically three options: Wired Equivalent Privacy (WEP), Wi-Fi Protected Access (WPA), and Wi-Fi Protected Access 2 (WPA2). The strongest of these options is WPA2 and is the option that you should select.
    - WEP–Designed to provide a wireless local area network (WLAN) with a level of security and privacy comparable to what is usually expected of a wired LAN.
    - WPA–Provides a more sophisticated data encryption and user authentication than WEP.
    - WPA2–Provides a higher level of assurance that only authorized users can access wireless networks.
- Turn off Wi-Fi Protected Setup (WPS). Your router may have a WPS button. If enabled, this button will allow the router to send out the network and password information automatically to anyone that wishes to connect to the network. This option presents a vulnerability in your network security and should be disabled.

Protect your Smart (Cell) Phone

Not only should you protect your computer, but you need to protect your cell phone. As you install new apps, you will typically get requests asking for permission to access the phone's camera, microphone, and/or pictures. Pay close attention to these requests and find out why they need access before saying yes. By giving access to your pictures, the app can use any picture you store on your phone. By giving permission to the microphone, the app can listen to what you are saying.

Your apps may also be running in the background collecting data from other sites that you visit. Be sure to check your app settings on your phone to ensure that you are not giving them an all-access pass to your data.

Avoid Oversharing

Another way to manage your data is to avoid oversharing. The lure of social media enables us to post our daily tasks, locations, and personal information. Hackers can piece together the random details posted online and target your information.

Let's look at an example. Your cousin has a baby and the family is very excited. Grandma posts the baby's full name, date of birth, and tags the location of the hospital in the post. Everyone in their social media network now has the information needed to request a birth certificate and Social Security number for that baby. Now let's say the post gets 50 "likes." The information is now visible to the social network of all 50 people that "liked" the original post. The baby that may only be days old could soon be a victim of identify fraud.

So how do we stay safe on social media? Keep the following tips in mind when you are on social media:

- Avoid posts with names, phone numbers, addresses, or locations.
- Turn off geotagging. Geotagging is *the process of adding geographical information to images and data posted on social media.*
- Only accept people you know in person into your social networks.
- Set your privacy settings to the most restrictive they can be.

Abstract city map, with pin pointers, geotag marker.

Courtesy of Maria Kazanova/Shutterstock

Get Creative with Passwords

The most important thing you can do to manage your data is to keep it password-protected. When selecting a password, be creative! A creative password is typically a strong password. Ensure your password is at least 12 characters long and includes numbers, capital letters, and symbols. The best passwords use a pass phrase. A pass phrase is *a string of random words strung together that make sense to you, but no one else.* For example, the pass phrase February#17PinkFi$h is an excellent pass phrase.

You should use a different password for each account. This is a lot to remember, so it's recommended that you use a password manager like 1Password or LastPass to help you manage your passwords.

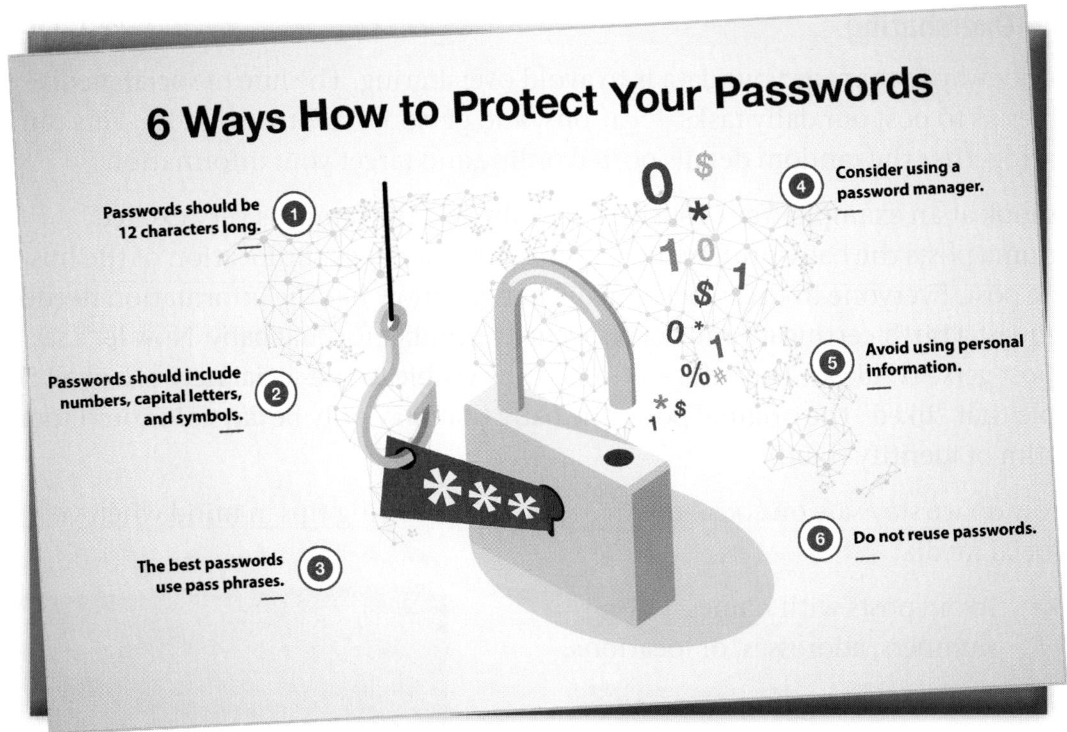

Six ways to protect your passwords.
CourteCourtesy of Tomas Knopp/Shutterstock

How to Safely Back Up Your Data

In addition to managing your data and keeping it secure, it is important that you identify a backup strategy for your data. Online cloud storage solutions are the most popular methods for data backup, but they are not the only options. Online data management tools provide free and affordable options to access your data from any location. However, the free accounts typically have limited storage and you cannot access the data without an internet connection. When researching options, make sure you choose an organization that is established. You don't want to lose access to your data when the organization goes under. Review Table 8.1 for information on common online data management tools and methods.

In addition to online storage options, several other data backup solutions are available:

- External hard drive. An external hard drive is *a portable storage device that can be connected to a computer via USB*. External hard drives are very easy to use, and backups can be scheduled through the software associated with the hard drive. However, the hard drive can fail or be destroyed; they can be expensive and should be stored off-site in case of theft or a fire.
- Thumb drive. As with the external hard drive, a thumb drive uses USB technology with one major difference. The thumb drive or flash drive is a small, solid-state devise that also allows users to store and transfer files to and from computers.

TABLE 8.1 *Online Data Management Tools and Methods*

Name	Devices	Cost
iCloud	iPhone and Apple products	5MB free, pay for additional storage
Google Drive	Android integration and desktop application for computers	15MB free, pay for additional storage
OneDrive	Browser access and apps for smartphones	5GB free, pay for additional storage
Dropbox	Personal and business subscriptions with browser access and apps for smartphones	2BG free on the personal subscription, pay for additional storage or business subscription

- CD, DVD, or Blu-ray. Burning data to a CD, DVD, or Blu-ray is very easy, but it's time consuming. It may be pricey depending on the storage capacity of the CD and the amount of data you have to burn. In addition, once the data is burned it cannot easily be updated. You would need to re-burn the updated data.
- Network attached storage (NAS). Network attached storage (NAS) is *a networked server used for saving data*. It can back up more than one computer automatically, but it's a pricey option for personal use.
- Secure digital (SD) memory card. A flash memory card is used primarily with cameras and cell phones to gain additional memory storage.

Laptop with thumb drive, SD card, CD, and portable hard drive. Concept of data storage.

Courtesy of Daniel Krason/Shutterstock

Plan to be Hacked

Everyone should plan to be hacked! After all, 1.76 billion records were leaked in January 2019 alone. Planning to be hacked means that you have taken measures to secure your data and that you are aware of the types of attacks that could occur. Let's review some of the most common attacks and how you can be prepared for them.

Phishing schemes can use fraudulent emails to get access to your credit cards and financial institutions.

Courtesy of wk1003mike/Shutterstock

Phishing

Phishing *occurs when cybercriminals send fraudulent emails designed to look like they are from a legitimate source.* Phishing is responsible for 91% of the cyberattacks in the United States. In fact, 9 out of 10 attacks can be traced to a phishing attempt. These emails aim to get you to click on a fraudulent link and coerce your personal information or username and passwords from you. For example, you get an email from your bank that you need to log into your account to receive an urgent message. The link is provided in the email. When you click, the link takes you to a criminal site that is masked to look like your bank's site. When you enter your username and password, the information is transferred to the criminals and your bank account can now be hacked.

To protect yourself from phishing schemes, be vigilant and suspicious of every email or social media request you receive. Phishing emails used to be unsophisticated and contain various spelling and grammar mistakes, but over time these cybercriminals have evolved, and phishing emails are more sophisticated. There are still red flags that you should look for:

- Ensure the email address looks valid. An email from your bank should not come from Gmail or another email provider.
- The email asks you to confirm personal information.
- The website link looks suspicious.
- The email content is designed to make you panic. For example, the email may inform you that you need to act immediately or the police will be sent to arrest you.

Malware and Ransomware

As we learned in Chapter 8 Lesson 1, malware is any software program designed to cause harm. The most common type of malware is ransomware. These programs attack your data and hold it hostage until a ransom is paid. Over 92% of malware is initiated via an email. Being suspicious of all email can also help protect you against malware.

Did You Know?

Many cybercriminals send email messages or use phone calls to imitate the IRS and inform you that you need to pay back taxes or they will notify the authorities. The IRS only communicates with taxpayers through US Postal Service mail; email and telephone calls are phishing schemes and should be avoided.

The best way to prevent malware on your computer is to use an antivirus software program. These programs are the first step in protecting your computer. Keep in mind these programs only protect against known malware. New malware is being developed every day, and antivirus software updates are essential for keeping yourself protected.

Cryptojacking

Cryptojacking is a relatively new cyberthreat. Cryptojacking *occurs when hackers hijack your computer and use its CPU power to mine cryptocurrencies*. Cryptocurrencies are *digital currencies in which encryption techniques are used to regulate the generation of units of currency and verify the transfer of funds, operating independently of a central bank*. Bitcoin is an example of a cryptocurrency. It is stored as a digital file on a computer. Today, one bitcoin is worth over $8,000.

Cryptojacking is similar to ransomware but runs in the background of your computer. So, you may not even notice that you've been affected. Cryptojacking is delivered through a phishing email or through ads on a third-party website. When you click the ad, you are giving permission to access your computer.

To prevent cryptojacking, install an ad-blocker on your web browsers, install an antivirus protection software, and be suspicious of all emails you receive.

Cryptojacking uses your computer to mine for and steal cryptocurrency.

Courtesy of igorstevanovic/Shutterstock

Help! I've been Hacked

If you discover that you've been hacked, there are several key steps that you should take immediately to minimize the damage.

1. Identify the type of attack. By knowing if you were attacked using a phishing scheme or ransomware, you're better prepared to deal with the situation.
2. Do damage control. Reset all your passwords to ensure access to your data is cut off. Remove any corrupt files on your computer. Your antivirus system can help identify corrupt files. You may need to take your computer offline to identify and remove the corrupt files.
3. Inform others who may be affected. Be sure to inform your social media network if you have been hacked through your social media accounts. Your social media network could be the next target if this is how the hacker got to your data.
4. Report the incident to the police. While it's difficult to trace the attack to the source, it's good to file a complaint and report the incident to the police.
5. Safeguard against future attacks. Once you have done damage control and contain the situation, prepare yourself for a future attack by securing your data.

Cyber Strategies against Cyberweapons

The US has many strategies that aim to establish the United States as the highest-standard in cybersecurity initiatives. One of the keys to a strong cyber strategy is to have highly trained personnel. The US has committed to investing in training and advocating for a strong cybersecurity workforce.

Currently, the number of available cybersecurity jobs greatly outweighs the number of qualified candidates available to fill the positions. The cybersecurity job sector is expected to grow 37% a year through 2022.

Cybersecurity jobs are in big demand.
Courtesy of Tupungato/Shutterstock

The pay for cybersecurity jobs averages about $100,000 per year, with an average entry level pay of approximately $67,000 annually--not bad pay for an entry level job. To help prepare candidates for cybersecurity jobs, CISA established the National Cybersecurity Workforce Framework. The initiative is aimed at increasing the size and capability of the US cybersecurity workforce. They are working with educators, students, employers, employees, and policy workers to ensure the curriculum is aligned to the skills and opportunities needed.

The top cybersecurity jobs are:

1. Cybersecurity engineer
2. Cybersecurity analyst
3. Network engineer/architect
4. Cybersecurity manager/administrator
5. Systems engineer
6. Software developer/engineer
7. Systems administrator
8. Vulnerability analyst/penetration tester
9. Cybersecurity consultant

Higher education is required for cybersecurity jobs. Approximately 89%of all cybersecurity jobs require higher education to earn needed certifications, with 67% requiring a bachelor's degree and 23% requiring a master's degree.

IT certifications are also available for cybersecurity professionals. The top certifications are:

- CompTIA Cybersecurity Career Pathway–This certification takes technicians from a beginner to mastery level of knowledge over a series of courses.
- Certified Information Systems Security Professional (CISSP)–This is an advanced-level certificate in cybersecurity.
- Cisco Certified Network Associate (CCNA)–This is a certificate available to those with an associate's degree in a variety of specialties.
- Global Information Assurance Certification (GIAC)–This is certification in cyber defense, penetration testing, digital forensics and incident response, developer, and information security management

Understand the Human Side of Cyber Weapons

An often-overlooked side effect of cybercrime is the human impact. Cyberattacks leave people feeling vulnerable and powerless. With limited capabilities to track down and punish the culprits, many victims are left with a lack of justice. Across the world, 65% of adults are victims of cybercrime.

Most people today are aware of and do think about cybercrime: about 28% expect they will be victims of cybercrime someday. Despite that information, the online behavior of most internet users hasn't change. People expect that cybercrime will happen and feel hopeless about preventing an attack.

Cybercrime statistics are thought to be lower than the number of crimes that actually occur because people are not reporting cybercrime to authorities. Cybercriminals are faceless criminals. As a result, many victims of cybercrime believe the perpetrators will not and cannot be found and punished for their actions, so they do not report incidents. In fact, less than half of all cybercrime victims report the incident to their financial institution or the police.

Did You Know?

Ninety percent of today's cybercrime is a result of organized crime.

When cybercrime occurs, victims go through a series of emotions. They may feel angry, annoyed, cheated, frustrated, and distrustful. Surprisingly, many victims also feel a sense of guilt—guilt because they previously had felt safe and protected online and afterward feel they had been naive or careless. In fact, many victims try to resolve cybercrime themselves by limiting the sites they visit online or by trying to identify the criminal themselves.

Cybercrime victims are often left with a lengthy and complicated process to resolve the effects of the attack. It takes approximately four weeks to resolve a cybercrime incident. That amount of time can cause a lot of emotional baggage and stress. It also costs the victim an average of $334 to resolve each attack.

It is imperative that all cybercrime be reported to the police and your financial institution. Criminals make a habit of stealing small amounts of money from many accounts. You may not think it's worth reporting for such a small amount, but if you don't report it you are helping the criminals steal from other accounts. That small amount from your account can grow quite large as the criminals add in the small amounts from thousands of other accounts.

Cybercrime and cybersecurity are complex and multifaceted.

Courtesy of design36/Shutterstock

Did You Know?

Almost half the people you meet are probably cybercriminals in some way or another. It's illegal to download a music track without paying for it. What about sharing or editing pictures that you didn't take? Or viewing someone else's email or browser history? These are all activities that many people see as legal activity, but they are not legal or ethical in nature.

So, what are some commonsense precautions that you can take to protect yourself online? Start by creating an email address that you use only for online purchases. And use a credit card with a small credit limit for all your online purchases. Use this credit card for only online purchases and nothing else. Report all incidents to the police and follow a cybersecurity community to keep up-to-date on new threats and ways to protect yourself.

Implementing National and International Cybersecurity Laws

Surprisingly, national and international cybersecurity laws are virtually nonexistent. The nation and world are still working out how to properly hold cybercriminals accountable for their actions. However, policies and regulations exist at state, national, and international levels.

National Cybersecurity Laws

In September 2018, the National Cyber Strategy was signed into law. This was the first major cybersecurity initiative for the nation in 15 years. The objectives of the National Cyber Strategy include:

1. Defend the homeland by protecting networks, systems, functions, and data.
2. Promote American prosperity by nurturing a secure, thriving digital economy and fostering strong domestic innovation.
3. Preserve peace and security by strengthening the United States' ability–in concert with allies and partners–to deter and, if necessary, punish those who use cyber tools for malicious purposes.
4. Expand American influence abroad to extend the key tenets of an open, interoperable, reliable, and secure internet.

Some of the key elements of the National Cyber Strategy include strengthening cybersecurity among government contractors, incentivizing organizations that invest in cybersecurity, building new cybersecurity approaches, and updating cybercrime laws.

The strategy also includes sections to enhance maritime and space cybersecurity. The strategy aims to establish international alliances for a Cyber Deterrence Initiative to build support and cooperation against cyberattacks.

Many new laws are aimed at prosecuting against cybercrime.
Courtesy of Ruslan Grumble/Shutterstock

The National Cyber Strategy provided the framework for the Department of Homeland Security (DHS) to establish the Cybersecurity and Infrastructure Security Agency (CISA) to secure all federal systems other than the Department of Defense (DoD) and Intelligence Community (IC). The DoD and IC maintain their own cyber policies, which will be discussed in Chapter 8 Lesson 3.

In addition, more than 35 states introduced about 256 bills to address cybersecurity in 2018. The legislative efforts aim to improve government security practices, provide funding for cybersecurity programs, and promote workforce training and development.

International Cybersecurity Laws

As we discussed in Chapter 8 Lesson 1, initiatives for cybersecurity policies must be applied internationally. Currently, we have an established set of laws that allow retaliation by military attacks against a foreign nation when a conflict occurs. But do these laws apply to cybercrime? Can a nation use military force to retaliate against cybercrime?

There is room for a lot of misinterpretation when applying the laws of war to cybercrime. Many gaps currently exist in how to deal with international cybercrime. To date, there is no major cybersecurity treaty among nations.

The flag of the United Nations.
Courtesy of Alexandros Michailidis/Shutterstock

The United Nations (UN) is an international organization founded in 1945 to guide the initiatives of its members. There are currently 193 member nations of the UN. The UN takes actions on humanitarian issues, such as human rights, climate change, terrorism, health emergencies, and gender equality. The International Telecommunication Union (ITU) is the UN's specialized agency for information and communication technologies. ITU established the Global Cybersecurity Index (GCI) to measure the commitment of countries to cybersecurity and raise awareness of the cybersecurity issues faced by many nations.

The GCI measures each country based on five categories:

- Legal–Measures the existence of legal institutions to deal with cybersecurity and cybercrime.
- Technical–Measures the existence of technical intuitions to deal with cybersecurity
- Organizational–Measures the coordination of policy and strategies at a national level
- Capacity building–Measures the research and development, education, and certified professionals of a nation
- Cooperation–Measures the partnerships and cooperative networks with other nations in addressing cybersecurity

In addition, the level of commitment of the various countries is evaluated in the assessment. The United States is considered a country that demonstrates a high commitment in all five of the areas. The 2018 results showed a large gap in the commitment and initiatives in addressing cybersecurity across the world.

Lesson 2 Review

Using complete sentences, answer the following questions on a sheet of paper.

1. What are the seven main components of a computer?
2. What are the advantages of virtualization?
3. What role do routers play in accessing the internet?
4. How can you protect your home Wi-Fi?
5. What are some ways to protect yourself from phishing schemes?
6. What should you do if you've been hacked?
7. What is the expected growth rate for cybersecurity jobs?
8. Why are cybercrime statistics thought to be lower than the actual number of cybercrimes committed?
9. What are the key elements of the National Cyber Strategy?

APPLYING YOUR LEARNING

10. If you could establish one cybersecurity policy to deter cybercrime, what would it be?

LESSON 3 National Cyber Policy

Quick Write

How do you think CyberCity and simulation exercises can help enhance cybersecurity for individual computer users like yourself?

Learn About

- US Department of Homeland Security
- US Department of Defense cyber strategy
- cyber warfare

CyberCity is a 48-square-foot miniature city that sits in an office building in New Jersey. The miniature city, built from parts from a hobby shop, is used to simulate cyber warfare attacks and identify systems to defend from attacks.

CyberCity was created by the SANS Institute, which provides security training to military, civilian, and government officials. The SANS Institute operates computer simulation games with CyberCity to teach cybersecurity techniques.

CyberCity has its own transportation system, hospital, bank, residential area, and retail area. The SANS Institute partnered with companies like Rockwell Automation, Siemens, and Phoenix Contact to create systems for CyberCity. The companies selected for partnership control a large majority of infrastructure systems in the market today.

Trainees are given simulated incidents, such as cyberattacks on the power grid. Their mission then is to resolve the situation and restore power to CyberCity. This realistic training method is providing hands-on experience to cyber professionals to learn defensive and offensive actions for managing cybersecurity. It's a new way of learning that can give cyber professionals the leg up on potential cyberattacks.

Miniature transportation system similar to the system in CyberCity.
Courtesy of Pernataya/Shutterstock

US Department of Homeland Security

The Department of Homeland Security (DHS) was established in 2002 after the 9/11 terror attacks. DHS combined 22 government agencies and departments under one entity and quickly grew to become the third largest government department. The DHS aims to strengthen security in the US and ensure that it is equipped to handle threats faced by the nation. The DHS is involved in many areas of security, but when it comes to cybersecurity, nothing is more important than monitoring and protecting the critical infrastructure of our nation.

Critical infrastructure describes *the physical and cyber systems and assets that are so vital to the United States that their incapacity or destruction would have a debilitating impact on our physical or economic security or public health or safety*. Our infrastructure provides essential services to society. In 2018, the Cybersecurity and Infrastructure Security Agency (CISA) was established under the DHS to provide cyber and physical protection for the nation's infrastructure. CISA's main goals are to:

- Provide free tools and resources for government and private sector partners
- Facilitate critical infrastructure vulnerability assessments
- Strengthen security and resilience across the chemical sector
- Provide training, encourage information sharing, and foster sector partnerships and international engagement

In addition, CISA combats cybercrime and provides cyber incident response services. It is responsible for protecting federal networks and critical infrastructure, and providing cybersecurity governance. CISA also promotes information sharing, training, and cybersafety information with the public.

Vocabulary

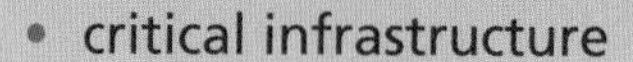

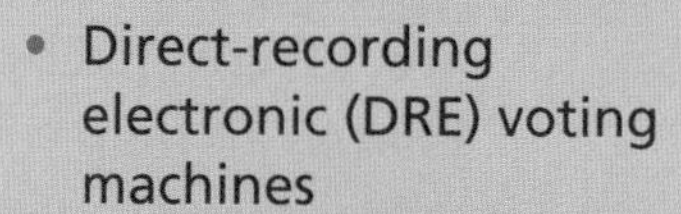

- critical infrastructure
- Direct-recording electronic (DRE) voting machines
- Optical/digital scan voting ballots
- Ballot-marking devices (BMD)
- system administrators
- cyber warfare
- virus
- computer worm
- Denial-of-service (DoS)
- Distributed denial-of-service (DDoS)
- zero-day vulnerabilities
- cyber espionage
- propaganda

Department of Homeland Security seal.
Courtesy of chrisdorney/Shutterstock

When a natural disaster occurs, the Federal Emergency Management Agency (FEMA) leads the response efforts on behalf of the federal government. Disaster assistance includes financial or direct assistance to residential and business owners who have lost property in a disaster and are not covered by insurance. FEMA can help with renting temporary properties and home repair and replacement.

Destruction in Mexico Beach, Florida, 16 days after Hurricane Michael.

Courtesy of Terry Kelly/Shutterstock

Just as important as physical security is the economic security of the nation. The DHS plays a role in identifying vulnerabilities in our nation's economic security and collaborating to secure global systems. The DHS works with international partners to ensure safe travel of residents and successful trade relationships. DHS monitors terrorist financing, fraud, counterfeiting, and intellectual property rights.

Ensuring the democratic nature of our nation is an additional responsibility of the DHS. The DHS ensures that our election process is secure and valid. We'll discuss additional details on election security later in this lesson.

Another aspect of the DHS is human trafficking. The DHS established its Blue Campaign to educate the public and law enforcement on human trafficking threats. Blue Campaign works to prevent human trafficking and protect exploited persons.

The DHS works with international partners, as well as state and local government agencies, to combat terrorism against the country. The goal is to detect and prevent terrorist attacks on American soil. Preventing terrorist attacks was the primary reason for establishing the DHS.

The DHS Privacy Office is responsible for evaluating the Department's programs, systems, and initiatives for potential privacy impacts, and providing strategies to reduce the privacy impact.

Some additional areas of DHS involvement include working with all levels of government, the private and nonprofit sectors, and individual citizens to make our nation more resilient to acts of terrorism, cyberattacks, pandemics, and natural disasters.

The DHS Science and Technology Directorate (S&T) is the Department's primary research and development arm and manages science and technology research, from development through transition, for the Department's operational components and first responders.

The final component of the DHS is the Transportation Security Administration (TSA). TSA works to secure transportation systems within the country, as well as in and out of the country. You may be familiar with TSA security in airports when you travel, but TSA also provides security for cargo screening and VISA screening.

Election Security

Are our elections secure? Over the last few years, this question has plagued the country. In 2016, the Illinois Board of Elections was hacked. The hackers quietly breached the network and spent weeks sifting through data while no one knew they were there. After a few weeks, the hackers crashed a server, alerting the Illinois Board of Elections to their presence. This was not the first or last attack against an election board. Clearly, policies needed to be put into place to ensure election security. In January 2017, election infrastructure security was moved under CISA's responsibilities.

Election Process

Let's first look at the process of having an election. An election always begins with a pre-election process. During the pre-election process, voters register for voting in their state, county, and/or district. Voters can register at in-person registration centers, such as their local driver's license center, via the mail, or online. It's imperative that voters register prior to an election because they will not be able to vote if they do not register. The voter registration process allows states to manage their voters, provide absentee voting, and establish voting locations convenient to registered voters.

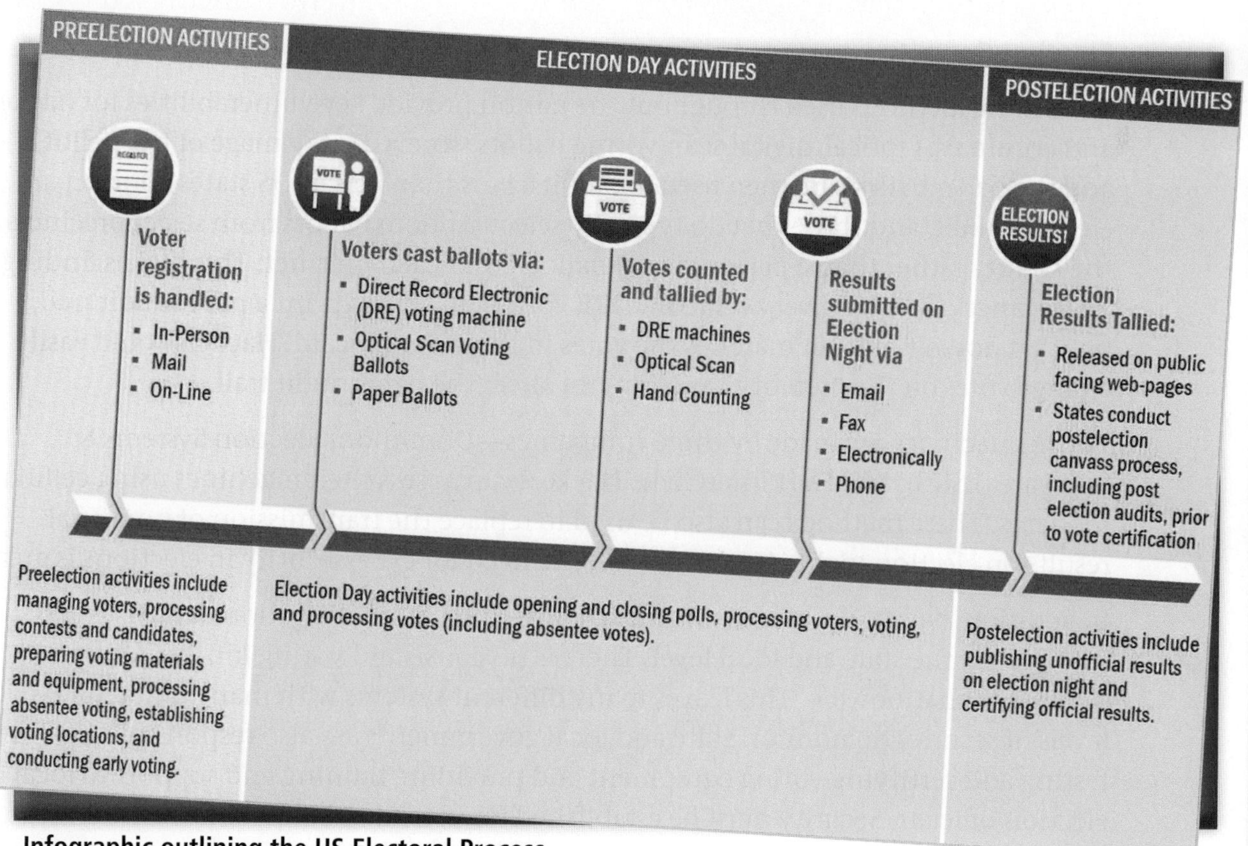

Infographic outlining the US Electoral Process.

Courtesy of Department of Homeland Security

A DRE voting machine.

Courtesy of CLS Digital Arts/Shutterstock

After voters are registered, election day is the next step in the electoral process. Voters can cast their ballots using Direct-recording electronic (DRE) voting machines, optical/digital scan voting ballots, or paper ballots.

Direct-recording electronic (DRE) voting machines are *portable monitors that allow voters to cast their ballot electronically*. When using a DRE voting machine, the votes get tallied into the memory of the machine. Some machines even provide a paper audit trail with the details of each vote to verify the votes manually.

An optical/digital scan voting ballot is *a paper ballot that can be read by optical or digital equipment to tally votes*. Some areas use ballot-marking devices (BMD) that *allow the voter to select their votes on a screen and then prints a paper ballot with their votes to be submitted*. This type of device does not store voting information into its memory. In rural areas, many voting places still use paper ballots that are manually counted.

As votes are counted by the various voting methods, unofficial results are submitted on election night via email, fax, and phone. After an election, postelection activities begin to publish official results, conduct audits, and verify the unofficial results presented on election night.

Election Security

The voting methods used throughout the nation provide key vulnerabilities for hackers and criminals. Optical/digital scan voting ballots store a digital image of the ballot, and the paper ballots are then used for audits. Less than half of US states conduct an election audit, and those that do typically select random ballots from select precincts in the county rather than a percentage of ballots from each precinct. This makes finding an instance of hacking very difficult. DRE voting machines print a paper audit trail, but that doesn't mean it matches the votes in the memory card. Hackers could easily change votes on the memory card, but not affect the paper audit trail.

Voting machines are made by three companies—Dominion, Election Systems & Software (ES&S), and Hart InterCivic. Hackers can access their machines using cellular modems. These methods can also be used to replace the transmission of unofficial results on election night. As you can see, the need for cybersecurity in elections is great.

In addition, day-to-day election infrastructure and voter registration databases are managed at the state and local level. There is no consistent voting infrastructure or database nationwide. This leaves many different systems with many different levels of security to monitor. State and local governments are also responsible for testing and certifying voting equipment and providing training and support to local election officials. Security must be established for voter databases, IT infrastructure, voting systems storage facilities, and polling places to ensure the security of elections.

Through CISA, the DHS provides infrastructure security to reduce cyber and physical threats. Through coordination with federal, state, and local agencies, DHS provides threat information, cybersecurity (by request), processes for coordination, and sensors to detect malicious activity (by request).

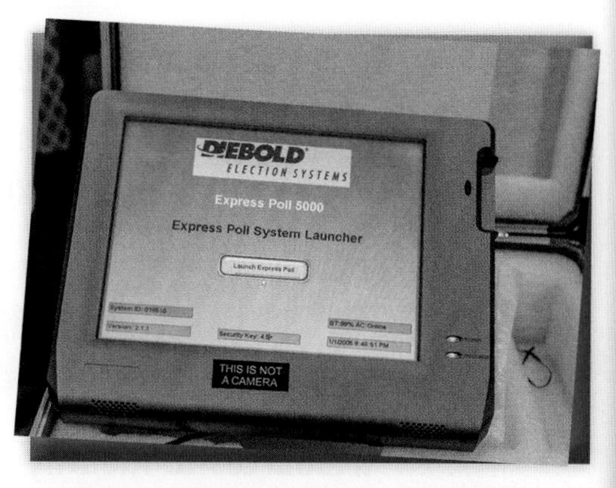

A hacked voting machine on display at the 25th DEF CON® hacker conference held at Las Vegas, Nevada, on July 29, 2017, showing splash screen after being booted up by hacker.

Courtesy of Bing Wen/Shutterstock

CISA services are available at no cost to state and local officials, but they must be requested by state and local officials. Services offered include cybersecurity advising, cybersecurity assessments, detection and prevention of threats, information sharing and awareness, incident response, and training or career development.

National Cyber Awareness System

CISA provides a National Cyber Awareness System to share information on potential threats. The National Cyber Awareness System provides technical information for those with a technical background on known and trending cyber threats. The information is provided via current activity reports, alerts, bulletins, tips, and analysis reports.

Did You Know?

Every state does not have a Chief Information Officer to oversee election security.

Current activity reports provide up-to-date information about high-impact types of cybersecurity activity affecting the community at large. The information is gathered from the incident reports received by CISA.

Alerts provide information about current security issues, vulnerabilities, and exploits that you should look out for. The alerts provide technical details for cybersecurity professionals who monitor for security issues.

Bulletins are weekly summaries of new vulnerabilities discovered by CISA. For any vulnerabilities that have a known solution, the patch information is also provided. A patch is *a set of changes to a computer program that are required to update, fix, or improve the program.* Each vulnerability listed in the weekly report is given a high, medium, or low score to identify how often the vulnerability is likely to occur.

Tips describe and offer advice about common security issues for nontechnical computer users. For example, there are articles on protecting against ransomware, dealing with cyberbullies, and properly disposing of your electronic devices. These articles are written for the average computer user and do not require technical expertise.

Analysis reports provide in-depth analysis on new or evolving cyber threats. These reports provide ample technical details and are geared to those with a technical expertise.

The information from the National Cyber Awareness System can be accessed online via its website or you can subscribe to receive the information via email or text message.

National Resources for Federal Government

As part of CISA's Critical Infrastructure Cyber Community (C3, pronounced "C-cubed"), Voluntary Program , resources are provided for the nation's agencies to identify, protect, detect, and respond to cyber threats.

Resources for Federal Government

Resources to Identify

Resource	Description
Cybersecurity Evaluation Tool (CSET®) and On-Site Cybersecurity Consulting	CSET® is a self-assessment tool to evaluate cyber-security risks. In addition, on-site cybersecurity is available.
Industrial Control Systems Computer Emergency Readiness Team (ICS-CERT) recommended practices	The ICS-CERT recommended practices are a list of practices aimed at helping the industry understand and prepare for cybersecurity issues.
National Cybersecurity Assessments & Technical Services (NCATS)	NCATS utilizes the best cybersecurity assessment techniques to provide risk management guidance and recommendations.
Federal Virtual Training Environment (FedVTE)	FedVTE is a library of prerecorded classroom cybersecurity training that is available for federal government employees and contractors, as well as state and local government personnel.
CyberChain Portal-Based Assessment Tool	The portal provides risk assessment tools, scenario simulations, and information sharing to identify vulnerabilities and risks.

Resources to Protect

Resource	Description
ICS-CERT Training	Cybersecurity training is provided via classroom and web-based formats for beginner, intermediate, and advanced levels.
ICS-CERT Recommended Practices	ICS-CERT's list of recommended practices helps the industry understand and prepare for cybersecurity issues.
US Computer Emergency Readiness Team (US-CERT) and ICS-CERT Alerts, Bulletins, Tips, and Technical Documents	This site provides access to alerts, bulletins, tips, and technical documents to combat cyber threats.
Federal Network Resilience (FNR)	FNR collaborates with other federal agencies to enhance the nation's cybersecurity policies and prevent long-term vulnerabilities.

Resources to Protect (continued)

Information Systems Security Line of Business Security and Awareness Training	This training provides security training provides and services for the federal government.
Network Security Deployment (NSD)	NSD works to improve the cybersecurity of federal government departments and agencies.
National Security Agency (NSA)/Information Assurance Directorate (IAD) National Security Cyber Assistance Program	The NSA/IAD has established a National Security Cyber Assistance Program through which commercial organizations can receive accreditation for cyber incident response services.
STOP. THINK. CONNECT.™ Campaign	This campaign was launched in 2010 to educate and empower Americans to prevent cyber threats and safely operate online.

Resources to Detect

Continuous Diagnostics and Mitigation (CDM)	CDM deploys sensors for hardware asset management, software asset management, vulnerability management, compliance management, and data management.

Resources to Respond

Cyber Incident Response and Analysis	Incident response services are offered to those who have experienced cyberattacks. Services include digital media and malware analysis, identification of the source of an incident, analyzing the extent of the compromise, and developing strategies for recovery and improving defenses.
National Security Agency (NSA)/Information Assurance Directorate (IAD) National Security Cyber Assistance Program	Through the accreditation offered by the NSA/IAD, commercial organizations can be utilized for incident response services.

US Department of Defense Cyber Strategy

Cyberspace is a critical asset to the Department of Defense (DoD) and all military branches. Cyberspace is used to gather intelligence, strike targets remotely, and work from anywhere in the world. Unfortunately, cyberspace is also under attack and the DoD handles cyber threats on a daily basis. In 2018, the DoD issued its cyber strategy. The DoD's cyberspace objectives are:

1. Ensure the Joint Force can achieve its missions in a contested cyberspace environment.
2. Strengthen the Joint Force by conducting cyberspace operations that enhance US military advantages.
3. Defend US critical infrastructure from malicious cyber activity that alone, or as part of a campaign, could cause a significant cyber incident
4. Secure DoD information and systems against malicious cyber activity, including DoD information on non-DoD-owned network.
5. Expand DoD cyber cooperation with interagency, industry, and international partners.

Five Things to Know About Military Cyberspace

So how will the DoD implement its cyber strategy and ensure that cyberspace is protected? There are five methods the DoD will use to implement its strategy.

1. The DoD needs to ensure that troops can accomplish their missions without the added worry of cyberattacks. To do this, the DoD will make processes more flexible. The DoD will use automation and data analysis to identify cyberattacks. It will also integrate commercial technology to ensure the maximum effectiveness in all DoD systems.
2. The DoD needs to prevent cyberattacks before they occur. To do this, it will need to strengthen cybersecurity across DoD systems. In addition, it will share information in a streamlined method with the public and private sectors. The DoD will upgrade critical infrastructure networks and systems to reduce the risk of attacks. In addition, standards for cybersecurity will be set and enforced. And, finally, the DoD must directly offer assistance to those outside the DoD when cyberattacks occur.
3. The DoD will expand its cooperation on cyber issues by building partnerships with commercial organizations. It will increase effectiveness by sharing information with other agencies and allies. In addition, they will utilize opportunities to identify and fix vulnerabilities.

4. DoD personnel must increase their cyber awareness by making sure that leaders and staff are knowledgeable in cyber terminology. They will hold personnel and contractors accountable for mistakes and procure services to keep pace with commercial information technology.
5. Finally, the DoD must cultivate the workforce and recruit cybersecurity talent to ensure success. The DoD will promote science, technology, engineering, and math classes in grade schools to grow cyber talent. They will also create competitions to identify top cyber specialists to help with the DoD's biggest challenges. In addition, computer-related jobs will be incentivized within the DoD. And a healthy mix of service members, civilians, and contractors will be employed within the DoD to best meet the needs of the mission.

Roles of Teams in Cybersecurity

In Chapter 8 Lesson 2, we discussed how you can protect yourself with cybersecurity, but how do organizations implement cybersecurity? Smaller organizations may outsource their cybersecurity and rely on a third party to monitor and secure their network. Larger organizations typically have a full-time cybersecurity team.

Hiring experienced cybersecurity professionals is a difficult task. There is a lack of talent in the cybersecurity field and an abundance of open positions. And people with experience come with a high price tag. Many companies are investing in their current employee base to create security experts. By training their employees they can create a pool of security experts who already know the company and understand the systems the company uses. Training existing employees is a big commitment. Once the proper employees are selected, a lot of time and money is required to provide the proper training.

The demand for cybersecurity jobs is high, but the pool of resources is low.

Courtesy of Akira Kaelyn/Shutterstock

Many companies give their system administrators (SA) dual duty and ask them to also be their security expert. System administrators are *information technology personnel who are responsible for ensuring that all system resources are available.* An SA role is a full-time position and does not have the time or expertise to also be the security expert. This is one of the biggest mistakes a company can make and can lead to a data breach.

A security expert is needed to identify vulnerabilities in the system and determine the best way to keep the network safe. A security expert knows how to find up-to-date information on vulnerabilities and knows how to address cyberattacks when they occur.

Training for Cybersecurity Positions

Many cyber professionals obtain the skills they need through certificate programs rather than degree programs. Many colleges are beginning to offer cybersecurity degrees, but this is a relatively new field of study.

There are two major certificate programs for cybersecurity professionals: SANS/GIAC and CISSP. SANS/GIAC (Global Information Assurance Certification) offers 11 certificates and is more appropriate for system engineers. The Certified Information Systems Security Professional (CISSP) certificate has been around a bit longer, but it's more appropriate for policy makers.

Common Cybersecurity Roles

After security experts are properly trained, they are typically employed on a security team. The roles on a security team include the following:

Role	Description
Security Analyst	A security analyst examines vulnerabilities and investigates countermeasures for each vulnerability. In addition, the security analyst recommends solutions and best practices. In the event of a data breach, the security analyst identifies damage and recommends recovery tools and strategies.
Security Engineer	A security engineer monitors system security and analyzes the data on the organizations systems. In the event of a data breach, the security engineer is responsible for performing forensics to determine the source of the incident. The security engineer also implements new technology and processes to prevent security attacks.
Security Architect	A security architect is responsible for designing security systems.
Security Administrator	A security administrator is responsible for installing and managing security systems.
Security Software Developer	A security software developer is responsible for developing new security software. Security software includes monitoring tools, traffic analyzers, intrusion detection, and antivirus software.
Security Consultant	A security consultant is a broad title for a security expert and can incorporate tasks from each of the other positions.

Cyber Warfare

Cyber warfare is *a digital attack by one country against another*. Cyber warfare is an attempt to damage another nation's computers or information networks. Cyberspace has introduced a new battlefield for the military. Nations are now using cyberattacks as a means for warfare against other nations. It is warfare that involves hackers attacking a nation's infrastructure.

Did You Know?

In 2018, the United States and Great Britain identified Russia as the mastermind behind the Petya cyberattack on the Ukraine. The Petya malware was hidden as ransomware and shut down power stations throughout the Ukraine in 2015. It took three years to determine the culprit behind the attack.

We rely on computer systems for almost all aspects of life. The computer systems attacked in cyber warfare are not the end target. These systems control a nation's infrastructure, such as power grids, banks, or airports. Accessing these types of systems allows hackers to virtually shut down essential infrastructure that results in chaos and destruction. In addition, the free press of many democratic nations and the rapid increased use of social media has opened up a gold mine of possibilities for cybercriminals.

Cyber warfare is used to "weaken, disrupt, or destroy the US," according to CISA. The threats faced by the US range from propaganda to espionage. The purpose of cyber warfare is to create damage and destruction in other nations.

Wars in the future will involve aspects of cyber warfare as a weapon.

Courtesy of Gorodenkoff/Shutterstock

Cyber Warfare or Cyberattack?

What is the difference between cyber warfare and a cyberattack? There are many factors that must be analyzed to determine if something is cyber warfare. We must first look at who initiated the attack. Cyber warfare occurs between nations, whereas cyberattacks occur between individuals. Next, we need to evaluate what the attackers are doing. Are they looking to bring down an entire system or steal sensitive information, or are they trying to trick people out of money? The amount of damage done by the attack must also be evaluated. Cyber warfare is aimed at being a large-scale attack that causes serious damage.

Did You Know?

Cybercrimes are estimated to cost the US economy over $100 million annually.

Types of Cyber Threats

Cyber threats from cyber warfare go beyond the typical threats that an individual faces. Typically, teams of developers and millions of dollars go into a cyber warfare attack. While there are many forms of cyber threats, the most common cyber warfare threats come from:

- Viruses and worms
- Denial-of-service (DoS)
- Zero-day exploits
- Cyber espionage
- Propaganda

Viruses and Worms

A virus is *computer code that is designed to spread from one computer to the next, infecting computers with malicious software.* A computer worm is *similar to a virus but can replicate itself automatically to infect other computers with malicious software.* Viruses require the user to send email, click on links, etc. to reach other systems. Viruses and worms can be fed into computers that run water systems, transportation systems, and power grids. After one computer is affected, the virus or worm can easily spread to other computers and bring down an entire network.

Denial-of-Service

Denial-of-service (DoS) is *an attack intended to shut down a computer or network.* DoS attacks basically lock authorized users out of a computer and/or network. DoS attacks occupy the resources of a network so it is no longer able to respond to any requests.

Distributed denial-of-service (DDoS) is *a more complex form of a DoS attack, in which large numbers of computers, devices, etc., are used to force the shutdown of a site.* A DoS attack typically uses a single computer to attack. With a DDoS attack, multiple devices are used to increase the severity of the attack. DDoS has become the weapon of choice for many attackers.

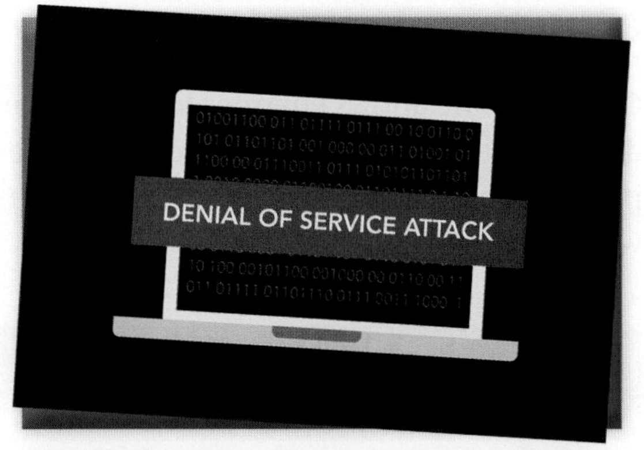

Denial of service attacks aim to overwhelm a network so it cannot respond to requests.
Courtesy of Eny Setiyowati/Shutterstock

Zero-Day Exploits

Every computer system has bugs or flaws in the code that the developers have yet to find. Zero-day vulnerabilities are *the bugs or flaws found in code that are used by hackers to exploit a system.* This type of attack is relatively easy to achieve. And many nations stockpile zero-day vulnerabilities as a weapon, in case they would need to attack. Once access is gained, hackers can steal sensitive data, disrupt the network, or corrupt files.

Did You Know?

One of the first instances of cyber warfare was Stuxnet. It was a computer worm designed to inflict physical damage. Suxnet is suspected to have been developed by the US and Israel to target the Iranian nuclear program. It used four zero-day exploits, cost millions of dollars, and lasted for months.

Cyber Espionage

Cyber espionage *occurs when hackers steal data and intellectual property.* With cyber espionage, hackers aim to steal information rather than do damage. Of course, depending on the information they steal there could be resulting damage. In 2015, the US Office of Personnel Management had 21 million US citizen records stolen in a case of cyber espionage. In this particular incident, the hackers also stole five million sets of fingerprints.

Propaganda

Propaganda is *misleading or inaccurate information used to publicize a particular political cause.* Propaganda can include theft of sensitive documents that get published or the sharing of false information. Propaganda is one of the easiest forms of cyber warfare and typically uses social media networks to spread.

Deterring Cyber Warfare

Cyber warfare allows enemies to attack from anywhere in the work and bypass traditional defense mechanisms. For this reason, government investment in cyber warfare prevention is imperative. In addition, by investing in cyber warfare we can also add a new weapon to the nation's arsenal.

The Right Stuff

According to international law, a nation is allowed to defend itself with force against an armed attack. If cyber warfare causes serious consequences and large-scale damage, this could be seen as an armed attack and trigger a physical attack or war. This attack would be within the rights of a nation, according to international law. Although this has not yet occurred, the rise of cyber warfare indicates that it could someday be a possibility.

The Tallinn Manual is a textbook prepared by a group of law scholars to explain how international law can apply to digital warfare. The Tallinn Manual is backed by NATO's Cooperative Cyber Defence Centre of Excellence (CCDCOE) in Tallinn, Estonia.

The initial release of the Tallinn Manual in 2014 clearly identified when an attack could be considered a violation of the international laws in cyberspace. It included 154 rules on how to apply the law to cyber warfare.

In 2017, the Tallinn Manual 2.0 was introduced to investigate the legal status for each type of hacking that occurs. In the three years since its initial release, the manual was greatly expanded and has become an influential resource for the law community.

Simple steps like setting complex passwords and changing passwords often can help deter a cyberattack.

Courtesy of designer491/Shutterstock

Surprisingly, basic cybersecurity practices are used to help defend against cyber warfare, such as changing passwords often, creating complex passwords, and keeping systems up-to-date. In addition, many nations are creating cyber deterrence programs to prevent digital attacks by making the cost of an attack too high. If the proper cybersecurity is in place, the cost of attacking the system will rise and cyber warfare would not be a feasible option.

Steps to Defend Against Cyber Warfare

There are four steps an organization must take in order to defend itself against cyber warfare.

1. Identify where your security team needs help. A security team may not be able to do everything needed to protect the organization. Identify where the organization needs help and where to find assistance.
2. Collaborate and share information. All organizations require cybersecurity. By sharing information with others, it enhances the network for all components of our infrastructure, creating a more secure nation.
3. Focus on security for key systems. Organizations may not need the best security for their online training system, but they do need top security to keep sensitive information secure. Ensure that the most needed systems receive the attention they deserve.
4. Establish contingency plans and incident response plans. Every organization should plan to be attacked at some point. Identify contingency plans and incident response plans so that you are ready when an attack occurs.

US Cyber Command (USCYBERCOM)

US service components that make up Cyber Command.

Courtesy of US Navy/US Cyber Command

The US Cyber Command is the military unit responsible for protecting all DoD networks. They monitor activity, block attacks, and defeat hackers on a daily basis. US Cyber Command (USCYBERCOM), located at Fort Meade, Maryland.

In 1972, DoD consultants warned of potential vulnerabilities to the government in computer and network security. This become a more pressing concern in 1995 after the end of the Cold War. A variety of initiatives were deployed to respond to the threat.

The US Cyber Command was established on November 12, 2008 and began formal operations in 2010 as a sub-command unit. In August 18, 2017, the US Cyber Command was promoted to a Unified Combatant Command to oversee cyberspace operations.

Active duty, reserve, and National Guard service members participate in the Cyber Guard and Cyber Flag exercises sponsored by US Cyber Command.
Courtesy of Chief Petty Officer Dennis Herring/US Cyber Command

As the nation's cyber warriors, USCYBERCOM operates daily in cyberspace against capable adversaries, some of whom are near-peer competitors in this domain. Our government has learned we must stop attacks before they penetrate our cyber defenses or impair our military forces; and through persistent, integrated operations, we can influence adversary behavior and introduce uncertainty into their decisions.

The USCYBERCOM has three main focus areas:

- Defend the DoD Information Network (DoDIN)
- Provide support to combatant commanders for execution of their missions around the world
- Strengthen our nation's ability to withstand and respond to cyberattacks

Each military service branch has its own cyber unit, that provides resources to USCYBERCOM. The 24th Air Force (Air Forces Cyber) is the organization that supports and protects Air Force networks. They are located at Joint Base San Antonio-Lackland, Texas.

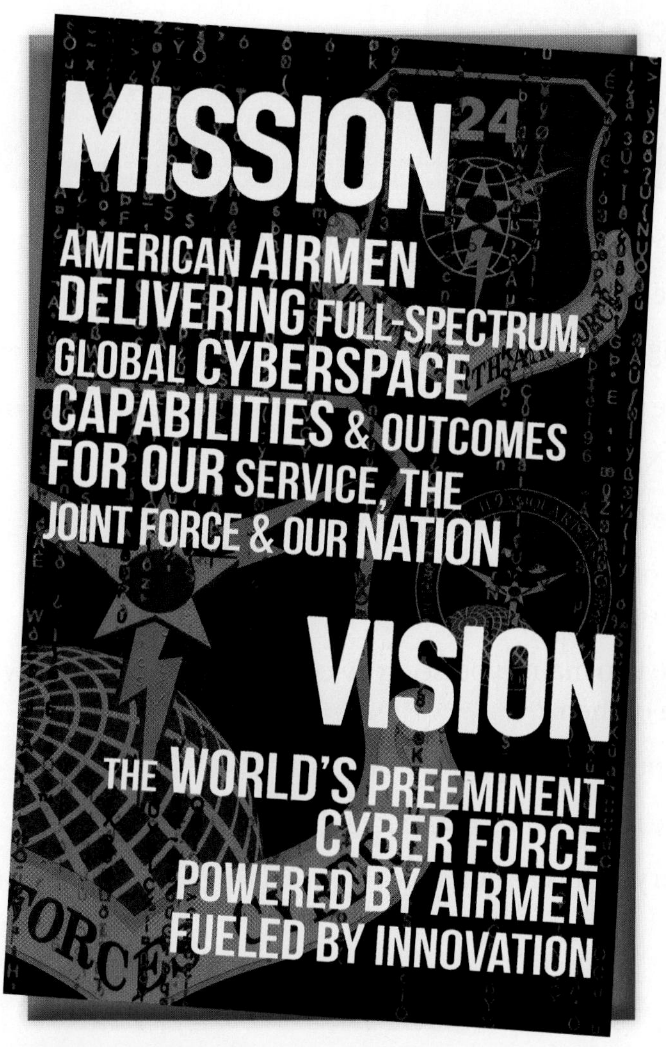

Mission of Air Forces Cyber.
Courtesy of US Air Force

Conclusion

Throughout this book, we have discussed the elements of space and humans' unwavering desire to explore this final frontier. As technology is enhanced and matures, new adventures in space will surely follow. We are living in the age of a new space race to see who can go farther than ever before—and with the added challenge of living on a foreign planet. There are new developments in space travel on a daily basis, and it's an exciting new time for space travel.

The technology required to travel through space brings along enhancements and new technology here on Earth. With that comes the risk of a different type of attack. This chapter has laid out some key threats and ways to protect yourself and our national infrastructure from cyberattacks. Stay on top of your cybersecurity and be prepared for an attack.

Lesson 3 Review

Using complete sentences, answer the following questions on a sheet of paper.

1. What are the main goals of CISA?
2. What DHS agency is responsible for natural disaster response efforts?
3. What are the three methods that voters can use to cast their votes?
4. What is the National Cyber Awareness System?
5. How will the DoD cultivate a cybersecurity work force?
6. What is one of the biggest mistakes a company can make with regard to cybersecurity?
7. What are the two main cybersecurity certifications and from whom would you get each of them?
8. What is the purpose of cyber warfare?
9. What are the most common cyber warfare threats?
10. What military unit is responsible for protecting all DoD networks?

APPLYING YOUR LEARNING

11. Cybersecurity knowledge was the desired outcome of this lesson. What can you do to help improve cybersecurity in your daily life?

References

CHAPTER 1 The History of Astronomy

LESSON 1 Prehistoric and Classical Astronomy

Cain, F. (2017, May 09). Do Stars Move? Tracking Their Movements Across the Sky. Retrieved from https://www.universetoday.com/135453/stars-move-tracking-movements-across-sky/

Canadian Museum of History; *Maya Civilization* https://www.historymuseum.ca/cmc/exhibitions/civil/maya/mmc07eng.shtml

The Celestial Sphere. (n.d.). Retrieved from http://planetary-science.org/astronomy/the-celestial-globe/

Johnson, V. (n.d.). Early Astronomers: Ptolemy, Aristotle, Copernicus, and Galileo. Retrieved from http://www.librarypoint.org/early_astronomers

Koupelis, T. (2014). *In quest of the universe* (7th ed.). Burlington, Massachusetts: Jones & Bartlett Learning.

Lippincott, K. (1994). Astronomy. New York: Dorling Kindersley.

Martyn Shuttleworth (Feb 4, 2010). Mayan Scientific Achievements. https://www.history.com/topics/mayan-scientific-achievements

Martyn Shuttleworth (Feb 4, 2010). Ancient Astronomy, Science And The Ancient Greeks. Retrieved ------- from Explorable.com: https://explorable.com/greek-astronomy

Martyn Shuttleworth (Feb 5, 2010). Egyptian Astronomy. Retrieved Jun 14, 2018 from Explorable.com: https://explorable.com/egyptian-astronomy

Sky Tellers - The Myths, the Magic and the Mysteries of the Universe. (n.d.). Retrieved from https://www.lpi.usra.edu/education/skytellers/constellations/

Smithsonian National Museum of the American Indian; *Living Maya Time*

https://maya.nmai.si.edu/calendar/calendar-system

Thompson, M. (2016). *A space travelers guide to the solar system*. New York: Pegasus Books.

LESSON 2 Astronomy and the Renaissance

Oskar Blakstad (Aug 18, 2011). Renaissance Astronomy. Retrieved Jul 09, 2018 from Explorable.com: https://explorable.com/renaissance-astronomy

Chaisson, E., & McMillan, S. (2017). *Astronomy: A beginner's guide to the universe* (8th ed.). Boston, MA: Pearson.

Darwinsbulldog. (2009, December 27). Shedding some light! [web log comment]. Retrieved from https://thonyc.wordpress.com/2009/12/27/shedding-some-light/

Koupelis, T. (2014). *In quest of the universe* (7th ed.). Burlington, Massachusetts: Jones & Bartlett Learning.

McClure, B. (2017, October 23). *What is an astronomical unit?* Retrieved from http://earthsky.org/space/what-is-the-astronomical-unit

Riebeek, H. (2009, July 7). *Planetary motion: The history of an idea that launched the scientific revolution*. Retrieved from https://earthobservatory.nasa.gov/Features/OrbitsHistory/

Shumueli, G. (2010). To explain or to predict? *Statistical Science*, 25(3), 1-22. doi: 10.1214/10-STS330

Williams, M. (2016, November 7). *What is the heliocentric model of the universe?* Retrieved from https://www.universetoday.com/33113/heliocentric-model/

LESSON 3 The Enlightenment and Modern Astronomy

Britannica, T. E. (2017, December 29). Equivalence principle. Retrieved from https://www.britannica.com/science/equivalence-principle

Galileo Galilei. (2014, September 14). Retrieved from https://www.famousscientists.org/galileo-galilei/

Galileo's Observations of the Moon, Jupiter, Venus and the Sun – Solar System Exploration: NASA Science. (2018, January 30). Retrieved from https://solarsystem.nasa.gov/news/307/galileos-observations-of-the-moon-jupiter-venus-and-the-sun/

Garner, R. (2017, July 31). An Earth-like Atmosphere May Not Survive Proxima b's Orbit. Retrieved from https://www.nasa.gov/feature/goddard/2017/an-earth-like-atmosphere-may-not-survive-proxima-b-s-orbit

Kepler's Laws and Newton's Laws. (n.d.). Retrieved from https://www.mtholyoke.edu/courses/mdyar/ast223/orbits/orb_lect.html

Koupelis, T. (2014). *In quest of the universe* (7th ed.). Burlington, Massachusetts: Jones & Bartlett Learning.

Lafferty, P. (2000). *Force & motion*. New York, NY: Dorling Kindersley.

Neil deGrasse Tyson. (n.d.). Retrieved from https://www.famousscientists.org/neil-degrasse-tyson/

Newton's Laws of Motion. (2015, May 5). Retrieved from https://www.grc.nasa.gov/www/K-12/airplane/newton.html

O'Neill, I. (2017, May/June). How a Total Solar Eclipse Helped Prove Einstein Right About Relativity. Retrieved from https://www.space.com/37018-solar-eclipse-proved-einstein-relativity-right.html

Planetary Motion: The History of an Idea That Launched the Scientific Revolution. (n.d.). Retrieved from https://earthobservatory.nasa.gov/Features/OrbitsHistory/page2.php

Redd, N. T. (2018, March 24). Stephen Hawking Biography (1942-2018). Retrieved from https://www.space.com/15923-stephen-hawking.html

CHAPTER 2 The Solar System

LESSON 1 The Earth and Moon

9 Things That Make Earth the Perfect Place for Life. https://curiosity.com/topics/9-things-that-make-earth-the-perfect-place-for-life-curiosity/

Chown, M. (2016). *Solar system: A visual exploration of the planets, moons, and other heavenly bodies that orbit our sun*. New York: Black Dog & Leventhal.

Dinwiddie, R. (2012). *Stars and planets*. London: DK Publishing.

Garner, R. (2018, January 30). Studying the Van Allen Belts 60 Years After America's First Spacecraft. Retrieved from https://www.nasa.gov/feature/goddard/2018/studying-the-van-allen-belts-60-years-after-america-s-first-spacecraft

How Far Away is the Moon? (2017, May 25). Retrieved from https://spaceplace.nasa.gov/moon-distance/en/

How the Moon was Formed: Retrieved from https://sservi.nasa.gov/articles/nasa-scientist-jen-heldmann-describes-how-the-earths-moon-was-formed/

Inglis-Arkell, E. (2015, December 16). How to measure the distance from the Earth to the Moon. Retrieved from https://io9.gizmodo.com/5688939/how-to-measure-the-distance-from-the-earth-to-the-moon

Koupelis, T. (2014). *In quest of the universe*. Burlington, Massachusetts: Jones & Bartlett Learning.

Mid-Atlantic Ridge. (n.d.). Retrieved from https://www.geolsoc.org.uk/Plate-Tectonics/Chap3-Plate-Margins/Divergent/Mid-Atlantic-Ridge

NASA's Goddard Space Flight Center: Astrophysics Science Division (ASD) https://heasarc.gsfc.nasa.gov/docs/rosat/gallery/display/saa.html

National Oceanic and Atmospheric Administration. (2014, August 01). What are Spring and Neap Tides? Retrieved from https://oceanservice.noaa.gov/facts/springtide.html

Redd, N. T. (2017, October 27). What is the Moon Made Of? Retrieved from https://www.space.com/19582-moon-composition.html

Structure of the Earth! (2017, October 04). Retrieved from https://www.natgeokids.com/uk/discover/geography/physical-geography/structure-of-the-earth/

Teitel, A. S. (2016, April 26). Do-It-Yourself Guide to Measuring the Moon's Distance. Retrieved from https://www.universetoday.com/91120/do-it-yourself-guide-to-measuring-the-moons-distance/

The Earth's Wobble – Precession. (2008, October 14). Retrieved from https://mydarksky.org/2008/10/14/the-earths-wobble-precession/

UCSB ScienceLine. (2015). Retrieved from http://scienceline.ucsb.edu/index.html

What is the South Atlantic Anomaly? (n.d.). Retrieved from https://image.gsfc.nasa.gov/poetry/ask/q525.html

Zell, H. (2015, March 02). Earth's Atmospheric Layers. Retrieved from https://www.nasa.gov/mission_pages/sunearth/science/atmosphere-layers2.html

LESSON 2 The Sun and Its Domain

Bahcall, J. (2000, June 29). How the Sun Shines. Retrieved August 8, 2018, from https://www.slac.stanford.edu/pubs/beamline/31/1/31-1-bahcall.pdf

Chaisson, E., & McMillan, S. (2017). *Astronomy: A beginner's guide to the universe* (8th ed.). Boston, MA: Pearson.

Dinwiddie, R., Will, G., Sparrow, G., & Stott, C. (2012). *Stars and planets*. London: DK Publishing.

Herridge, L. (2018, August 12). NASA's Parker Solar Probe Begins Journey to the Sun [Web log post]. Retrieved August 14, 2018, from NASA's Parker Solar Probe Begins Journey to the Sun

Koupelis, T. (2014). *In quest of the universe* (7th ed.). Burlington, Massachusetts: Jones & Bartlett Learning.

Lippincott, K. (1994). *Astronomy*. New York: Dorling Kindersley.

McClure, B. (2017, February 18). *Star brightness versus star luminosity*. Retrieved from http://earthsky.org/astronomy-essentials/stellar-luminosity-the-true-brightness-of-stars

Planets. (2018, February 06). Retrieved from https://solarsystem.nasa.gov/planets/overview/

Sharp, T. (2017, November 02). Atmosphere of the Sun: Photosphere, Chromosphere & Corona. Retrieved August 8, 2018, from https://www.space.com/17160-sun-atmosphere.html

Solar - Energy Explained, Your Guide To Understanding Energy - Energy Information Administration. (2017, December 15). Retrieved August 08, 2018, from https://www.eia.gov/energyexplained/index.php?page=solar_home

Young, C. (n.d.). Solar Structure. Retrieved from http://www.thesuntoday.org/the-sun/solar-structure/

LESSON 3 The Solar System

Chown, M. (2011). *Solar system: A visual exploration of the planets, moons, and other heavenly bodies that orbit our sun*. New York, NY: Black Dog & Leventhal.

Dinwiddie, R., Will, G., Sparrow, G., & Stott, C. (2012). *Stars and planets*. London: DK Publishing.

Fesenmaier, K. (2016, January 20). Caltech Researchers Find Evidence of a Real Ninth Planet. Retrieved September 27, 2018, from http://www.caltech.edu/news/caltech-researchers-find-evidence-real-ninth-planet-49523

Koupelis, T. (2014). *In quest of the universe* (7th ed.). Burlington, Massachusetts: Jones & Bartlett Learning.

Lippincott, K. (1994). *Astronomy*. New York: Dorling Kindersley.

Mars Pathfinder. https://marsmobile.jpl.nasa.gov/programmissions/missions/past/pathfinder/

Planets. (2018, February 06). Retrieved from https://solarsystem.nasa.gov/planets/overview/

Planet X. (2018, January 23). Retrieved from https://solarsystem.nasa.gov/planets/hypothetical-planet-x/in-depth/

Planetary Fact Sheet (2018, July 18). https://nssdc.gsfc.nasa.gov/planetary/factsheet/planet_table_british.html

Rao, J. (2017, December 31). When, Where and How to See the Planets in the 2018 Night Sky. Retrieved September 27, 2018, from https://www.space.com/39240-when-to-see-planets-in-the-sky.html

University of Central Florida. New research suggest Pluto should be reclassified as a planet. (2018, September 7). Retrieved September 27, 2018, from https://phys.org/news/2018-09-pluto-reclassified-planet.html

LESSON 4 Deep Space

'Monster' black hole consuming the Sun's mass every two days. (2018, May 15). Retrieved from https://astronomynow.com/2018/05/15/monster-black-hole-consuming-the-suns-mass-every-two-days/

Administrator, N. C. (2017, November 15). What Are Black Holes? Retrieved from https://www.nasa.gov/vision/universe/starsgalaxies/black_hole_description.html

Andrei, M. (2017, October 27). Our solar system seems to be inside a "bubble" of interstellar medium. Retrieved from https://www.zmescience.com/space/astrophysics-space/solar-system-interstellar-05042011/

Black Holes. (n.d.). Retrieved from https://science.nasa.gov/astrophysics/focus-areas/black-holes

Black Holes, Explained. (2018, September 25). Retrieved from https://www.nationalgeographic.com/science/space/universe/black-holes/

Brumfield, B. (2014, August 28). Solar system in big gas bubble, scientists corroborate. Retrieved from https://www.cnn.com/2014/08/27/tech/innovation/space-local-bubble/index.html

Cain, F. (2015, December 25). How Do Galaxies Get Named? Retrieved from https://www.universetoday.com/30765/how-do-galaxies-get-named/

Carlisle, C. M. (2016, August 26). Why Galaxies Have Spiral Arms. Retrieved from https://www.skyandtelescope.com/astronomy-news/galaxies-spiral-arms-2908201623/

Density wave theory. (2018, July 27). Retrieved from https://en.wikipedia.org/wiki/Density_wave_theory

Dinwiddie, R. (2012). *Stars and planets*. New York: DK Pub.

Dupzyk, K. (2018, August 21). How to Make a Scale Model of the Universe. Retrieved from https://www.popularmechanics.com/space/deep-space/a22757912/how-to-make-a-scale-model-of-the-universe/

Exoplanets 101. (2018, May 01). Retrieved from https://exoplanets.nasa.gov/the-search-for-life/exoplanets-101/

Garner, R. (2017, October 06). Messier 51 (The Whirlpool Galaxy). Retrieved from https://www.nasa.gov/feature/goddard/2017/messier-51-the-whirlpool-galaxy

Garner, R. (2017, October 06). Messier 31 (The Andromeda Galaxy). Retrieved from https://www.nasa.gov/feature/goddard/2017/messier-31-the-andromeda-galaxy

Garner, R. (2017, October 06). Messier 82 (The Cigar Galaxy). Retrieved from https://www.nasa.gov/feature/goddard/2017/messier-82-the-cigar-galaxy

Greicius, T. (2015, June 03). Charting the Milky Way From the Inside Out. Retrieved from https://www.nasa.gov/jpl/charting-the-milky-way-from-the-inside-out

Hodge, P. W. (2018, June 15). Milky Way Galaxy. Retrieved from https://www.britannica.com/place/Milky-Way-Galaxy/The-structure-and-dynamics-of-the-Milky-Way-Galaxy

Howell, E. (2017, November 07). What Is the Big Bang Theory? Retrieved from https://www.space.com/25126-big-bang-theory.html

Howell, E. (2018, August 25). Cosmic Microwave Background: Remnant of the Big Bang. Retrieved from https://www.space.com/33892-cosmic-microwave-background.html

HubbleSite - Discovering Planets Beyond - How Do Planets Form? (n.d.). Retrieved from http://hubblesite.org/hubble_discoveries/discovering_planets_beyond/how-do-planets-form

Redd, N. T. (2017, November 14). Milky Way Galaxy: Facts About Our Galactic Home. Retrieved from https://www.space.com/19915-milky-way-galaxy.html

Redd, N. T. (2018, January 10). The Andromeda Galaxy (M31): Location, Characteristics & Images. Retrieved from https://www.space.com/15590-andromeda-galaxy-m31.html

The Milky Way. (n.d.). Retrieved from https://stardate.org/astro-guide/milky-way

Wethington, N. (2016, April 26). The Milky Way Spiral. Retrieved from https://www.universetoday.com/24199/the-milky-way-spiral/

What Is Dark Matter? (2012, February 22) Retrieved from https://www.nasa.gov/audience/forstudents/9-12/features/what-is-dark-matter.html

What is the Big Bang? (2016, October 12). Retrieved from https://spaceplace.nasa.gov/big-bang/en/

CHAPTER 3 Space Exploration

LESSON 1 Why Explore Space?

Benefits Stemming from Space Exploration [PDF]. (2013, September). International Space Exploration Coordination Group.

Brenner, L. (2018, March 20). Why Are Private Companies in Space? Retrieved from https://sciencing.com/why-are-private-companies-in-space-13710338.html

Fishman, C. (2016, December 01). Is Jeff Bezos' Blue Origin the Future of Space Exploration? Retrieved from https://www.smithsonianmag.com/innovation/rocketeer-jeff-bezos-winner-smithsonians-technology-ingenuity-award-180961119/

Griffin, M. (2007, July 01). The Real Reasons We Explore Space. Retrieved from https://www.airspacemag.com/space/the-real-reasons-we-explore-space-18816871/

Kiger, P. J. (2018, March 08). 10 Reasons Why Space Exploration Matters to You. Retrieved from https://science.howstuffworks.com/10-reasons-space-exploration-matters.htm

Klesman, A. (2018, March 23). The demise of (the) Humanity (Star). Retrieved from http://www.astronomy.com/news/2018/03/the-demise-of-the-humanity-star

Life Signs | The Search for Life – Exoplanet Exploration: Planets Beyond our Solar System. (2017, January 24). Retrieved from https://exoplanets.nasa.gov/the-search-for-life/life-signs/

Masunaga, S. (2018, September 18). Meet the man who's paying SpaceX to fly him around the moon and back: Billionaire Yusaku Maezawa. Retrieved from http://www.latimes.com/business/la-fi-spacex-moon-passenger-20180917-story.html

NASA Strategic Plan 2018 [PDF]. (2018). NASA.

Savov, V. (2018, July 13). A ticket on Jeff Bezos' space tourism rocket will cost at least $200,000. Retrieved from https://www.theverge.com/2018/7/13/17567872/jeff-bezos-blue-origin-space-tourism-price-ticket

Sheetz, M. (2018, May 29). Richard Branson says Virgin Galactic is '2 or 3' flights away from taking people to space. Retrieved from https://www.cnbc.com/2018/05/29/richard-branson-virgin-galactic-is-2-or-3-flights-away-from-space.html

Stromberg, J. (2014, September 17). How did private companies get involved in space? Retrieved from https://www.vox.com/cards/private-space-flight/private-spaceflight-history-NASA

Wall, M. (2016, March 08). NASA's Shuttle Program Cost $209 Billion - Was it Worth It? Retrieved from https://www.space.com/12166-space-shuttle-program-cost-promises-209-billion.html

Why Is STEM Education So Important? (2016, February 02). Retrieved from https://www.engineeringforkids.com/about/news/2016/february/why-is-stem-education-so-important-/

Wiles, J. (2013, June 13). Why We Explore. Retrieved from https://www.nasa.gov/exploration/whyweexplore/why_we_explore_main.html#.W7t_jPZRdhF

Williams, M. (2017, January 06). What is Low Earth Orbit? Retrieved from https://www.universetoday.com/85322/what-is-low-earth-orbit/

LESSON 2 Assembling a Space Mission

Canright, S. (Ed.) (2009, April 10). The People Behind the Astronauts. Retrieved from https://www.nasa.gov/audience/foreducators/k-4/features/F_People_Behind_the_Astronauts.html

Canright, S. (Ed.) (2004, August 05). Your Body in Space: Use It or Lose It. Retrieved from https://www.nasa.gov/audience/forstudents/5-8/features/F_Your_Body_in_Space.html

Dunbar, B. (Ed.) (2004, May 02). Astronauts Take a Dive. Retrieved from https://www.nasa.gov/audience/foreducators/9-12/features/F_Astronauts_Take_Dive.html

Greenfieldboyce, N. (2018, August 3). NASA Announces Crew For First Commercial Space Flights. Retrieved from https://www.npr.org/2018/08/03/635344671/nasa-announces-crew-for-first-commercial-space-flights

Howell, E. (2018, February 26). How Astronaut Leland Melvin Went from the NFL to Space (Exclusive Video). Retrieved from https://www.space.com/39818-astronaut-leland-melvin-nfl-space-video.html

Martin, A. & Woollaston, V. (2018, October 11). SpaceX, Blue Origin, Virgin Galactic: Who's who in private space travel? Retrieved from http://www.alphr.com/space/1003058/private-space-travel-spacex-blue-origin-virgin-galactic

May, S. (2015, June 05). A Day in the Life Aboard the International Space Station. Retrieved from https://www.nasa.gov/audience/foreducators/stem-on-station/dayinthelife

NASA. (2018). NASA Strategic Plan 2018 (Rep.). Retrieved October 22, 2018, from https://www.nasa.gov/sites/default/files/atoms/files/nasa_2018_strategic_plan.pdf

NASA (2011). Astronaut Selection and Training (Rep). Retrieved October 31, 2018 from https://www.nasa.gov/centers/johnson/pdf/606877main_FS-2011-11-057-JSC-astro_trng.pdf

Selection and Training of Astronauts. (n.d.). Retrieved October 22, 2018, from https://science.ksc.nasa.gov/mirrors/msfc/crew/astronaut.html

Stuart, C. (2018). *How to live in space: Everything you need to know for the not-so-distant future*. Washington, DC: Smithsonian Books.

Weitering, H. (2017, May 23). Astronaut Leland Melvin Celebrated for Work Championing Women. Retrieved from https://www.space.com/36961-leland-melvin-makers-men.html

Wild, F. (Ed.) (2017, June 21). Astronaut Requirements. Retrieved October 22, 2018, from https://www.nasa.gov/audience/forstudents/postsecondary/features/F_Astronaut_Requirements.html

Wong, C. (2017, July 18). How to Shower in Space. Retrieved October 22, 2018, from https://airandspace.si.edu/stories/editorial/how-shower-space

Ziv, S. (2017, June 07). How NASA picks its new astronauts. Retrieved October 22, 2018, from https://www.newsweek.com/how-nasa-chooses-new-astronauts-621235

LESSON 3 The Hazards for Spacecraft

Braga, M. (2014, December 05). How a Spacecraft Like Orion Survives the Harsh Radiation of Orbit. Retrieved from https://motherboard.vice.com/en_us/article/pgam3y/orion-radiation-survival

Canright, S. (Ed.) (2004, August 26). Understanding Space Radiation. Retrieved from https://www.nasa.gov/audience/foreducators/postsecondary/features/F_Understanding_Space_Radiation.html

Garcia, M. (2015, April 14). Space Debris and Human Spacecraft. Retrieved from https://www.nasa.gov/mission_pages/station/news/orbital_debris.html

Granath, B. (2015, June 30). NASA Met Unprecedented Challenges Sending Spacecraft to Pluto. Retrieved from https://www.nasa.gov/feature/nasa-met-unprecedented-challenges-sending-spacecraft-to-pluto

Harbaugh, J. (2016, September 19). Autonomous Landing Hazard Avoidance Technology (ALHAT). Retrieved from https://www.nasa.gov/mission_pages/tdm/alhat/index.html

Hoffpauir, D. (2016, May 26). Space Debris: Understanding the Risks to NASA Spacecraft. Retrieved from https://www.nasa.gov/offices/nesc/articles/space-debris

Howell, E. (2018, March 01). New Horizons: Exploring Pluto and Beyond. Retrieved from https://www.space.com/18377-new-horizons.html

Keeter, B. (2018, June 19). ISS Daily Summary Report – 6/19/2018. Retrieved from https://blogs.nasa.gov/stationreport/2018/06/19/iss-daily-summary-report-6192018/

Manual, B. (2017, April 26). Meet the Space Custodians: Debris Cleanup Plans Emerge. Retrieved from https://www.space.com/36602-space-junk-cleanup-concepts.html

Mars, K. (2017, April 13). Why Space Radiation Matters. Retrieved from https://www.nasa.gov/analogs/nsrl/why-space-radiation-matters

New Horizons The First Mission to Pluto and the Kuiper Belt: Exploring Frontier Worlds. (n.d.). http://pluto.jhuapl.edu/common/content/missionGuide/NH_MissionGuide.pdf

Pultarova, T. (2018, June 21). 1st Satellite Built to Harpoon Space Junk for Disposal Begins Test Flight. Retrieved from https://www.space.com/40960-removedebris-space-junk-cleanup-test-flight.html

Ryba, J. (Ed.) (2004, April 1). Fire Prevention in Space. Retrieved from https://www.nasa.gov/missions/shuttle/f_fireprevention.html

Staff, S. (2016, March 08). Fire Burns Differently in Space, Space Station Experiment Shows. Retrieved from https://www.space.com/13766-international-space-station-flex-fire-research.html

Space Debris and Human Spacecraft. https://www.nasa.gov/mission_pages/station/news/orbital_debris.html

CHAPTER 4 Space Programs

LESSON 1 Strategic Significance of Space Programs

Cold War: A Brief History. (n.d.). Retrieved from http://www.atomicarchive.com/History/coldwar/page20.shtml

Commission to Assess United States National Security Space Management and Organization [PDF]. (2001, January 11).

David W. McFaddin, Lt Col, USAF, Can the U.S. Air Force Weaponize Space? (research paper, Air War College, Maxwell AFB, AL, 1998), 19.

SP-4012 NASA Historical Data Book: Volume IV NASA Resources 1969-1978, http://history.nasa.gov/SP-4012/vol4/ch1.htm (accessed December 12, 2011).

Presidential Directive/NSC-37, "National Space Policy," May 11, 1978. Retrieved from https://www.hq.nasa.gov/office/pao/History/nsc-37.html

Hayden, D. (2004-01). *The International Development of Space and Its Impact on U.S. National Space Policy*. (Airpower Research Institute, 2004). The White House, "Presidential Directive on National Space Policy," 11 February 1988, 1.

Hayden, D. (2004-01). *The International Development of Space and Its Impact on U.S. National Space Policy*. (Airpower Research Institute, 2004). The Commission to Assess United States National Security Space Management and Organization, "The Report of the Commission to Assess United States National Security, Space Management and Organization," May 2001, 99-100.

Hayden, D. (2004-01). *The International Development of Space and Its Impact on U.S. National Space Policy*. (Airpower Research Institute, 2004). The White House, Presidential Directive re. National Space Policy Review, NSPD-15, June 28, 2002.

Wikipedia, The Free Encyclopedia. *Space policy of the George W. Bush administration*

Retrieved January 24 2019 from: https://en.wikipedia.org/wiki/Space_policy_of_the_George_W._Bush_administration

President Donald Trump is Unveiling an America First National Space Strategy

March 23, 2018. Retrieved from:

https://www.whitehouse.gov/briefings-statements/president-donald-j-trump-unveiling-america-first-national-space-strategy/

Department of Space Indian Space Research Organisation. (n.d.). Retrieved from https://www.isro.gov.in/about-isro

Dickerson, K. (2015, October 19). Here's why NASA won't work with China to explore space. Retrieved from https://www.businessinsider.com/nasa-china-collaboration-illegal-2015-10

Dickerson, K. (2015, October 19). Here's why NASA won't work with China to explore space. Retrieved from https://www.businessinsider.com/nasa-china-collaboration-illegal-2015-10

Dujmovic, J. (2018, January 28). China is becoming a space superpower, just like the U.S. Retrieved from https://www.marketwatch.com/story/china-is-becoming-a-space-superpower-just-like-the-us-2018-01-24

ESA. (n.d.). History of Europe in space. Retrieved from https://www.esa.int/About_Us/Welcome_to_ESA/ESA_history/History_of_Europe_in_space

IndiaToday.in. (2017, December 29). These 5 achievements by ISRO proves that the space agency had a record-breaking 2017. Retrieved from https://www.indiatoday.in/education-today/gk-current-affairs/story/achievements-by-isro-in-2017-1118417-2017-12-29

JAXA Japan Aerospace Exploration Agency. (n.d.). Retrieved from http://global.jaxa.jp/article/special/solar-b/3/index_e.html

JAXA | JAXA History. (n.d.). Retrieved from http://global.jaxa.jp/about/history/index.html

Jones, A. (2018, June 20). China appears to be preparing to deorbit its Tiangong-2 space lab. Retrieved from https://spacenews.com/china-appears-to-be-preparing-to-deorbit-its-tiangong-2-space-lab/

Kobie, N. (2018, September 29). The epic tale of China's out of this world plan for space domination. Retrieved from https://www.wired.co.uk/article/china-space-moon-base-mars-landing

Launius, R. D. (2018). *The Smithsonian history of space exploration: From the ancient world to the extraterrestrial future*. Washington, DC: Smithsonian Books.

Leonov, A. (2005, January 01). The Nightmare of Voskhod 2. Retrieved from https://www.airspacemag.com/space/the-nightmare-of-voskhod-2-8655378/

May, S. (Ed.) (2018, December 18). What Was Project Mercury? Retrieved from https://www.nasa.gov/audience/forstudents/5-8/features/nasa-knows/what-was-project-mercury-58.html

May, S. (Ed.) (2011, March 16). What Was the Gemini Program? Retrieved from https://www.nasa.gov/audience/forstudents/5-8/features/nasa-knows/what-was-gemini-program-58.html

NASA. Glenn Orbits the Earth. (2012, February 16). Retrieved from https://www.nasa.gov/centers/glenn/about/bios/mercury_mission.html

Petersen, C. C. (2018, May 3). How China Became a Leader in Space Exploration. Retrieved from https://www.thoughtco.com/chinese-space-program-4164018

Pike, J. (n.d.). China's Taikonauts, "Yuhangyuan" (Astronauts) Program. Retrieved from https://www.globalsecurity.org/space/world/china/taikonauts.htm

Presidential Directive on National Space Policy [PDF]. (1988, February 11). The White House.

Wade, M. (n.d.). Shenzhou. Retrieved from http://www.astronautix.com/s/shenzhou.html

Wall, M. (2018, April 02). Farewell, Tiangong-1: Chinese Space Station Meets Fiery Doom Over South Pacific. Retrieved from https://www.space.com/40101-china-space-station-tiangong-1-crashes.html

Wall, M. (2017, December 27). Presidential Visions for Space Exploration: From Ike to Trump. Retrieved from https://www.space.com/11751-nasa-american-presidential-visions-space-exploration.html

Williams, D. R. (2018, November 20). Retrieved from https://nssdc.gsfc.nasa.gov/planetary/lunar/cnsa_moon_future.html

Williams, M. (2018, January 11). The surprising scale of China's space program. Retrieved from https://phys.org/news/2018-01-scale-china-space.html

Xinhua. (2018, January 22). China to select astronauts for its space station (Pengying, Ed.). Retrieved from http://www.xinhuanet.com/english/2018-01/22/c_136914938.htm

LESSON 2 US Manned Space Program

Allen, K. (2017, February 08). Soccer Ball from Challenger Explosion Makes It to Space 31 Years Later. Retrieved November 27, 2018, from https://abcnews.go.com/US/soccer-ball-challenger-explosion-makes-space-31-years/story?id=45341034

Bray, N. (2013, June 19). Shuttle Operations at Kennedy. Retrieved from https://www.nasa.gov/centers/kennedy/shuttleoperations/

Challenger: A Management Failure. (2014, September 08). Retrieved November 24, 2018, from http://www.spacesafetymagazine.com/space-disasters/challenger-disaster/challenger-management-failure/

Grady, D. (2012, July 23). American Woman Who Shattered Space Ceiling. Retrieved November 24, 2018, from https://www.nytimes.com/2012/07/24/science/space/sally-ride-trailblazing-astronaut-dies-at-61.html

Hall, J. L. (2003). Columbia and Challenger: Organizational failure at NASA. Retrieved November 24, 2018, from https://josephhall.org/papers/nasa.pdf

Heroes of Space: Sally Ride | Space Facts – Astronomy, the Solar System & Outer Space | All About Space Magazine. (n.d.). Retrieved November 24, 2018, from https://www.spaceanswers.com/space-exploration/heroes-of-space-sally-ride/

Howell, E. (2017, February 08). Guion Bluford: First African-American in Space. Retrieved November 24, 2018, from https://www.space.com/25602-guion-bluford-biography.html

Howell, E. (2017, November 15). Columbia Disaster: What Happened, What NASA Learned. Retrieved from https://www.space.com/19436-columbia-disaster.html

Launius, R. D. (2018). *The Smithsonian history of space exploration: From the ancient world to the extraterrestrial future*. Washington, DC: Smithsonian Books.

May, S. (Ed.) (2018, December 18). What Was Project Mercury? Retrieved from https://www.nasa.gov/audience/forstudents/5-8/features/nasa-knows/what-was-project-mercury-58.html

May, S. (Ed.) (2011, March 16). What Was the Gemini Program? Retrieved from https://www.nasa.gov/audience/forstudents/5-8/features/nasa-knows/what-was-gemini-program-58.html

NASA. Glenn Orbits the Earth. (2012, February 16). Retrieved from https://www.nasa.gov/centers/glenn/about/bios/mercury_mission.html

NASA. (2004, October 22). The Crew of the Challenger Shuttle Mission in 1986. Retrieved November 27, 2018, from https://history.nasa.gov/Biographies/challenger.html

NASA. (2009, September 17). NASA Biographies. Retrieved November 27, 2018, from https://history.nasa.gov/columbia/Biographies.html

NASA. (2018). NASA Space Shuttle. Retrieved November 24, 2018, from https://www.nasa.gov/mission-pages/shuttle

NASA. (n.d.). Retrieved November 24, 2018, from https://er.jsc.nasa.gov/seh/explode.html

Pruitt, S. (2016, January 28). 5 Things You May Not Know About the Challenger Shuttle Disaster. Retrieved November 24, 2018, from https://www.history.com/news/5-things-you-might-not-know-about-the-challenger-shuttle-disaster

Spaceflight Now (2019, January 7). SpaceX about one month away from first commercial crew test flight, retrieved from https://spaceflightnow.com

The Space Shuttle Columbia Disaster. (n.d.). Retrieved November 24, 2018, from http://www.spacesafetymagazine.com/space-disasters/columbia-disaster/

Wall, M. (2016, March 08). How the Space Shuttle Was Born. Retrieved November 24, 2018, from https://www.space.com/12085-nasa-space-shuttle-history-born.html

LESSON 3 Making Space People-Friendly

Bisphosphonates as a Countermeasure to Space Flight Induced Bone Loss. (2014, October 4). Retrieved from https://www.nasa.gov/mission_pages/station/research/experiments/239.html

Blinding Flashes. (2004, October 22). Retrieved from https://science.nasa.gov/science-news/science-at-nasa/2004/22oct_cataracts

Brief, J. (Ed.) (2013, June 06). National Space Biomedical Research Institute. Retrieved from https://www.nasa.gov/exploration/humanresearch/HRP_NASA/research_at_nasa_NSBRI.html

Callini, C. (2015, February 23). Biomedical Engineering for Exploration Space Technology. Retrieved from https://www.nasa.gov/content/biomedical-engineering-for-exploration-space-technology

Fessenden, M. (2015, April 02). Houston, We Might Have Some Major Problems Making Babies in Space. Retrieved from https://www.smithsonianmag.com/smart-news/houston-we-might-have-some-major-problems-making-babies-space-180954828/

Gushanas, T. (2017, November 07). Human Factors and Behavioral Performance (HFBP). Retrieved from https://www.nasa.gov/hrp/elements/hfbp

Imster, E. (2017, October 18). Traveling to Mars? Top 6 health challenges. Retrieved from https://earthsky.org/space/human-health-dangers-mars-travel

Kelly, S., & Dean, M. L. (2018). *Endurance: My year in space, a lifetime of discovery*. New York: Vintage Books, a division of Penguin Random House, LLC.

Mann, A. (2017, September 27). How Will We Deal With The Psychological Problems Of Spaceflight? Retrieved from https://medium.com/@adammann930/how-will-we-deal-with-the-psychological-problems-of-spaceflight-e191196006df

Mars, K. (2015, April 14). Twins Study. Retrieved from https://www.nasa.gov/twins-study

Mars, K. (2016, March 30). The Human Body in Space. Retrieved from https://www.nasa.gov/hrp/bodyinspace

Mars, K. (2017, June 07). Space Radiation (HRP Elements). Retrieved from https://www.nasa.gov/hrp/elements/radiation

Mars, K. (2017, June 19). About Space Radiation. Retrieved from https://www.nasa.gov/hrp/elements/radiation/about

NASA. (n.d.). Muscle Atrophy. NASA Information.

PhD, Y. H. (2017, November 03). Space travel: Beware of the health risks. Retrieved from https://www.medicalnewstoday.com/articles/319971.php

Radiation Sickness. (n.d.). Retrieved from https://www.mayoclinic.org/diseases-conditions/radiation-sickness/symptoms-causes/syc-20377058

CHAPTER 5 Space Stations and Beyond

LESSON 1 From Salyut to the International Space Station

Hanes, E. (2012, July 11). The Day Skylab Crashed to Earth: Facts About the First U.S. Space Station's Re-Entry. Retrieved from https://www.history.com/news/the-day-skylab-crashed-to-earth-facts-about-the-first-u-s-space-stations-re-entry

Harbaugh, J. (2016, February 18). Biography of Wernher Von Braun. Retrieved from https://www.nasa.gov/centers/marshall/history/vonbraun/bio.html

Salyut 7: retrieved from http://www.astronautix.com/s/salyut.html

History and Timeline of the ISS. (n.d.). Retrieved from https://www.iss-casis.org/about/iss-timeline/

Hollingham, R. (2014, November 18). Future - The rise and fall of artificial gravity. Retrieved from http://www.bbc.com/future/story/20130121-worth-the-weight

Howell, E. (2018, February 08). International Space Station: Facts, History & Tracking. Retrieved from https://amp.space.com/16748-international-space-station.html

Kitmacher, G., Miller, R., Pearlman, R., & Stott, N. (2018). Space stations: The art, science, and reality of working in space. Washington, DC: Smithsonian Books.

Martin, P. K. (2018, May 16). Examining the Future of the International Space Station [PDF].

May, S. (2015, June 05). A Day in the Life Aboard the International Space Station. Retrieved from https://www.nasa.gov/audience/foreducators/stem-on-station/dayinthelife

Stuart, C. (2018). How to live in space: Everything you need to know for the not-so-distant future. Washington, DC: Smithsonian Books.

LESSON 2 The Future in Space

Anderson, P. S. (2018, November 18). Lunar Outpost unveils small, exploratory moon rovers. Retrieved from https://earthsky.org/space/lunar-outpost-rovers-aka-lunar-resource-prospector

Atkinson, N. (2015, December 24). What Are Asteroids Made Of? Retrieved from https://www.universetoday.com/37425/what-are-asteroids-made-of/

Bartels, M. (2018, September 18). SpaceX's Lunar BFR Mission Isn't the 1st Private Moon Tourist Plan. Not Even for SpaceX. Retrieved from https://www.space.com/41855-spacex-private-moon-flight-commercial-lunar-plans.html

Boyle, A. (2018, October 11). Blue Origin resets schedule: First crew to space in 2019, first orbital launch in 2021. Retrieved from https://www.geekwire.com/2018/blue-origin-resets-schedule-first-crew-space-2019-first-orbital-launch-2021/

Choi, C. Q. (2017, September 21). Asteroids: Fun Facts and Information About Asteroids. Retrieved from https://www.space.com/51-asteroids-formation-discovery-and-exploration.html

Desjardins, J. (2016, November 02). Infographic: There's Big Money to Be Made in Asteroid Mining. Retrieved from https://www.visualcapitalist.com/theres-big-money-made-asteroid-mining/

Devlin, H., & Lyons, K. (2019, January 03). Far side of the moon: China's Chang'e 4 probe makes historic touchdown. Retrieved from https://www.theguardian.com/science/2019/jan/03/china-probe-change-4-land-far-side-moon-basin-crater

Dunbar, B. (Ed.) (2018, June 29). Moon to Mars Overview. Retrieved from https://www.nasa.gov/topics/moon-to-mars/overview

NASA, Lunar Orbital Platform-Gateway. (2018, May 2). Gateway Memorandum for the Record [Press release]. Retrieved from https://www.nasa.gov/sites/default/files/atoms/files/gateway_domestic_and_international_benefits-memo.pdf

NASA, Lyndon B. Johnson Space Center. (n.d.). Lunar Living - The Next Giant Leap [Press release]. Retrieved from https://www.nasa.gov/centers/johnson/pdf/160405main_lunar_living_fact_sheet.pdf

Mars Desert Research Station. (n.d.). Retrieved from http://mdrs.marssociety.org/

Potter, S. (2018, November 26). NASA InSight Lander Arrives on Martian Surface. Retrieved from https://www.nasa.gov/press-release/nasa-insight-lander-arrives-on-martian-surface-to-learn-what-lies-beneath

Robotic Exploration. (n.d.). Retrieved from http://exploration.esa.int/mars/

Roberts, J. (2015, October 01). The Next Generation of Suit Technologies. Retrieved from https://www.nasa.gov/feature/the-next-generation-of-suit-technologies

Snowden, S. (2018, October 2). A colossal elevator to space could be going up sooner than you ever imagined. Retrieved from https://www.nbcnews.com/mach/science/colossal-elevator-space-could-be-going-sooner-you-ever-imagined-ncna915421

SpaceX. (2016, September 20). Mars. Retrieved from https://www.spacex.com/mars

Stuart, C. (2018). How to live in space: Everything you need to know for the not-so-distant future. Washington, DC: Smithsonian Books.

Tate, K. (2014, May 09). NASA's Futuristic Z-2 Spacesuit: How It Works (Infographic). Retrieved from https://www.space.com/25708-how-nasa-z2-spacesuit-works-infographic.html

Today, U. (2017, May 01). China and Europe May Build A "Moon Village" in the 2020s. Retrieved from https://futurism.com/china-and-europe-may-build-a-moon-village-in-the-2020s

SpaceNews. Virgin Galactic achieves space on SpaceShipTwo test flight. (2018, December 14). Retrieved from https://spacenews.com/virgin-galactic-achieves-space-on-spaceshiptwo-test-flight/

Waldek, S. (2018, April 20). How to become a space tourist: 8 companies (almost) ready to launch. Retrieved from https://www.popsci.com/how-to-become-a-space-tourist#page-3

Warner, C. (2018, February 13). NASA's Lunar Outpost will Extend Human Presence in Deep Space. Retrieved from https://www.nasa.gov/feature/nasa-s-lunar-outpost-will-extend-human-presence-in-deep-space

LESSON 3 Space in Your Daily Life

Air Education and Training Command. (2009, October 16). AFJROTC cadets track NASA's moon spacecraft. Retrieved February 06, 2019, from https://www.aetc.af.mil/News/Article/262421/afjrotc-cadets-track-nasas-moon-spacecraft/

Anti-satellite weapon. (2019, May 12). Retrieved May 14, 2019, from https://en.wikipedia.org/wiki/Anti-satellite_weapon

ASAT program of China. (2019, May 13). Retrieved May 14, 2019, from https://en.wikipedia.org/wiki/ASAT_program_of_China

Brown, G., & Harris, W. (2018, March 08). How Satellites Work. Retrieved from https://science.howstuffworks.com/satellite.htm/printable

Brumfiel, G. (2013, April 26). Can You Hear Me Now? Cellphone Satellites Phone Home. Retrieved from https://www.npr.org/2013/04/26/178846158/can-you-hear-me-now-cellphone-satellites-phone-home

GPS device critical factor in saving lives in backcountry hiking. (2018, August 1). Retrieved January 29, 2019, from

https://www.verdenews.com/news/2018/aug/01/gps-device-critical-factor-saving-lives-backcountr/

Heaven, D. (n.d.). The first detailed look at how Elon Musk's space internet could work. Retrieved from https://www.newscientist.com/article/mg24032033-300-the-first-detailed-look-at-how-elon-musks-space-internet-could-work/

How do GPS devices work? (2005, November 28). Scientific American, Inc. Retrieved January 29, 2019, from https://www.scientificamerican.com/article/how-do-gps-devices-work/

Institute of Physics. (n.d.) How does GPS work? (n.d.) Retrieved January 29, 2019, from http://www.physics.org/article-questions.asp?id=55

Intelsat. (2018). Intelsat History. Retrieved January 29, 2019 from http://www.intelsat.com/about-us/history/

NASA. (2018). 20 Inventions We Wouldn't Have Without Space Travel. Retrieved January 29, 2019 from https://www.jpl.nasa.gov/infographics/infographic.view.php?id=11358

NASA. (2017). Commercial Space Transportation. Retrieved January 29, 2019, from https://www.nasa.gov/exploration/commercial/index.html

Track Me. (2016, April 08). Retrieved May 14, 2019, from https://www.youtube.com/watch?v=2AnSq8qQXbo

NASA. (2015). How Does GPS Work? Retrieved January 29, 2019, from https://spaceplace.nasa.gov/gps/en/

NASA (2019). Technology Transfer Portal (T2P). Retrieved January 29, 2019, from https://technology.nasa.gov/

National Oceanic and Atmospheric Administration (2018, November 05). NESDIS News & Articles. Retrieved February 06, 2019, from https://www.nesdis.noaa.gov/content/here's-how-satellite-data-helps-forecasters-issue-snow-squall-alerts

Origins of the Commercial Space Industry [PDF file]. (n.d.) Retrieved January 29, 2019, from https://www.faa.gov/about/history/milestones/media/Commercial_Space_Industry.pdf

Satellite television. (n.d.). Retrieved from https://www.esa.int/kids/en/learn/Technology/Mission_control/Satellite_television

Schultz, C. (2012, July 10). Fifty Years Ago, Lyndon Johnson Answered the First Satellite Phone Call. Retrieved from https://www.smithsonianmag.com/smart-news/fifty-years-ago-lyndon-johnson-answered-the-first-satellite-phone-call-1392634/

Smithsonian Institute (n.d.). Model, Communications Satellite, Marisat. Retrieved January 29, 2019, from https://airandspace.si.edu/collection-objects/model-communications-satellite-marisat

Smithsonian Institute (n.d.). Telemetry Module, Communications Satellite, Satcom 1. Retrieved January 29, 2019, from https://airandspace.si.edu/collection-objects/telemetry-module-communications-satellite-satcom-1

The Future of Space Commercialization [PDF file]. Retrieved from https://republicans-science.house.gov/sites/republicans.science.house.gov/files/documents/TheFutureofSpaceCommercializationFinal.pdf

Weather Questions & Answers. (n.d.). Retrieved from http://www.weatherquestions.com/How_do_weather_satellites_work.htm

Whalen, D.J. (n.d.) Communications Satellites: Making the Global Village Possible. Retrieved January 29, 2019, from https://history.nasa.gov/satcomhistory.html

CHAPTER 6 6 Space Probes and Robotics

LESSON 1 Space Probe Missions

About the Mission | Mission – Solar System Exploration: NASA Science. (2019, January 04). Retrieved from https://solarsystem.nasa.gov/missions/cassini/mission/about-the-mission/

ARTEMIS - The First Earth-Moon Libration Orbiter. (2015, March 25). Retrieved from https://www.nasa.gov/mission_pages/themis/news/artemis-orbit.html

Bartels, M. (2018, October 02). Should We Land on Venus Again? Scientists Are Trying to Decide. Retrieved from https://www.space.com/41986-venus-landing-mission-venera-d.html

Boyle, A. (2019, January 25). New Horizons probe sends a sharper view of space snowman ... with a bashed-in face. Retrieved from https://www.geekwire.com/2019/new-horizons-probe-sends-sharper-view-space-snowman-bashed-face/

Galileo. (n.d.). Retrieved from https://www.jpl.nasa.gov/missions/galileo/

Garner, R. (2017, May 30). Parker Solar Probe. Retrieved from https://www.nasa.gov/content/goddard/parker-solar-probe

Garner, R. (2018, November 07). Parker Solar Probe Reports Good Status After Close Solar Approach. Retrieved from https://www.nasa.gov/feature/goddard/2018/parker-solar-probe-reports-good-status-after-close-solar-approach

Greicius, T. (2015, March 13). Juno Overview. Retrieved from https://www.nasa.gov/mission_pages/juno/overview/index.html

Howell, E. (2012, November 19). Venera 13: First Color Pictures From Venus. Retrieved from https://www.space.com/18551-venera-13.html

Howell, E. (2016, March 08). A Brief History of Mars Missions. Retrieved from https://www.space.com/13558-historic-mars-missions.html

Howell, E. (2018, July 21). Mars Reconnaissance Orbiter: Mapping Mars in High Definition. Retrieved from https://www.space.com/18320-mars-reconnaissance-orbiter.html

In Depth | Helios 1 – Solar System Exploration: NASA Science. (2018, January 25). Retrieved from https://solarsystem.nasa.gov/missions/helios-1/in-depth/

In Depth | Ulysses – Solar System Exploration: NASA Science. (2018, January 26). Retrieved from https://solarsystem.nasa.gov/missions/ulysses/in-depth/

Kruse, R. (n.d.). Space Probes to the Outer Planets. Retrieved from https://historicspacecraft.com/Probes_Outer_Planets.html

Lunar Reconnaissance Orbiter. (n.d.). Retrieved from https://lunar.gsfc.nasa.gov/

Major NASA Lunar Probes History. (n.d.). Retrieved from https://history.nasa.gov/luneprob.html

Mariner 2. (n.d.). Retrieved from https://www.jpl.nasa.gov/missions/mariner-2/

Mars Observer. (n.d.). Retrieved from https://mars.nasa.gov/programmissions/missions/past/observer/

Mars Phoenix Lander Finishes Successful Work on Red Planet. Retrieved from https://www.nasa.gov/mission_pages/phoenix/news/phoenix-20081110.html

NASA Missions Uncover the Moon's Buried Treasures. (n.d.). Retrieved from https://www.nasa.gov/centers/ames/news/releases/2010/10-89AR.html

NASA Space Science Data Coordinated Archive: Lunar Reconnaissance Orbiter (LRO). (n.d.). Retrieved from https://nssdc.gsfc.nasa.gov/nmc/spacecraft/display.action?id=2009-031A

NASA Space Science Data Coordinated Archive: Pioneer 6. (n.d.). Retrieved from https://nssdc.gsfc.nasa.gov/nmc/spacecraft/display.action?id=1965-105A

NASA Space Scient Data Coordinated Archive: Magellan. (n.d.). Retrieved from https://nssdc.gsfc.nasa.gov/nmc/spacecraft/display.action?id=1989-033B

New Horizons: NASA's Mission to Pluto and the Kuiper Belt. (2019). Retrieved from http://pluto.jhuapl.edu/Pluto/Why-Pluto.php

Phoenix Mars Mission. (n.d.). Retrieved from http://phoenix.lpl.arizona.edu/science.php

Pioneer Venus 1, Orbiter and Multiprobe spacecraft (included NASA Ames partnership). Retrieved from https://www.nasa.gov/centers/ames/missions/archive/pioneer-venus.html

Pioneer-Venus. Retrieved from https://www.nasa.gov/mission_pages/pioneer-venus/

The Pioneer Missions. (2015, March 03). Retrieved from https://www.nasa.gov/centers/ames/missions/archive/pioneer.html

Ulysses. (n.d.). Retrieved from https://www.jpl.nasa.gov/missions/ulysses/

Ulysses Gets A New Partner In The Hunt For The Source Of Gamma-Ray Bursts – NASA's Mars Exploration Program. (2009, September 15). Retrieved from https://mars.nasa.gov/news/235/ulysses-gets-a-new-partner-in-the-hunt-for-the-source-of-gamma-ray-bursts/

Venera Program. (n.d.). Retrieved from https://imagine.gsfc.nasa.gov/science/toolbox/missions/venera.html

Venus Climate Orbiter AKATSUKI web site. (n.d.). Retrieved from http://akatsuki.isas.jaxa.jp/en/

Venus Express: Mission overview. (n.d.). Retrieved from http://sci.esa.int/venus-express/33010-summary/

Voyager: Fact Sheet. (n.d.). Retrieved from https://voyager.jpl.nasa.gov/frequently-asked-questions/fact-sheet/

Wall, M. (2018, December 21). New Horizons May Make Yet Another Flyby

After Ultima Thule. Retrieved from https://www.space.com/42808-nasa-new-horizons-possible-third-flyby.html

What Is a Space Probe? (n.d.). Retrieved from https://www.nasa.gov/centers/jpl/education/spaceprobe-20100225.html

LESSON 2 Robotics in Space

About Canadarm2. (2018, June 15). Retrieved from http://www.asc-csa.gc.ca/eng/iss/canadarm2/about.asp

About Dextre. (2018, July 30). Retrieved from http://asc-csa.gc.ca/eng/iss/dextre/about.asp

Brabaw, K. (2015, December 16). Do the Robot! NASA's Valkyrie Gets Down in New Video. Retrieved from https://www.space.com/31393-nasa-valkyrie-robot-dance-video.html

Burton, B. (2014, September 02). Japan's ISS Kirobo robot is lonely in space. Retrieved from https://www.cnet.com/news/japans-iss-kirobo-robot-is-lonely-in-space/

Campbell, L. (2018, March 20). Astrobotic wins NASA award to produce small lunar rover. Retrieved from https://www.spaceflightinsider.com/missions/commercial/astrobotic-wins-nasa-award-produce-small-lunar-rover/

David, L. (2018, March 16). This Tiny Private CubeRover Could Reach the Moon by 2020. Retrieved from https://www.space.com/40000-astrobotic-cuberover-moon-launch-2020.html

Dodge, A., Dodge, A., & Dodge, A. (2018, April 12). Robots in Space: How Technology Helps Us Explore the Solar System. Retrieved from https://blog.ozobot.com/2018/04/12/robots-space-technology-helps-us-explore-solar-system/

Howell, E. (2014, September 16). NASA's Robonaut 2 Droid Gets Its Legs on Space Station. Retrieved from https://www.space.com/27161-space-station-robonaut-legs.html

Howell, E. (2017, October 16). Real-Life Replicants: 6 Humanoid Robots Used for Space Exploration. Retrieved from https://www.space.com/amp/38460-humanoid-robots-for-space-exploration.html

Kanis, S. (2015, June 03). The History of SPHERES. Retrieved from https://www.nasa.gov/spheres/the-history-of-spheres

Kanis, S. (2016, November 08). Astrobee. Retrieved from https://www.nasa.gov/astrobee

Landau, E. (2015, June 09). RoboSimian Drives, Walks and Drills in Robotics Finals. Retrieved from https://www.jpl.nasa.gov/news/news.php?feature=4617

Pultarova, T. (2018, November 29). AI Robot CIMON Debuts at International Space Station. Retrieved from https://www.space.com/42574-ai-robot-cimon-space-station-experiment.html

Spice, B. (2014, November 24). Carnegie Mellon Unveils Lunar Rover "Andy". Retrieved from

https://www.cmu.edu/news/stories/archives/2014/november/november24_lunarroverandy.html

Tabor, A. (2017, December 18). Robotic Assistants on the International Space Station. Retrieved from https://www.nasa.gov/feature/ames/robotic-assistants-on-the-international-space-station

Wall, M. (2018, June 29). Meet CIMON, the 1st Robot with Artificial Intelligence to Fly in Space. Retrieved from https://www.space.com/amp/41041-artificial-intelligence-cimon-space-exploration.html

Why do we send robots to space? (2017, September 17). Retrieved from https://spaceplace.nasa.gov/space-robots/en/

Wilcox, B., Ambrose, R., & Kumar, V. (n.d.). Assessment of International Research and Development in Robotics, Chapter 3: Space Robotics [PDF]. National Science Foundation.

LESSON 3 The Mars Rover Expedition

Howell, E. (2018, July 17). Mars Curiosity: Facts and Information. Retrieved March 07, 2019, from https://www.space.com/17963-mars-curiosity.html

Ma, C., & Ma, C. (2012, November 16). At Age 11, This Girl Named the Curiosity Rover. Retrieved March 07, 2019, from https://mashable.com/2012/11/16/clara-ma-curiosity-rover/#h64mGc.JNgqW

NASA Radioisotope Power Systems. (2014, June 03). Retrieved March 07, 2019, from https://rps.nasa.gov/

NASA. (2018, July 17). Gravity Assist Podcast: Exploring Mars, with Steve Squyres. Retrieved March 07, 2019, from https://www.space.com/41176-gravity-assist-podcast-exploring-mars-with-steve-squyres.html

NASA. (2019). Mars 2020 Mission. Retrieved March 07, 2019, from https://mars.nasa.gov/mars2020/

NASA. (2019). Mars Exploration Rovers. Retrieved March 07, 2019, from https://mars.nasa.gov/mer/

NASA. (2019). Mars Science Laboratory. Retrieved March 07, 2019, from https://mars.nasa.gov/msl/

NASA. (n.d.). Mars: Mars Science Laboratory: Alpha Particle X-Ray Spectrometer (APXS). Retrieved March 07, 2019, from https://marsmobile.jpl.nasa.gov/msl/mission/instruments/spectrometers/apxs/

O'Brien, M. (2018, October 29). Steve Squyres – On Exploring Mars, and Other Celestial Objects. Retrieved March 07, 2019, from https://milesobrien.com/steve-squyres-on-exploring-mars-and-other-celestial-objects

Pyle, R. (2014). *Curiosity: An inside look at the Mars rover mission and the people who made it happen*. New York (N.Y.): Prometheus Book.

Wall, M. (2011, November 25). Girl Who Named NASA's Mars Rover Excited to Watch It Blast Off. Retrieved March 07, 2019, from https://www.space.com/13735-nasa-mars-rover-curiosity-clara-ma.html

CHAPTER 7 Orbiting, Space Travel, and Rockets

LESSON 1 Orbits and How They Work

Brown, G., & Harris, W. (2018, March 08). How Satellites Work. Retrieved from https://science.howstuffworks.com/satellite.htm/printable

Elliptical Orbits. (n.d.). Retrieved from https://www.windows2universe.org/physical_science/physics/mechanics/orbit/ellipse.html

ESA. (n.d.). Types of orbits. Retrieved from https://www.esa.int/Our_Activities/Space_Transportation/Types_of_orbits

NASA. (2015, June 01). What Is an Orbit? Retrieved from https://www.nasa.gov/audience/forstudents/5-8/features/nasa-knows/what-is-orbit-58.html

Space Environment. (n.d.). Retrieved from http://www.qrg.northwestern.edu/projects/vss/docs/space-environment/1-what-causes-an-orbit.html

I. (n.d.). Types of Orbits. Retrieved from http://www.polaris.iastate.edu/EveningStar/Unit4/unit4_sub3.htm

Williams, M. (2016, April 26). How Satellites Stay in Orbit. Retrieved from https://www.universetoday.com/93077/how-satellites-stay-in-orbit/

LESSON 2 How to Travel in Space

Basics of Space Flight - Solar System Exploration: NASA Science. (n.d.). Retrieved from https://solarsystem.nasa.gov/basics/chapter4-1/

Basics of Space Flight - Solar System Exploration: NASA Science. (n.d.). Retrieved from https://solarsystem.nasa.gov/basics/chapter13-1/

Dawn. (n.d.). Retrieved from https://www.jpl.nasa.gov/missions/dawn/

Deep Space 1. (n.d.). Retrieved from https://www.jpl.nasa.gov/missions/deep-space-1-ds1/

Dunbar, B. (Ed.). (2004, December 7). Ion Propulsion: Farther, Faster, Cheaper. Retrieved from https://www.nasa.gov/centers/glenn/technology/Ion_Propulsion1.html

George, L. E., & Kos, L. D. (1998, July). *Interplanetary Mission Design Handbook: Earth-to-Mars Mission Opportunities and Mars-to-Earth Return Opportunities 2009-2024* [PDF]. NASA - Marshall Space Flight Center.

Gravity Assist. (n.d.). Retrieved from http://www.gravityassist.com/

How does solar electric propulsion (ion propulsion) work? (n.d.). Retrieved from http://www.qrg.northwestern.edu/projects/vss/docs/propulsion/zoom-solar-ion.html

Mohon, L. (2015, May 20). Deep Space Atomic Clock (DSAC) Overview. Retrieved from https://www.nasa.gov/mission_pages/tdm/clock/overview.html

Moving in Space. (n.d.). Retrieved from http://howthingsfly.si.edu/flight-dynamics/moving-space

NASA. (n.d.). Deep Space Navigation. Retrieved from https://scienceandtechnology.jpl.nasa.gov/research/research-topics-list/communications-computing-software/deep-space-navigation

Trajectories and Orbits. (n.d.). Retrieved from https://history.nasa.gov/conghand/traject.htm

LESSON 3 How Rockets Work

Berger, E., & Utc. (2017, August 10). NASA's plasma rocket making progress toward a 100-hour firing. Retrieved from https://arstechnica.com/science/2017/08/nasas-plasma-rocket-making-progress-toward-a-100-hour-firing/

Berger, E. (2019, February 26). Europe unveils design of reusable rocket that looks a lot like a Falcon 9. Retrieved from https://arstechnica.com/science/2019/02/europe-unveils-design-of-reusable-rocket-that-looks-a-lot-like-a-falcon-9/

Brain, M. (2018, June 28). How Rocket Engines Work. Retrieved from https://science.howstuffworks.com/rocket.htm

Canright, S. (Ed). (n.d.). The Engine That Does More. Retrieved from https://www.nasa.gov/audience/foreducators/k-4/features/F_Engine_That_Does_More.html

Cassini Launch [PDF]. (1997, October). NASA.

Choi, C. Q. (2017, June 09). Will Mini Fusion Rockets Provide Spaceflight's Next Big Leap? Retrieved from https://www.space.com/37146-nuclear-fusion-rockets-interstellar-spaceflight.html

Dhitt. (2016, August 02). All Roads Lead to the Pad. Retrieved April 12, 2019, from https://blogs.nasa.gov/Rocketology/2016/08/02/all-roads-lead-to-the-pad/

Doody, D. (2017, February). *Basics of Space Flight*. Jet Propulsion Laboratory: California Institute of Technology, Pasadena, CA.

Dr. Wernher von Braun. (n.d.). Retrieved April 12, 2019, from https://www.nasa.gov/centers/marshall/history/vonbraun/bio.html

Esa. (n.d.). Solid and liquid fuel rockets. Retrieved from https://www.esa.int/Education/Solid_and_liquid_fuel_rockets

Exploring space: The high frontier. (2011). Sudbury, MA: Jones & Bartlett Learning.

Garner, R. (2019, March 08). NASA's Clean Room: Last Stop for New Hubble Hardware. Retrieved from https://www.nasa.gov/mission_pages/hubble/servicing/series/cleanroom.html

Gilman, I. (Ed.). (2010, December 10). Rockets of the Future. Retrieved from https://www.nasa.gov/audience/forstudents/nasaandyou/home/rockets_bkgd_en.html

Hall, N. (Ed.). (n.d.). Solid Rocket Engine. Retrieved from https://www.grc.nasa.gov/www/k-12/airplane/srockth.html

Hall, N. (Ed.). (n.d.). Liquid Rocket Engine. Retrieved from https://www.grc.nasa.gov/www/k-12/airplane/lrockth.html

Harbaugh, J. (2016, February 18). Biography of Wernher Von Braun. Retrieved April 12, 2019, from https://www.nasa.gov/centers/marshall/history/vonbraun/bio.html

Heiney, A. (2012, February 23). Aiming for an Open Window. Retrieved from https://www.nasa.gov/centers/kennedy/launchingrockets/launchwindows.html

Heiney, A. (Ed.). (2018, June 19). The Crawlers. Retrieved from https://www.nasa.gov/content/the-crawlers

How does a rocket work? (n.d.). Retrieved from https://www.esa.int/kids/en/learn/Technology/Rockets/How_does_a_rocket_work

How Rockets Work [PDF]. (n.d.). NASA.

Howell, E. (2018, April 20). Delta IV Heavy: Powerful Launch Vehicle. Retrieved April 07, 2019, from https://www.space.com/40360-delta-iv-heavy.html

Howell, E. (2018, October 25). The History of Rockets. Retrieved from https://www.space.com/amp/29295-rocket-history.html

Inside the LEO Doghouse: Nuclear Thermal Engines. (2014, June 30). Retrieved from https://blogs.nasa.gov/J2X/tag/nuclear-fission/

Johnson, S. (2019, March 02). What Are the Different Kinds of Rockets? Retrieved from https://sciencing.com/different-kinds-rockets-8552176.html

Kruse, R. (n.d.). Miscellaneous American Rockets and Missiles. Retrieved from https://historicspacecraft.com/rockets.html

Launch a rocket from a spinning planet. (n.d.). Retrieved April 12, 2019, from https://spaceplace.nasa.gov/launch-windows/en

Launius, R. D., & Johnston, A. K. (2009). *Smithsonian atlas of space exploration.* Piermont, N.H: Bunker Hill Publ.

Lethbridge, C. (n.d.). History of Rocketry: Goddard. Retrieved July 16, 2019, from https://www.spaceline.org/history/22.html

Lewis, D. (2016, November 16). Why the U.S. Government Brought Nazi Scientists to America After World War II. Retrieved from https://www.smithsonianmag.com/smart-news/why-us-government-brought-nazi-scientists-america-after-world-war-ii-180961110/

Looking Closer at the Saturn V. (2019, April 08). Retrieved April 12, 2019, from https://airandspace.si.edu/stories/editorial/looking-closer-saturn-v

May, S. (Ed.). (2015, June 02). What Was the Saturn V? Retrieved from https://www.nasa.gov/audience/forstudents/5-8/features/nasa-knows/what-was-the-saturn-v-58.html

Mohon, L. (2015, January 21). Meet the Rocket. Retrieved from https://www.nasa.gov/exploration/systems/sls/index.html

NASA (2013, June 07). Launch Schedule 101. Retrieved from https://www.nasa.gov/missions/highlights/schedule101.html

NASA. (2019). Space Launch System (SLS) Overview. Retrieved April 34, 2019, from https://www.nasa.gov/exploration/systems/sls/overview.html

On Atlas' Shoulders. (n.d.). Retrieved April 11, 2019, from https://www.lockheedmartin.com/en-us/news/features/history/atlas.html

Reneke, D. (2018, July 7). Rockets Of The Future? Retrieved from https://www.davidreneke.com/rockets-of-the-future/

RussianSpaceWeb.com. (n.d.). Retrieved April 10, 2019, from http://www.russianspaceweb.com/

Steigerwald, B. (n.d.). New and Improved Antimatter Spaceship for Mars Missions. Retrieved from https://www.nasa.gov/exploration/home/antimatter_spaceship.html

Thompson, A. (2019, February 09). This Year SpaceX Made Us All Believe in Reusable Rockets. Retrieved from https://www.wired.com/story/this-year-spacex-made-us-all-believe-in-reusable-rockets/

United Launch Alliance. (2019, March 16). Retrieved from https://en.wikipedia.org/wiki/United_Launch_Alliance

Vostok and Voskhod. (n.d.). Retrieved April 11, 2019, from https://airandspace.si.edu/exhibitions/space-race/online/sec300/sec330.htm

Vostok-K. (2019, January 24). Retrieved from https://en.wikipedia.org/wiki/Vostok-K

Wall, M. (2013, October 07). Quick Fusion-Powered Trips to Mars No Fantasy, Scientists Say. Retrieved from https://www.space.com/23084-mars-exploration-nuclear-fusion-rocket.html

Wheeler, R. (2009). Apollo lunar landing launch window: The controlling factors and constraints. Retrieved from https://history.nasa.gov/afj/launchwindow/lw1.html

CHAPTER 8 Cyber Security

LESSON 1 Foundations of Cyber Security

Crane, C. (2019, April 30). 80 Eye-Opening Cyber Security Statistics for 2019. Retrieved from https://www.thesslstore.com/blog/80-eye-opening-cyber-security-statistics-for-2019/

Cybersecurity. (2019, March 01). Retrieved from https://www.dhs.gov/topic/cybersecurity#

Dunbar, B. (2015, May 21). What Is the Hubble Space Telescope? Retrieved from https://www.nasa.gov/audience/forstudents/k-4/stories/nasa-knows/what-is-the-hubble-space-telecope-k4.html

Fidler, D. (2018, April 3). Cybersecurity and the New Era of Space Activities. Retrieved from https://www.cfr.org/report/cybersecurity-and-new-era-space-activities

How to do satellite jamming. (2015, April 27). Retrieved from https://iicybersecurity.wordpress.com/2015/04/27/how-to-do-satellite-jamming/

Livingstone, D., & Lewis, P. (2016, September). *Space, the Final Frontier for Cybersecurity?* [PDF]. Chatham House.

Lord, N. (2019, April 26). What is Cyber Security? Definition, Best Practices & More. Retrieved from https://digitalguardian.com/blog/what-cyber-security

Lunden, I. (2015, June 3). 6.1B Smartphone Users Globally By 2020, Overtaking Basic Fixed Phone Subscriptions. Retrieved from https://techcrunch.com/2015/06/02/6-1b-smartphone-users-globally-by-2020-overtaking-basic-fixed-phone-subscriptions/

May, S. (Ed.). (2015, May 21). What Is the Hubble Space Telescope? Retrieved from https://www.nasa.gov/audience/forstudents/k-4/stories/nasa-knows/what-is-the-hubble-space-telecope-k4.html

Number of smartphone users worldwide 2014-2020. (n.d.). Retrieved from https://www.statista.com/statistics/330695/number-of-smartphone-users-worldwide/

Popkin, G. (2018, December 27). China's quantum satellite achieves 'spooky action' at record distance. Retrieved from https://www.sciencemag.org/news/2017/06/china-s-quantum-satellite-achieves-spooky-action-record-distance

Whitwam, R. (2019, March 08). Hacking Satellites Is Surprisingly Simple. Retrieved from https://www.extremetech.com/extreme/287284-hacking-satellites-is-probably-easier-than-you-think

LESSON 2 Principles of Cybersecurity and Computer Basics

About International Telecommunication Union (ITU). (n.d.). Retrieved from https://www.itu.int/en/about/Pages/default.aspx

Armerding, T. (2018, October 09). Cybersecurity: Not Just "A" Job - Many Jobs Of The Future. Retrieved from https://www.forbes.com/sites/taylorarmerding/2018/10/09/cybersecurity-not-just-a-job-many-jobs-of-the-future/#557245de3f2b

Cooper, S. (2018, June 06). How to secure your home wireless network from hackers - Best practices. Retrieved from https://www.comparitech.com/blog/information-security/secure-home-wireless-network/

Global Cybersecurity Index (GCI) [PDF]. (2018). International Telecommunication Union (ITU).

Hoffman, C. (2018, February 16). How Does the Internet Work? Retrieved from https://www.howtogeek.com/341866/how-does-the-internet-work/amp/

Hoffman, C. (2018, August 08). What is an Operating System? Retrieved from https://www.howtogeek.com/361572/what-is-an-operating-system/

Hurwitz, S. (2019, May 21). WiFi passwords hacked at local coffee shop, security compromised. Retrieved from https://www.kptv.com/news/wifi-passwords-hacked-at-local-coffee-shop-security-compromised/article_88e645ae-7c49-11e9-9aae-5386cbe68e0b.html

James, M. (2018, August 22). 5 Ways to Spot a Phishing Email. Retrieved from https://staysafeonline.org/blog/5-ways-spot-phishing-emails/

Johnston, L. (2019, May 09). 5 Ways to Back up Your Data and Keep It Safe. Retrieved from https://www.lifewire.com/ways-to-back-up-your-data-2640426

Lowmaster, K. (2018, June 1). The Ten Commandments for Data Management: Personal Data Management Fundamentals. Retrieved from https://oneworldidentity.com/research/personal-data-management-fundamentals/

Mardisalu, R. (2019, May 15). 14 Most Alarming Cyber Security Statistics in 2019. Retrieved from https://thebestvpn.com/cyber-security-statistics-2019/

Microsoft. (n.d.). What Is Cloud Computing? A Beginner's Guide | Microsoft Azure. Retrieved from https://azure.microsoft.com/en-us/overview/what-is-cloud-computing/

Nadeau, M. (2018, December 13). What is cryptojacking? How to prevent, detect, and recover from it. Retrieved from https://www.csoonline.com/article/3253572/what-is-cryptojacking-how-to-prevent-detect-and-recover-from-it.html?page=2

National Conference of State Legislatures. (2019, February 8). Cybersecurity Legislation 2018. Retrieved from http://www.ncsl.org/research/telecommunications-and-information-technology/cybersecurity-legislation-2018.aspx

National Cyber Strategy of the United States of America [PDF]. (2018, September).

Porup, J. (2017, December 27). What is cyber security? How to build a cyber security strategy. Retrieved from https://www.csoonline.com/article/3242690/what-is-cyber-security-how-to-build-a-cyber-security-strategy.html

Spiro, S. (2018, June 28). You've Been Hacked: 5 Ways to Minimize the Damage. Retrieved from https://staysafeonline.org/blog/hacked-5-ways-minimize-damage/

Symantec. (n.d.). Norton Cybercrime Report: The Human Impact [PDF].

Symantec. (n.d.). Why Hackers Love Public WiFi. Retrieved from https://us.norton.com/internetsecurity-wifi-why-hackers-love-public-wifi.html

Tauchman, E. R. (2018, August 10). The Top 9 Cybersecurity Jobs and What You Need to Get One. Retrieved from https://certification.comptia.org/it-career-news/post/view/2018/08/10/the-top-9-jobs-in-cybersecurity

Tips for Strong, Secure Passwords & Other Authentication Tools. (n.d.). Retrieved from https://www.connectsafely.org/tips-to-create-and-manage-strong-passwords/

United Nations. (n.d.). Retrieved from https://www.un.org/en/sections/about-un/overview/index.html

What is a Virtual Machine and How Does it Work | Microsoft Azure. (n.d.). Retrieved from https://azure.microsoft.com/en-us/overview/what-is-a-virtual-machine/

What is a Virtual Machine and How Does it Work? - Definition from WhatIs.com. (n.d.). Retrieved from https://searchservervirtualization.techtarget.com/definition/virtual-machine

Woodford, C. (2018, December 25). How do computers work? A simple introduction. Retrieved from https://www.explainthatstuff.com/howcomputerswork.html

LESSON 3 National Cyber Policy

Cybersecurity. (2019, March 01). Retrieved from https://www.dhs.gov/topic/cybersecurity

Cybersecurity Roles and Job Titles. (n.d.). Retrieved from https://www.cs.seas.gwu.edu/cybersecurity-roles-and-job-titles

Department of Defense. (2018). Summary: Department of Defense Cyber Strategy 2018 [PDF].

Election Security. (2019, March 05). Retrieved from https://www.dhs.gov/topic/election-security

Garber, M. (2013, January 04). The Future of Cybersecurity Could Be Sitting in an Office in New Jersey. Retrieved from https://www.theatlantic.com/technology/archive/2013/01/the-future-of-cybersecurity-could-be-sitting-in-an-office-in-new-jersey/266849/

Lange, K. (2018, October 2). DOD's Cyber Strategy: 5 Things to Know. Retrieved from https://www.defense.gov/explore/story/Article/1648425/dods-cyber-strategy-5-things-to-know/

Melnick, J. (2018, May 15). Top 10 Most Common Types of Cyber Attacks. Retrieved from https://blog.netwrix.com/2018/05/15/top-10-most-common-types-of-cyber-attacks/

National Cyber Awareness System. (n.d.). Retrieved from https://www.us-cert.gov/ncas

NCSL. (2018, August 20). Voting Equipment. Retrieved from http://www.ncsl.org/research/elections-and-campaigns/voting-equipment.aspx

NetWars: CyberCity. (n.d.). Retrieved from https://www.sans.org/netwars/cybercity

O'Hara, K. (n.d.). The Future of Cybersecurity Jobs. Retrieved from https://www.monster.com/career-advice/article/future-of-cybersecurity-jobs

Porche, I. R., III. (2018, March 03). Getting Ready to Fight the Next (Cyber) War. Retrieved from https://www.rand.org/blog/2018/03/getting-ready-to-fight-the-next-cyber-war.html

RAND. (n.d.). Cyber Warfare. Retrieved from https://www.rand.org/topics/cyber-warfare.html

Ranger, S. (2018, December 04). What is cyberwar? Everything you need to know about the frightening future of digital conflict. Retrieved from https://www.zdnet.com/article/cyberwar-a-guide-to-the-frightening-future-of-online-conflict/

Rouse, M., Ferguson, K., & Rosencrance, L. (2019, May). What is cyberwarfare? - Definition from WhatIs.com. Retrieved from https://searchsecurity.techtarget.com/definition/cyberwarfare

Schwartz, E. (2015, February). Security Think Tank: Four steps to defend against cyber attack . Retrieved from https://www.computerweekly.com/opinion/Security-Think-Tank-Four-steps-to-defend-against-cyber-attack

Stevens, M. (2017, July 10). Cybersecurity Team Structure: 7 Important Roles & Responsibilities. Retrieved from https://www.bitsight.com/blog/cybersecurity-teams

Tallinn Manual 2.0. (n.d.). Retrieved from https://ccdcoe.org/research/tallinn-manual/

Topics. (2018, June 08). Retrieved from https://www.dhs.gov/topics

Torrisi, J. (2014, May 16). Cyberwarfare: Protecting 'soft underbelly' of USA. Retrieved from https://www.cnbc.com/2014/05/15/defending-against-cyberwarfare-inside-americas-soft-underbelly.html

US Cyber Command. (n.d.). Retrieved from https://www.cybercom.mil/

Weisman, S. (2019). What are Denial of Service (DoS) attacks? DoS attacks explained. Retrieved from https://us.norton.com/internetsecurity-emerging-threats-dos-attacks-explained.html

Zetter, K. (2018, September 26). The Crisis of Election Security. Retrieved from https://www.nytimes.com/2018/09/26/magazine/election-security-crisis-midterms.html

Glossary

A

acceleration—the act of increasing speed or velocity. (p. 41)

aerobraking—a maneuver used to slow down a spacecraft by flying through a planet's atmosphere to produce aerodynamic drag. (p. 309)

aeronautics—the study of the science of flight. (p. 159)

aeroshell—protects the lander, with the rover inside, from the intense heating of entry into the thin Martian atmosphere. (p. 339)

airlock—last area of the flight deck. (p. 218)

aft—The rear part of the orbiter is the region where the main engines and maneuvering systems are located to propel and guide the orbiter. (p. 219)

alpha Particle X-Ray Spectrometer (APXS)—calculated the amount of chemical elements in rocks and soils by measuring the way different materials respond to two kinds of radiation: x-rays and alpha particles. (p. 338)

alluvial fan—a fan-shaped deposit of sediment crossed and built up by streams. (p. 345)

angular momentum—the quantity of rotation of a body. (p. 365)

antimatter—a substance composed of subatomic particles that have the mass, electric charge, and magnetic moment of the electrons, protons, and neutrons of ordinary. (p. 388)

antimatter rockets—a class of rockets that uses antimatter as the energy source. (p. 388)

aphelion—the planet's farthest distance from the Sun. (p. 31)

apogee—position where the Moon is the farthest from Earth. (p. 65); the farthest point away in an elliptical orbit. (p. 356)

artificial intelligence (AI)—the development of computer systems that are able to perform tasks that normally require human intelligence. (p. 332)

asteroid belt—randomly scattered asteroids between the orbits of Mars and Jupiter. (p. 106)

asteroids—small rocky objects orbiting the Sun mostly between Mars and Jupiter, although some have passed close to Earth. (p. 83)

astrology—the interpretation of the influence of the heavenly bodies on human affairs. (p. 4)

astronomical unit (AU)—the average distance between the Earth and the Sun. (p. 26)

astronomy—the study of the universe beyond the earth's atmosphere. (p. 7)

astrophysics—science that applies the laws of physics and chemistry to explain the birth, life, and death of stars, planets, galaxies, and nebulae. (p. 158)

atmospheric drag—drag from Earth's atmosphere, which will eventually decay the satellite. (p. 355)

B

ballot-marking devices (BMD)—allow the voter to select their votes on a screen and then prints a paper ballot with their votes to be submitted. (p. 434)

biomedicine—the medical study of principles of the natural sciences, especially biology and biochemistry. (p. 242)

black hole—an area of intense gravitational pull. (p. 128)

booster—the first stage of a multistage launch vehicle. (p. 374)

bus—the metal or composite frame and body of a satellite. (p. 285)

C

cassini Division—the space between the second and third rings of Saturn. (p. 101)

cataracts—occur when the lens over the eye becomes cloudy. (p. 241)

celestial sphere—this imaginary shell formed by the sky in which the center is the observer's position. (p. 8)

cellular communications—use on-land towers to send and receive signals. (p. 289)

centrifugal force—a force that makes objects move outwards when they are spinning around something. (p. 248)

charge exchange—the passing solar wind stealing electrons and emitting radiation. (p. 119)

chasms—channels that appear to have been made by water or glaciers. (p. 96)

chemCam—fires a laser and reviews the elemental composition of vaporized materials from areas smaller than 0.04 inch on the surface of Mars. (p. 343)

chemistry and Mineralogy Instrument (CHEMIN)—instrument that uses x-ray beams to examine soil and drilled rock to identify minerals. (p. 344)

chromosphere—the layer just above the photosphere, and below the corona. (p. 77)

chronicles—a factual written account of important or historical events in the order of their occurrence. (p. 5)

clean room—a sterile environment, completely free of contaminants, with a strong ventilation system and highly advanced air filtration. (p. 391)

cloud computing—internet-based computing where large groups of remote servers are networked to allow for virtual data processing, data storage, and online access. (p. 398); uses a network of servers hosted on the internet rather than a local server. (p. 410)

cloud services—allow you to easily access data and information stored on the internet. (p. 410)

collection Handling for Interior Martian Rock Analysis (CHIMRA)—collects rock powder for deposit into Curiosity's body for examination. (p. 343)

coma—a large cloud of gas and dust. (p. 108)

combustion chamber—an enclosed space where combustion takes place. (p. 386)

comets—objects made of ice and dust and having a "tail" that points away from the Sun. (p. 83)
computer programs—used to give the computer instructions on how to process the data. (p. 409)
computer worm—similar to a virus but can replicate itself automatically to infect other computers with malicious software. (p. 442)
concentric—circles, arcs, or other shapes that share the same center. (p. 8)
conduction—occurs in solids as energy transfers from atom to atom or molecule to molecule. (p. 75)
constellation—a group of stars forming a recognizable pattern named after its apparent form. (p. 7)
constellation satellites—clusters of smaller satellites working together. (p. 285)
continental drift—the gradual movement of the Earth's surface. (p. 54)
convection—the generation of heat. (p. 59); occurs in liquids or gases when atoms move from one location to another. (p. 72)
copernican Revolution—a shift from Ptolemy's geocentric model to the heliocentric model with the Sun at the center of the Solar System. (p. 24)
corona—The highest part of the Sun's atmosphere. (p. 78)
cosmic microwave background (CMB)—the radiation leftover from the Big Bang. (p. 136)
cosmological—a branch of philosophy dealing with the origin and general structure of the universe. (p. 16)
cosmonauts—individuals trained and certified by the Russian Space Agency. (p. 156)
crawler-transporter—used to transport the launch vehicle, payload, and spacecraft to the launch pad. (p. 392)
critical infrastructure—the physical and cyber systems and assets that are so vital to the United States that their incapacity or destruction would have a debilitating impact on our physical or economic security or public health or safety. (p. 431)
cryogenic propellants—gasses chilled to subfreezing temperatures to create very combustible liquids. (p. 382)
cryptocurrencies—digital currencies in which encryption techniques are used to regulate the generation of units of currency and verify the transfer of funds, operating independently of a central bank. (p. 421)
cryptojacking—occurs when hackers hijack your computer and use its CPU power to mine cryptocurrencies. (p. 421)
cyber espionage—occurs when hackers steal data and intellectual property. (p. 443)
cyber warfare—a digital attack by one country against another. (p. 441)
cyberattack—an attack that attempts to damage or destroy a computer network or system. (p. 396)
cybersecurity—the precautions taken to guard against crime that involves the internet, especially unauthorized access to computer systems and data. (p. 398)
cyclical—occurring in cycles or recurrent. (p. 5)

D

dark matter—invisible matter that can be detected through its gravitational pull. (p. 117)

dark nebulae—nebulae that do not emit or reflect light. (p. 132)

data management—the process of acquiring, storing, and protecting information. (p. 413)

denial-of-service (DoS)—an attack intended to shut down a computer or network. (p. 442)

density—a measure of how much material (or mass) is packed into a given volume. (p. 52)

differential rotation—is seen when a rotating object moves with different periods of rotation. (p. 98)

direct-recording electronic (DRE) voting machines—portable monitors that allow voters to cast their ballot electronically. (p. 434)

distributed denial-of-service (DDoS)—a more complex form of a DoS attack, in which large numbers of computers, devices, etc., are used to force the shutdown of a site. (p. 443)

domain name system (DNS) server—contains a list of the domains' alphabetic or alphanumeric sequence followed by a suffix. (p. 411)

downlink—the signal transmitted from the satellite to Earth. (p. 283)

cwarf planet—orbits the Sun, is large enough to be a spherical shape, but not large enough to have cleared away other objects of a similar size near its orbit. (p. 85)

dwarf star—a star with average or low luminosity, mass, and size. (p. 117)

dwell time—the time a satellite sits over one part of Earth. (p. 357)

dynamic albedo of neutrons (DAN)—instrument that measures water content below the surface. (p. 344)

earth-moon libration orbit—an orbit that relies on balancing the Sun, Earth, and Moon's gravity so that a spacecraft can orbit a virtual location. (p. 306)

eccentricity—used to describe the shape of the ellipse. (p. 356)

eclipse—occurs when the Moon's rotation is very close to or on the ecliptic line. (p. 13)

ecliptic—the sun's apparent path during the year. (p. 10)

elliptical galaxy—elliptical in shape and does not have spiral arms or gas and dust particles as seen in the Milky Way. (p. 121)

ellipse—a geometrical shape where each point on the ellipse is at the same total distance from two fixed points, or foci. (p. 30)

elliptic—the circle formed by the Earth's orbit with the celestial sphere. (p. 10)

emission nebulae—nebulae clouds lit up by interstellar gas. (p. 132)

epicycle—a small circle in which the center of the circle moves around a larger circle. (p. 19)

equinox—the points where the Sun crosses the celestial equator. (p. 58)

escape velocity—the lowest velocity an object must have in order to escape the gravitational pull of another object. (p. 354)
event horizon—an object is pulled into a black hole to the point of no return. (p. 129)
exoplanets—planets that orbit a star outside of our solar system. (p. 119)
expendable launch vehicles (ELVs)—used only once to carry payloads into space. (p. 376)
exploitation—the action of making use of and benefiting from resources such as new technology. (p. 188)
external hard drive—a portable storage device that can be connected to a computer via USB. (p. 418)
extravehicular activity (EVA)—when an astronaut leaves the protective environment of a spacecraft and enters outer space. (p. 219)

F

Fission rockets—an efficient method of using nuclear power to propel a rocket. (p. 388)
flight simulators—safe and cost-effective alternatives to actual flights to gather data, and provide facilities for practice and training. (p. 164)
flyby—the close approach of a spacecraft to a planet, Sun, or Moon for observation. (p. 304)
fusion—happens when the nuclei of two or more atoms combine. (p. 388)
fusion rockets—small rockets that use fusion propulsion. (p. 388)

G

g-force—the gravitational force that is put on a body during acceleration. (p. 164)
galaxy—a group of stars, gas, and dust bound together by gravity. (p. 113)
galaxy clusters—galaxies form groups. (p. 128)
gamma—ray burst-a short-lived burst of energetic explosives in distance galaxies. (p. 302)
galactic cosmic rays—high-energy protons and heavy ions that originate outside our solar system. (p. 172)
galactic radiation—energetic particles produced in solar storms and the solar winds. (p. 302)
galactic year—the galaxy is so big that an entire orbit takes 250 million years. (p. 114)
galilean satellites—Jupiter's four largest moons named because they were first observed by Galileo in 1610. (p. 99)
geocentric—where the Earth is the center of the universe. (p. 25)
gene expression—how your body reacts to your environment. (p. 239)
geostationary orbit—puts the satellite in the same plane as the equator. (p. 284)
geosynchronous orbit—an orbit that puts the satellite in the same place in the sky over a specific point on the surface each day. (p. 282)

geotagging—the process of adding geographical information to images and data posted on social media. (p. 417)
globular clusters—large compact spherical star clusters, typically of old stars in the outer regions of a galaxy. (p. 113)
goldilocks zone—the habitable zone around a star in which planets can sustain life. (p. 53)
gravity assist—occurs when a planet's gravity is used to give a spacecraft additional momentum. (p. 365)
gravitational contraction—the shrinking and compression of gases, caused by gravity. (p. 72)
gravitational force—the force that attracts any two objects with mass. (p. 42)
gravitational lensing—a technique used to study dark matter. (p. 131)
greenhouse effect—is the exchange of incoming and outgoing radiation that warms a planet. (p. 93)
gypsum—a salt that precipitates from liquid water. (p. 341)

H

hacking—occurs when unauthorized access to data or a computer system is obtained. (p. 402)
heavy lift launch vehicles—launch larger payloads into space. (p. 376)
heliocentric—where the Earth and other planets orbit the Sun. (p. 24)
heliopause—the boundary between the end of the Sun's magnetic influence and the beginning of interstellar space. (p. 317)
heliophysics—the study of the Sun and its effects on space. (p. 158)
hematite—made up of iron oxide and usually forms in water. (p. 340)
highly elliptical orbit—an orbit in which the satellite spends the majority of its time in apogee, moving very slowly. (p. 358)
humanoid robots—robots that look like humans. (p. 330)
hydrostatic equilibrium—occurs when compression due to gravity is balanced by outward pressure from the Sun's core. (p. 74)
hypergolic fuels—can be stored as liquids at room temperature and are easily ignited. (p. 381)

I

inertia—the tendency of an object to resist a change in motion. (p. 40)
intercontinental ballistic missiles (ICBMs)—missiles with a flight capability of over 3,400 miles. (p. 186)
international space station (ISS)—a large spacecraft in orbit around Earth. (p. 156)
internet—a global computer network using a standardized communication protocol. (p. 411)

internet of things (IoT)—any device that is connected to the internet. (p. 413)
interstellar medium (ISM)—the term used to describe the matter in galaxies that exists between solar systems. (p. 118)
in vitro fertilization (IVF)—scientists inject the seminal fluid of the animal into the egg to create an embryo. (p. 243)
ion propulsion—uses magnetism and electricity to propel a spacecraft through space. (p. 368)
irregular galaxy—galaxies that do not have a clear shape. (p. 122)

J

jamming—occurs when the signal from a satellite is degraded or disrupted by interference. (p. 401)
jovian planets—much larger planets and made up of gases. (p. 85)

K

kuiper belt— a ring of icy bodies outside of Neptune's orbit. (p. 88)

L

launch pad—the circular structure from which the launch takes place. (p. 392)
launch sites—the physical locations from which rockets take off. (p. 389)
launch vehicle—a rocket that is used to propel the spacecraft into orbit. (p. 160)
launch vehicle stage adapter (LVSA)—connects two major sections of the launch vehicle's stages. (p. 391)
launch window—the time period in which a particular mission must be launched. (p. 355)
lenticular galaxy—a spherically shaped galaxy with a disk of stars and gas around a nucleus. (p. 122)
light year—the distance traveled by light in one year (approximately 5,880,000,000,000 miles). (p. 113)
local area network—a system of connecting devices within a building or cluster of buildings. (p. 411)
long-period comets—Comets that take more than 200 years to orbit the Sun. (p. 108)
low earth orbit (LEO)—objects orbit at between 99 to 1,200 miles above the Earth's surface. (p. 151)
luminosity—the amount of energy produced per unit of time by a star. (p. 74)
lunar eclipse—occurs when the Moon passes directly behind Earth and into its shadow. (p. 14)

M

macroinstructions (macros)—a programmable pattern that translates a sequence of input into a preset sequence of output. (p. 409)

magnetic field—magnetic forces are communicated. (p. 59)

magnetosphere—the region surrounding a planet, star, or other body, in which the body's magnetic field traps charged particles and dominates their behavior. (p. 307)

malware—software that is designed to disrupt, damage, or gain unauthorized access to a computer system. (p. 396)

margins—edges of the plates. (p. 54)

mars hand lens imager (MAHLI)—gives close-up views of the Martian rocks, debris and dust. (p. 343)

mass-weight. (p. 44)

max-Q—when aerodynamic forces reach their maximum. (p. 221)

mercury—ion trap technology-uses electrically charged mercury atoms created in an electromagnetic trap for better control and accuracy. (p. 362)

meteor—a very small meteoroid, usually smaller than a marble, which burns up as it enters a planet's atmosphere. (p. 109)

meteorite—a meteor that makes it through the atmosphere and strikes the surface of the planet or moon. (p. 109)

meteoroids—similar to asteroids but smaller. (p. 83); a tiny rocky object in space usually the debris from a comet or an asteroid. (p. 109)

methane—a colorless, odorless flammable gas on Earth is produced mostly by microorganisms. (p. 347)

microgravity—the condition in which people or objects appear to be weightless. (p. 162)

micrometeoroids—small pieces of asteroids and comets that cannot be tracked. (p. 174)

microscopic imager (MI)—attached to the robotic arm of each rover provided high-resolution, close-up views rocks and soils. (p. 338)

mid-atlantic ridge—an area in the Atlantic Ocean where lava flows upward creating new mountains underwater. (p. 54)

miniature thermal emission spectrometer (Mini-TES)—used infrared scans to help identify different types of compounds based on different temperatures. (p. 336)

mission directorates—groups that study space from Earth, and in space. (p. 158)

molecular diffusion—the thermal motion of all particles at temperatures above absolute zero. (p. 178)

momentum—forward motion. (p. 353)

monopolies—the exclusive possession or control of something. (p. 268)

mössbauer spectrometer (MB)—attached to the robotic arm, looked for the makeup of iron-bearing rocks and soils. (p. 337)

MSL entry, descent, and landing instrument (MEDLI)—instrument that measures pressure and temperature during entry and landing. (p. 344)

multi-factor authentication (MFA)—an authentication process that uses two or more factors for authentication. (p. 413)
multistage—refers to an ELV having two or more engines, stacked one on top of another and firing in succession. (p. 379)
muscle atrophy—when muscles waste away from lack of use. (p. 233)

N

neap tides—occur when the difference between the high tide and low tide is the smallest. (p. 66)
near-earth asteroids—asteroids that orbit closer to Earth than to the Sun. (p. 265)
nebulae—dust and gas are very dense and create clouds. (p. 132)
network attached storage (NAS)—a networked server used for saving data. (p. 419)
neutron star—ball of neutrons. (p. 134)
north celestial pole—the point on the celestial sphere directly above the North Pole. (p. 8)
northern hemisphere—the half of the Earth that lies north of the equator. (p. 57)
nuclear-electric rockets—use a nuclear reactor to generate electricity, then use the electricity to make high-velocity ions. (p. 388)
nuclear-thermal rockets—use a reactor to create heat to generate turbine gas and thrust. (p. 388)
nuclear fusion—hydrogen atoms compress and fuse into helium. (p. 72)
nucleus—the center of the comet's head made mostly of ice and frozen carbon dioxide (dry ice). (p. 108)

O

o-ring—a type of gasket, used in the field joints between each fuel segment of the SRBs. (p. 221)
oblate—somewhat flattened at its poles. (p. 98)
oort cloud—a spherical shell surrounding our Sun with a distance of up to 100,000 astronomical units (AU). (p. 108)
opposition—on the opposite side of the Earth from the Sun. (p. 94)
optical/digital scan voting ballot—a paper ballot that can be read by optical or digital equipment to tally votes. (p. 434)
orbit—a regular, repeating path of one object around another object. (p. 353)
orbital debris—man-made particles in space, like trash from spent rockets and broken satellites. (p. 174)
orbital flight—when an object is placed on a flight trajectory that keeps it in space for at least one orbit around a planet. (p. 187)
orbital velocity—the velocity needed to achieve a balance between gravity and momentum. (p. 354)

orbiter—what NASA calls a space shuttle. It is a space plane that is one of the four components of what most people think of as the space shuttle. (p. 217)

osteoporosis—a medical condition in which bones become brittle and fragile from loss of tissue. (p. 232)

oxidizer—a type of chemical used to burn fuel in space where an atmosphere containing oxygen is absent. (p. 382)

ozone layer—protects us from solar ultra-violet (UV) rays. (p. 55)

P

parallax—the apparent displacement of an observed object due to a change in the position of the observer. (p. 17)

pass phrase—a string of random words strung together that make sense to you, but no one else. (p. 417)

patch—a set of changes to a computer program that are required to update, fix, or improve the program. (p. 435)

perigee—position where the Moon is closest to the Earth. (p. 65); the closest point in an elliptical orbit. (p. 356)

perihelion—the planet's closest approach to the Sun. (p. 31)

phishing—occurs when cybercriminals send fraudulent emails designed to look like they are from a legitimate source. (p. 420)

photosphere—the deepest layer of the Sun's atmosphere. (p. 77)

photovoltaic (PV)—refers to the production of energy from light. (p. 71)

planet—a celestial body moving in an elliptical orbit around a star. (p. 15)

planetary nebulae—occurs when a star reaches the end of its life. (p. 132)

plasma rockets—accelerate by using plasma that is produced by stripping negative ions from hydrogen atoms in a magnetic field and pushing it out of the engine. (p. 387)

plate tectonics—the theory used to explain the structure of the Earth's crust and its movement over time. (p. 54)

plates—broken pieces. (p. 54)

plutonium-238—a material that emits steady heat due to its natural radioactive decay. (p. 344)

polar-orbiting satellites—satellites that orbit across the Earth's poles. (p. 288)

population II stars—older stars that are less luminous. (p. 115)

potassium perchlorate candle—a chemical candle that contains sodium chlorate and iron, which when burned produces oxygen. (p. 237)

powder acquisition drill system (PADS)—drills rock into powder. (p. 343)

precession—the slow and gradual shift of the Earth's axis. (p. 67)

predictive power—the ability of a given theory to allow predictions to be made. (p. 26)

principle of equivalence—the fundamental law of physics that states that gravitational and inertial forces are of a similar nature and often indistinguishable. (p. 45)

prominence—loop or arc of glowing gas extending from the photosphere into the corona. (p. 79)

propaganda—misleading or inaccurate information used to publicize a particular political cause. (p. 443)

propellant—a substance, usually a mixture of fuel and oxidizer, used to accelerate an object. (p. 381)

protoplanet—a large body of matter in orbit around the sun or a star and thought to be developing into a planet. (p. 369)

pulsar—a rapidly spinning neutron star that emits radio waves in pulses. (p. 134)

R

radiation—the movement of energy as waves through great distances without having the particles closely packed together. (p. 76)

radiation assessment detector (RAD)—instrument that measures radiation levels. (p. 344)

radiation sickness—the damage to the body caused by a large amount of radiation over a short amount of time. (p. 240)

radioisotope power system (RPS)—a type of nuclear energy technology that uses heat to produce electric power for operating spacecraft. (p. 344)

radiometers—instruments for detecting or measuring the intensity or force of radiation. (p. 287)

reconnaissance—the military observation of a region to locate an enemy or ascertain strategic features. (p. 294)

red planet—because of the rust in the iron-filled Martian soil. (p. 94)

reflection nebulae—scattered nebula clouds that reflect light from the nearby stars. (p. 132)

remote manipulator system (RMS)—a 50-foot long mechanical arm that deploys and retrieves things on the outside of the orbiter. (p. 218)

retrorocket—a small auxiliary rocket on a spacecraft that is designed to slow down the craft. (p. 204)

reusable launch system (RLS)—allows for the recovery of all or part of the system for reuse of launch components, including rockets. (p. 385)

rock abrasion tool (RAT)—attached to the robotic arm of each rover was simply a wire brush for dusting off rocks prior to examination. (p. 338)

router—used to transfer or route information between computers. (p. 411)

rover—a space exploration vehicle that moves across the surface of the planet. (p. 314)

S

sample analysis at Mars (SAM)—instrument that is used to study carbon, hydrogen, oxygen and nitrogen. (p. 344)
satellite—an object that orbits around a bigger object. (p. 279)
satellite cellular phones—use satellites orbiting the Earth to send and receive signals. (p. 289)
scientific revolution—refers to changes in thoughts and beliefs in modern science that occurred from 1550-1700. (p. 24)
security token—a small device typically attached to a key chain that randomly generates a unique code at regular intervals. (p. 414)
seismic waves—waves of energy that travel through the Earth during an earthquake. (p. 54)
silica—a compound made of crystals that occur in quartz on Earth. (p. 340)
short-period comets—comets that take less than 200 years to orbit the Sun. (p. 108)
smart device—an electronic device that is connected to other devices or networks. (p. 413)
solar eclipse—occurs when the Moon blocks the sun's path to Earth. (p. 13)
solar energy—the solar radiation that reaches Earth. (p. 70)
solar flare—an eruption on the Sun's surface. (p. 80)
solar maximum—the point in the 11-year solar cycle when the sun is most active. (p. 302)
solar wind—the emission of high-speed protons and electrons. (p. 78)
solid-fuel rocket—a fuel and oxidant mixed together as fine powders and then pressed into a solid "cake." (p. 385)
solid rocket boosters (SRBs)—large solid propellant motors that provide 80% of the thrust needed during the first two minutes of launch. (p. 219)
solstice—occurs twice a year when the sun reaches the greatest distance from the equator. (p. 4)
south celestial pole—the point on the celestial sphere directly above the South Pole. (p. 8)
southern hemisphere—the half of the Earth that lies south of the equator. (p. 57)
space debris—natural and man-made particles in space. (p. 174)
space launch system (SLS)—a heavy lift launch ELV rocket that is part of NASA's deep space exploration plans. (p. 383)
space probe—a spacecraft that travels through space to collect science information. (p. 301)
space race—the competition between the United States and the Soviet Union to prove their superiority with the technology and man power required to send a man to space. (p. 186)
space radiation—contains atoms that have been stripped of their electrons as they accelerate through space at speeds close to the speed of light. (p. 172)

space robots—general purpose machines that are capable of surviving the rigors of the space environment and performing exploration, assembly, construction, maintenance, servicing, or other tasks in space. (p. 324)
space shuttle—a reusable spacecraft designed to transport people and cargo between Earth and space. (p. 212)
space shuttle main engines (SSMEs)—three large engines located on the rear of the orbiter. (p. 220)
space-time—a combination of space and time. (p. 45)
spinoffs—are the technologies, products, and processes based on NASA innovations. (p. 295)
spiral galaxy—spiral in shape and have long spiral arms with young stars and star formation areas. (p. 122)
spring tides—occur when the Sun and Moon are aligned and cause the oceans to bulge more than usual, producing a higher tide than usual. (p. 66)
spoofing—occurs when the signal from a satellite is replaced with a false signal. (p. 402)
starburst—a high rate of star formation. (p. 126)
sublimate—the direct change from a solid to a gas. (p. 106)
suborbital flight—is a flight trajectory that does not complete a full orbit of the Earth. (p. 187)
sunspots—these are dark regions that appear when hot gases cool during the convection process. (p. 79)
superclusters—galaxy clusters can also group together. (p. 128)
supernova—a massive explosion of a star. (p. 118)
supernova remnant—the remnants of a star that has expanded to create a nebula. (p. 132)
synchronous rotation—the Moon spins very little and takes 27 days to complete one rotation, so we only see one side of the Moon. (p. 63)
system administrators—information technology personnel who are responsible for ensuring that all system resources are available. (p. 439)

T

terrestrial planets—small and dense, with rocky surfaces and a great deal of metals in their core. (p. 85)
theory—an educated explanation of something but without real proof. (p. 83)
thor missile—an intermediate range ballistic missile (IRBM). (p. 381)
thrust—the strength of a rocket measured in pounds. (p. 374)
thrusters—small rockets on a spacecraft that are used to make alterations to the spacecrafts flight path. (p. 366)
tides—the reaction of gravity that causes the sea levels to rise and fall. (p. 66)
tracking and data relay satellite system (TDRSS)—provides near-constant communication links between the ground and the orbiting satellites. (p. 160)

trajectory—the path of a body in space. (p. 364)
transponder—receives a radio signal and immediately transmits a different signal. (p. 283)
trilateration—when a GPS receiver calculates the distances to satellites by measuring angles. (p. 292)
trojan asteroids—asteroids that share orbits with larger planets. (p. 107)

U

underwater training—simulates conditions similar to the weightlessness in space. (p. 163)
universal serial bus (USB)—a common interface that enables communication between devices and a host controller, such as a personal computer. (p. 414)
uplink—the signal transmitted from Earth. (p. 283)

V

virtual machine—an operating system or application that is installed on software that imitates hardware. (p. 409)
virus—computer code that is designed to spread from one computer to the next, infecting computers with malicious software. (p. 442)

Z

zero-day vulnerabilites—the bugs or flaws found in code that are used by hackers to exploit a system. (p. 443)
zodiac—an imaginary belt in the sky that includes the paths of the planets and is divided into twelve constellations. (p. 16)

Index

B

C

D

E

F

G

H

I

J

K

L

M

N

O

P

Q

R

S

T

U

V

W

X

Y

Z

TERMS OF USE AND WARRANTY DISCLAIMERS
READ BEFORE OPENING THIS PACKAGE